Synthesis Lectures on Biomedical Engineering

Series Editor

John Enderle, Storr, USA

This series consists of concise books on advanced and state-of-the-art topics that span the field of biomedical engineering. Each Lecture covers the fundamental principles in a unified manner, develops underlying concepts needed for sequential material, and progresses to more advanced topics and design. The authors selected to write the Lectures are leading experts on the subject who have extensive background in theory, application, and design. The series is designed to meet the demands of the 21st century technology and the rapid advancements in the all-encompassing field of biomedical engineering.

Farzad Salehpour · Saeed Sadigh-Eteghad ·
Javad Mahmoudi · Farzin Kamari ·
Paolo Cassano · Michael Richard Hamblin

Photobiomodulation for the Brain

Photobiomodulation Therapy in Neurology and Neuropsychiatry

 Springer

Farzad Salehpour
College for Light Medicine
and Photobiomodulation
Starnberg, Germany

Saeed Sadigh-Eteghad
Neurosciences Research Center
Tabriz University of Medical Sciences
Tabriz, Iran

Javad Mahmoudi
Neurosciences Research Center
Tabriz University of Medical Sciences
Tabriz, Iran

Farzin Kamari
Department of Neurophysiology
Institute of Physiology, University of Tübingen
Tübingen, Germany

Paolo Cassano
Division of Neuropsychiatry
and Neuromodulation
Massachusetts General Hospital
Boston, MA, USA

Michael Richard Hamblin
Faculty of Health Science
Laser Research Centre
University of Johannesburg
Johannesburg, Gauteng, South Africa

ISSN 1930-0328   ISSN 1930-0336 (electronic)
Synthesis Lectures on Biomedical Engineering
ISBN 978-3-031-36230-9   ISBN 978-3-031-36231-6 (eBook)
https://doi.org/10.1007/978-3-031-36231-6

This Springer imprint is published by the registered company Springer Nature Switzerland AG
The registered company address is: Gewerbestrasse 11, 6330 Cham, Switzerland

To my beloved family, and most notably Melissa and Anissa—your support has made this journey possible.

—Farzad Salehpour

Contents

Introduction

1.1 A Brief History of Photobiomodulation

A photon is a fundamental packet of energy created by the perpendicular oscillation of the electric and magnetic fields according to Maxwell's equations, and like all quantum phenomena, exhibits both wave-like and particle-like behavior. In 1905, Albert Einstein introduced the concept of "*quanta*" as particles of light [1]. Shortly thereafter, in 1917, Einstein formulated the theoretical principle of stimulated emission, the quantum mechanical process that underlies the production of lasers [2]. However, the building of the first working laser had to wait more than 40 years until Theodore Maiman invented the Ruby laser in 1960 [3].

Traditionally, Endre Mester is believed to be the first physician to use a laser in medical science, with the goal of stimulating tissue rather than destroying it [4]. In 1967, Mester was studying the possible effects of the Ruby laser (694 nm) on tumoral tissue in a mouse model and, surprisingly, he found that the laser not only failed to destroy cancer or to cause cancer but instead stimulated skin wound healing and hair growth in the implanted tumor site [5]. Thus, the first "*photobiostimulation*" phenomenon using monochromatic light was unintentionally discovered. Experimental investigations into laser biostimulation have continued since that time in an attempt to determine the action mechanisms and possible biological effects. In 1982, Tiina Karu, a well-known photobiology scientist, started to explore the biological action mechanisms of visible light [6]. In 1991, Glen Calderhead introduced the term "low reactive-level laser therapy" or LLLT, which was defined as a "dose" of light that was under the damage threshold for tissue [7]. During the last three decades, a number of other terms have been used in the scientific literature for photobiostimulation processes, such as "low-level light therapy", "low-power laser therapy", "low-intensity laser therapy", "high-power laser therapy", "cold laser therapy", "cold low-level laser therapy", "soft laser therapy", "light therapy", "phototherapy", "biostimulation

F. Salehpour et al., *Photobiomodulation for the Brain*, Synthesis Lectures on Biomedical Engineering, https://doi.org/10.1007/978-3-031-36231-6_1

laser therapy", "Class 3 laser therapy", and "Class 4 laser therapy". Apart from these, in 1997, the first formal use of the term "photobiomodulation" was described in the PubMed literature by Yu et al. [8]. "Photo" means light, "Bio" means living tissues, and "Modulation" means alteration. Thereafter, during the 2000s and the first half of the 2010s, an international effort was launched by many pioneers in the field to use the term "photobiomodulation" (PBM) as a common language [9]. In 2014, a nomenclature consensus meeting was held and co-chaired by Juanita Anders and Jan Bjordal. In this meeting, the terms "Photobiomodulation" and "Photobiomodulation Therapy" were accepted by the majority of the international participants to describe the mechanistic and curative applications of low-level light therapy [9]. Finally, in 2016, after significant efforts by Michael R. Hamblin and Praveen Arany to establish the universal terminology, "photobiomodulation therapy" was added to the Medical Subject Headings (MeSH) database as an indexing term [10]. So far, more than 4000 scientific peer-reviewed articles on "photobiomodulation" have been published according to the NCBI PubMed database [11]. Nowadays, PBM is a growing field in photomedicine, so that about 40 new papers are published every month in this field, and hopefully more PBM-related papers will be published in high-impact factor journals.

1.2   What is Photobiomodulation?

Photobiomodulation (PBM), previously known as low-level light/laser therapy (LLLT), refers to the non-thermal application of the non-ionizing form of electromagnetic radiation to stimulate biological processes [11]. PBM therapy is a specific term for the therapeutic application of PBM in various disease conditions. Almost all PBM procedures are conducted using visible and/or near-infrared (NIR) light in the wavelength range between 400 and 1100 nm from various light sources (e.g., lasers, light-emitting diodes (LEDs), or broadband light sources) [12]. PBM delivers photons at lower power as compared to other forms of laser application in biomedicine such as high power surgical lasers that can coagulate, ablate, and cut biological tissue due to a macroscopic temperature increase [13]. It should be noted that as mentioned above, PBM therapy (as with LLLT) is not defined by the emitted power level of the light source used but by the level of tissue reaction to the dose ("high" versus "low") of light delivered [14].

Although the action mechanisms underlying all PBM effects are not yet completely understood, several mechanisms for the biostimulatory effects of PBM have been proposed [15, 16]. Considerable research has suggested that mitochondrial photoacceptors for specific wavelengths are responsible for different cellular responses connected with PBM in the visible to NIR spectrum [17, 18]. As a primary action mechanism, it was proposed that mitochondria contain photoacceptor molecules that can absorb light and thereby increase the production of adenosine triphosphate (ATP). Following the primary photochemical effects induced by photon absorption by mitochondrial enzymes, a cascade of secondary photosignal transduction occurs in the dark phase (or illumination off period). These responses are linked with the changes in the cellular redox state and redox homeostasis, which in turn drive other cellular reactions such as proliferation, differentiation, growth, and survival [19].

Today, the applications of PBM therapy are quite diverse in modern medicine, and this light-based modality is gaining growing acceptance and importance among other therapeutic options. PBM therapy has been shown to be an effective strategy to promote microcirculation, tissue repair, proliferation, and regeneration, and to relieve pain, edema, oxidative stress, and inflammation in several traumatic, acute, and chronic diseases [20]. It is noteworthy that although PBM is now applied mostly "off-label" in a wide range of medical conditions, its use in medicine is still somewhat controversial for the following reasons. First of all, the exact underlying cellular and molecular mechanisms are not completely elucidated. As the action mechanisms become progressively understood, studies have suggested that there are a number of different photoacceptors in cells that might be responsible for the observed beneficial effects [16]. Secondly, PBM is based on biophysical principles, which involve a great number of dosimetry parameters (e.g., wavelength, irradiance, fluence, total power, total energy, spot area, pulsing mode, coherence, and polarization). The selection of less-than-optimal physical and treatment parameters is likely to result in unsatisfactory therapeutic outcomes [21]. Thirdly, PBM displays a biphasic dose–response pattern, also known as the Arndt-Schulz law, suggesting that there is an optimal therapeutic dose of light, and doses/fluences lower or higher than this optimal range have fewer positive effects and even at times can have negative effects [22].

1.3 Photobiomodulation and the Brain

There are several well-known areas of medicine where PBM has a key role to play as a treatment, including wound healing, dentistry, dermatological conditions, muscle and tendon repair, and neurogenic pain [13]. However, in recent years, there has been special and growing interest in the use of PBM as a neuroprotective and neurorestorative therapy for the treatment of central nervous system (CNS) injuries and diseases [23, 24]. It is believed that low levels of visible and NIR light can stimulate the functions of neuronal cells resulting in prevention of neuronal death, hypoxia, trauma, or neurotoxicity [25].

Publications in the 1980s had already begun to disclose the direct effects of PBM on the CNS. Historically, the earliest in vitro PBM study in the CNS was the investigation of the effects of PBM on cultured slices of cortical tissue in 1988 when Wade et al. [26] found that low levels of visible light (1.3 mW/cm^2) enhanced K$^+$-induced release of gamma-aminobutyric acid (GABA) from the brain slices. They suggested that photoreceptive molecules or processes might exist in neuronal tissue, which could be responsible for the observed phenomenon. The first in vivo study, in 1982, was by Shen-Zeng et al. [27] who compared the effects of two types of low power lasers, namely, He–Ne (632.8 nm) and nitrogen (337.1 nm), on rat brain concentrations of monoamines and amino acids. They reported that laser beams guided by optical fibers to the caudate nucleus or frontal cortex altered striatal concentrations of serotonin, dopamine, GABA, and aspartic acid, most likely via acting on enzymes involved in neurotransmitter metabolism. Shimon Rochkind can be considered a pioneer in the use of PBM for treatment of CNS injuries because of his extensive research during the 1980s and 1990s [28, 29]. Accumulated evidence from his laboratory experiments suggested that irradiation of low-power red (632.8 nm) and NIR (780 nm) lasers can improve neuronal metabolism and promote CNS tissue recovery and repair [30, 31]. Although the preclinical investigations of efficacy and safety of PBM in neuroprotection and neuroregeneration were initiated more than 40 years ago, the history of the first application of PBM in clinical research for neurorehabilitation purposes goes back to almost 15 years ago [32]. In 2007, in the first multicenter, double-blind, placebo-controlled study, Lampl et al. [33] reported that shining NIR laser (808 nm) light onto the entire surface of the head was safe and could be effective in the treatment of ischemic stroke in humans.

According to the vast evidence accumulated over the past four decades, the main neurobiological action mechanisms responsible for the neuroprotective efficacy of PBM include improvement of neuronal metabolic activity and cerebral blood flow, promotion of anti-neuroinflammatory, anti-apoptotic, and antioxidant responses, stimulation of neurogenesis and synaptogenesis via some neurotrophic mechanisms, and regulation of neurotransmitters [34]. Nowadays, the PBM research field consists of preclinical and clinical investigations for a wide range of neuroscience applications, ranging from neuro-trauma [19, 35, 36], neurodegeneration [37–40], and neuropsychiatric disorders [41–45], as well as enhancing brain function in healthy subjects [46]. However, it should be noted that PBM has not been extensively adopted as a neurotherapeutic strategy in clinical practice for multiple reasons: **(I)** its neurobiological effects and action mechanisms are not fully understood; **(II)** the delivery of a therapeutic amount of light (dosage) into the brain through the scalp/skull is considered challenging; **(III)** due to different types and severity of neuropathology, the best timing for the application of PBM for each particular disease is likely to vary; **(IV)** researchers have employed a large number of dosimetry and treatment parameters, making inter-trial comparisons of study findings challenging; **(V)** the lack of extensive randomized controlled trials to date for each neurologic and neuropsychiatric disorders.

1.4   Research in Brain Photobiomodulation

There is a great deal of hope that PBM, as a promising strategy, will continue to provide further breakthroughs in the field of neuroscience. It must be clarified that PBM using low-level light/laser for stimulation of brain tissues or neuronal cells differs from the other forms of light application in the neuroscience and neurology such as transcranial optical imaging, optogenetics, bright light therapy working through the eyes, photodynamic inactivation of neurons, laser interstitial thermal therapy, or neural photostimulation [47, 48]. Among the aforementioned techniques, neural photostimulation is the closest to PBM and is sometimes hard to distinguish from neuronal PBM. Indeed, the biophysical mechanism of neural photostimulation is based on evoked action potentials by a transient thermally mediated mechanism and generally, short light pulses (femtosecond to millisecond) with longer wavelengths (800–2000 nm) compared to PBM applications (generally below 1100 nm). In addition, differences between PBM and other neuromodulation techniques such as transcranial direct current stimulation (tDCS) include the fact that there is little evidence to date that PBM can produce direct neural activity. Also, to the best of our knowledge, there have not been any researches that have shown that PBM induces long-term potentiation (LTP) or long-term depression (LTD) in ex vivo brain slices [49].

In a scientometric evaluation regarding PBM and brain/neuron, we searched Web of Science and PubMed databases from 1967 to December 2020 using 234 relevant keywords of "phototherapy, low-level laser therapy, low-level light therapy, LLLT, low energy laser, low intensity laser, low power laser, photobiomodulation, photomodulation, photoneuromodulation, photostimulation, photobiostimulation, photoneurostimulation, light irradiation, laser irradiation, light-emitting diode, LED irradiation, laser treatment, light treatment, laser therapy, light therapy, LED therapy, LED treatment, transcranial laser, transcranial LED, transcranial photobiomodulation, transcranial low-level laser therapy, transcranial low-level light therapy, intranasal laser, intranasal LED, red light, red laser, NIR light, NIR laser, visible light, near-infrared light, near-infrared laser, near infrared light, near infrared laser, infrared light, infrared laser, He–Ne laser, cold laser, laser acupuncture, laser acupoints, wavelength, central nervous system, hippocampus, hippocampal, hippocampi, cortex, frontal, parietal, occipital, temporal, orbitofrontal, prefrontal, cingulate gyrus, thalamus, amygdala, entorhinal, olfactory bulb, hypothalamus, pineal gland, pituitary gland, substantia nigra, substantia nigra pars compacta, motor cortex, epithelium, epithelial, optic, neuropathy, skull, skin, scalp, penetration, transmission, transmittance, optical, optical properties, attenuation, propagation, absorption, scatter, scattering, atrophy, blood–brain barrier, default mode network, gamma wave, alpha wave, beta wave, oscillations, EEG, coherency, synchrony, GABA, cognitive, cognition, learning, memory, attention, executive function, concentration, psychomotor, insomnia, aggression, mild cognitive impairment, dementia, anhedonia, Alzheimer's disease, AD, Parkinson's disease, PD, Huntington's disease, HD, depression, depressive, MDD, seasonal affective disorder, SAD, anxiety, anxious, major depressive disorder, autism, traumatic brain injury, TBI, trauma, post-traumatic stress disorder, PTSD, addiction, attention deficit hyperactivity disorder, ADHD, obsessive compulsive disorder, OCD, stroke, ischemia, cerebral infarction, encephalopathy, neurodegeneration, neurodegenerative, neuropsychiatry, psychiatric, schizophrenia, bipolar, manic, psychosis, consciousness, sleep, aging, age, aged, elderly, adult, young, ageing, senescent, cerebral blood flow, CBF, preconditioning, astrocytes, microglia, circadian rhythm, inflammation, neuroinflammation, inflammatory, apoptosis, apoptotic, caspase, oxidative stress, antioxidant, neurogenesis, synaptogenesis, neurotoxicity, vascular dysfunction, cerebrovascular, BDNF, GDNF, NGF, CREB, ERK, BrdU, GV20, SVZ, amyloid, amyloid β-peptide, Amyloid-β peptide, β-amyloid, senile plaque, tau, phosphorylated tau, neurofibrillary tangles, iNOS, eNOS, nNOS, akt, acetylcholinesterase, heat-shock protein, mesenchymal stem cells, mitochondrial dysfunction, neurotransmitter, serotonin, dopamine, neuron, neural cell, neuronal cell, neuronal tissue, neural tissue, brain, cerebral, neuroprotection, cortical neuron, primary cortical neuron, cultured cortical neuron, calcium channel, mitochondria, cytochrome c oxidase, mitochondrial enzyme, ion channel, heat-gated ion channel, light-gated ion channel, intracellular calcium, ATP, nitric oxide, reactive oxygen species, NO, ROS, transcription factor, signal transduction, retrograde mitochondrial signaling, microcirculation".

The samples were reduced to peer-reviewed publications (e.g., articles, articles in press, reviews, editorials, letters, perspectives, hypothesis, opinions, and notes) as well as book chapters and books. In the searches, conference papers, dissertations and theses, and patents were not included and results were limited to publications written in English. Figures 1.1 and 1.2 show graphs of the numbers of in vitro, in vivo, and clinical studies, with year-by-year evolution, conducted in the field of PBM and brain/neuron. The initial search yielded more than 16,560 unique articles, of which 419 were eligible for inclusion.

Of these, 66 were in vitro, 161 were in vivo, and 95 were original clinical articles. The remaining 97 studies were other types of articles (reviews, editorials, letters, perspectives, opinions, hypotheses, and notes). In addition, 25 chapter books and one book have been published to date in this field of research.

Of note, in this book, only the effects and applications of PBM to the brain (CNS) (not the spinal cord or retina) are discussed. For both in vivo and clinical reports, the effects of both direct and indirect/systemic (remote tissue irradiation) PBM approaches on brain

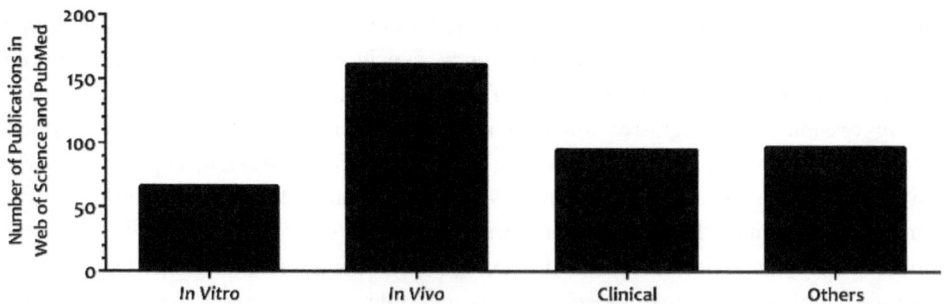

Fig. 1.1 Total number of articles published in the field of brain PBM therapy from 1967 to December 2020

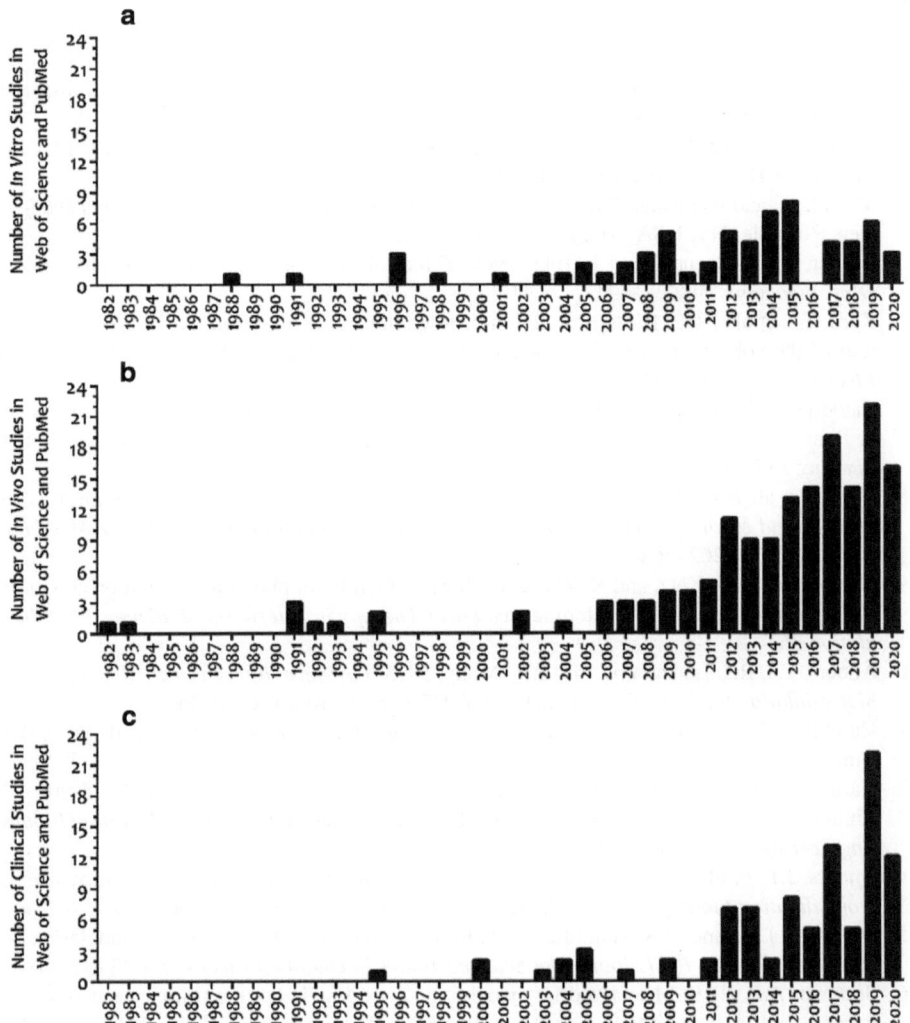

Fig. 1.2 Number of **a** in vitro, **b** in vivo, and **c** clinical studies published in the field of brain PBM therapy from 1967 to December 2020

functions were included. The beneficial effects of laser acupuncture or intravenous laser blood irradiation therapy on the brain are also reviewed. It should be mentioned that the neurotherapeutic applications of bright light therapy using broadband wavelengths via the eyes or ear canal, which are based on the phototransduction mechanisms through the light-sensitive opsins, are not discussed.

References

1. Einstein, A. 1905. On a heuristic point of view concerning the production and transformation of light. *Annalen der Physik* 1–18.
2. Einstein, A. 1917. Zur quantentheorie der strahlung. *Physikalishce Zeitschrift* 18: 121–128.
3. Maiman, T.H. 1960. *Stimulated optical radiation in ruby.*
4. Mester, A., and A. Mester. 2017. *The history of photobiomodulation: Endre Mester (1903–1984).* New Rochelle, NY, USA: Mary Ann Liebert, Inc.
5. Mester, E., B. Szende, and J. Tota. 1967. Effect of laser on hair growth of mice. *Kísérletes Orvostudomány* 19: 628–631.
6. Karu, T.Ĭ, et al. 1982. Biological action of low-intensity visible light on HeLa cells as a function of the coherence, dose, wavelength, and irradiation regime. *Soviet Journal of Quantum Electronics* 12 (9): 1134.
7. Calderhead, R.G. 1991. Watts a joule: On the importance of accurate and correct reporting of laser parameters in low reactive-level laser therapy and photobioactivation research. *Laser Therapy* 3 (4): 177–182.
8. Yu, W., et al. 1997. Improvement of host response to sepsis by photobiomodulation. *Lasers in Surgery and Medicine: The Official Journal of the American Society for Laser Medicine and Surgery* 21 (3): 262–268.
9. Anders, J.J., A.K. Ketz, and X. Wu. 2017. Basic principles of photobiomodulation and its effects at the cellular, tissue, and system levels. *Laser Therapy in Veterinary Medicine: Photobiomodulation* 36.
10. Anders, J.J., R.J. Lanzafame, and P.R. Arany. 2015. *Low-level light/laser therapy versus photobiomodulation therapy.* New Rochelle, NY, USA: Mary Ann Liebert, Inc.
11. Hamblin, M.R., et al. 2018. *Low-level light therapy: Photobiomodulation.* SPIE Press Bellingham.
12. Karu, T.I. 2003. Low-power laser therapy. *Biomedical Photonics Handbook* 48: 1–20.
13. Chung, H., et al. 2012. The nuts and bolts of low-level laser (light) therapy. *Annals of Biomedical Engineering* 40 (2): 516–533.
14. Anders, J.J., et al. 2019. *Light-emitting diode therapy and low-level light therapy are photobiomodulation therapy.* New Rochelle, NY, USA: Mary Ann Liebert, Inc.
15. de Freitas, L.F., and M.R. Hamblin. 2016. Proposed mechanisms of photobiomodulation or low-level light therapy. *IEEE Journal of Selected Topics in Quantum Electronics* 22 (3): 348–364.
16. Sommer, A.P. 2019. Mitochondrial cytochrome c oxidase is not the primary acceptor for near infrared light—It is mitochondrial bound water: The principles of low-level light therapy. *Annals of Translational Medicine.*
17. Adar, F., and T. Yonetani. 1978. Resonance Raman spectra of cytochrome oxidase. Evidence for photoreduction by laser photons in resonance with the Soret band. *Biochimica et Biophysica Acta (BBA)-Bioenergetics* 502 (1): 80–86.
18. Karu, T. 1988. Molecular mechanism of the therapeutic effect of low-intensity laser radiation. *Lasers in the Life Sciences* 2 (1): 53–74.
19. Hamblin, M.R. 2018. Photobiomodulation for traumatic brain injury and stroke. *Journal of Neuroscience Research* 96 (4): 731–743.
20. Mussttaf, R.A., D.F. Jenkins, and A.N. Jha. 2019. Assessing the impact of low level laser therapy (LLLT) on biological systems: A review. *International Journal of Radiation Biology* 1–24.
21. Zein, R., W. Selting, and M.R. Hamblin. 2018. Review of light parameters and photobiomodulation efficacy: Dive into complexity. *Journal of Biomedical Optics* 23 (12): 120901.

22. Huang, Y.-Y., et al. 2009. Biphasic dose response in low level light therapy. *Dose-Response* 7 (4).
23. Salehpour, F., et al. 2018. Brain photobiomodulation therapy: A narrative review. *Molecular Neurobiology* 55 (8): 6601–6636.
24. Ramezani, F., et al. 2021. Mechanistic aspects of photobiomodulation therapy in the nervous system. *Lasers in Medical Science* 1–8.
25. Hamblin, M.R. 2016. Shining light on the head: Photobiomodulation for brain disorders. *BBA Clinical* 6: 113–124.
26. Wade, P.D., J. Taylor, and P. Siekevitz. 1988. Mammalian cerebral cortical tissue responds to low-intensity visible light. *Proceedings of the National Academy of Sciences* 85 (23): 9322–9326.
27. Shen-Zeng, X.-J., S. Lin, and L. Wang. 1982. Effects of a low power laser beam guided by optic fiber on rat brain striatal monoamines and amino acids. *Neuroscience Letters* 32: 203–208.
28. Rochkind, S. 2009. Review of 30-years experience: Laser phototherapy in neuroscience and neurosurgery part II-nerve cells, brain and spinal cord. *Laser Therapy* 18 (3): 127–136.
29. Rochkind, S. 2009. Review of 30-years experience: Laser phototherapy in neuroscience and neurosurgery part I-muscle and nerve. *Laser Therapy* 18 (1): 27–38.
30. Rochkind, S., and G.E. Ouaknine. 1992. New trend in neuroscience: Low-power laser effect on peripheral and central nervous system (basic science, preclinical and clinical studies). *Neurological Research* 14 (1): 2–11.
31. Rochkind, S. 2017. Photobiomodulation in neuroscience: A summary of personal experience. *Photomedicine and Laser Surgery* 35 (11): 604–615.
32. Hashmi, J.T., et al. 2010. Role of low-level laser therapy in neurorehabilitation. *Pm&r* 2: S292–S305.
33. Lampl, Y., et al. 2007. Infrared laser therapy for ischemic stroke: A new treatment strategy: Results of the NeuroThera Effectiveness and Safety Trial-1 (NEST-1). *Stroke* 38 (6): 1843–1849.
34. Hennessy, M., and M.R. Hamblin. 2016. Photobiomodulation and the brain: A new paradigm. *Journal of Optics* 19 (1): 013003.
35. Thunshelle, C., and M.R. Hamblin. 2016. Transcranial low-level laser (light) therapy for brain injury. *Photomedicine and Laser Surgery* 34 (12): 587–598.
36. dos Santos, J.G.R.P., W.S. Paiva, and M.J. Teixeira. 2018. Transcranial light-emitting diode therapy for neuropsychological improvement after traumatic brain injury: A new perspective for diffuse axonal lesion management. *Medical Devices (Auckland, NZ)* 11: 139.
37. Johnstone, D.M., et al. 2016. Turning on lights to stop neurodegeneration: The potential of near infrared light therapy in Alzheimer's and Parkinson's disease. *Frontiers in Neuroscience* 9: 500.
38. Hamblin, M.R. 2019. Photobiomodulation for Alzheimer's disease: Has the light dawned? In *Photonics*. Multidisciplinary Digital Publishing Institute.
39. Jack, C. 2017. Treating cognitive impairment with transcranial low level laser therapy. *Journal of Photochemistry and Photobiology B: Biology* 168: 149–155.
40. Litscher, D., and G. Litscher. 2014. Laser therapy and dementia: A database analysis and future aspects on LED-based systems. *International Journal of Photoenergy* 2014.
41. Caldieraro, M.A., and P. Cassano. 2018. Transcranial and systemic photobiomodulation for major depressive disorder: A systematic review of efficacy, tolerability and biological mechanisms. *Journal of Affective Disorders*.
42. Fonseca, M., et al. 2019. Transcranial photobiomodulation in the treatment of major depression. *Clinical Psychiatry* 5 (1): 1.
43. Askalsky, P., and D.V. Iosifescu. 2019. Transcranial photobiomodulation for the management of depression: Current perspectives. *Neuropsychiatric Disease and Treatment* 15: 3255.

44. Cassano, P., et al. 2016. Review of transcranial photobiomodulation for major depressive disor-
    der: Targeting brain metabolism, inflammation, oxidative stress, and neurogenesis. *Neuropho-
    tonics* 3 (3): 031404.
45. Salehpour, F., and S.H. Rasta. 2017. The potential of transcranial photobiomodulation therapy
    for treatment of major depressive disorder. *Reviews in the Neurosciences* 28 (4): 441–453.
46. Salehpour, F., et al. 2019. Transcranial photobiomodulation improves cognitive performance
    in young healthy adults: A systematic review and meta-analysis. *Photobiomodulation, Pho-
    tomedicine, and Laser Surgery.*
47. Vo-Dinh, T. 2014. *Biomedical photonics handbook, 3 volume set.* CRC Press.
48. Pierroz, V., and M. Folcher. 2018. From photobiolumination to optogenerapy. Recent advances
    in NIR light photomedicine applications. *Journal of Molecular and Genetic Medicine* 2 (10).
49. Giordano, J., et al. 2017. Mechanisms and effects of transcranial direct current stimulation.
    Dose-Response 15 (1): 1559325816685467.

Biophysical and Safety Aspects of Brain Photobiomodulation

2.1 Light Sources

A procedure called heliotherapy, or treatment with sunlight, is one of the oldest therapeutic interventions to treat many diseases, including psoriasis, infections, rickets, etc. The spectrum of solar radiation on the Earth's surface contains a wide range of wavelengths ranging from 315 to 2500 nm. Sunlight is a non-coherent and multidirectional light source and because its peak emission is in the green spectrum (around 500 nm), most of its energy does not penetrate more than 1 mm into the dermis. Photons at 400 nm and 600 nm have maximum penetration depths of 0.1 mm and 0.5 mm into the body skin, respectively [1]. In addition, almost 2% of the harmful ultraviolet-B radiation (280–315 nm) reaches the Earth's surface, so delivering optimum PBM fluences of red/near-infrared (NIR) light to an individual body, may also simultaneously cause erythema or sunburn [2]. Given this, polychromatic sunlight irradiation is not thought to be an effective treatment strategy for modern photobiomodulation (PBM) procedures. Nevertheless, it has been postulated that the onset of human hairlessness at 2 million years ago might have provided evolutionary benefits by allowing more red/NIR energy of solar radiation to penetrate the scalp/skull and stimulate human brain function [3]. In 1893, Niels Ryberg Finsen developed monochromatic red-filtered sunlight and realized that smallpox could be treated with pure red light [4]. Apparently, this was the first evidence of using monochromatic PBM therapy in modern medicine.

Nowadays, light sources designed for PBM purposes can be coherent sources (lasers), or non-coherent sources (light-emitting diodes, LEDs), or even broad-spectrum filtered lamps. The first modern PBM treatment was conducted using lasers, which were based on gases or crystals as the lasing medium [2]. Currently, the most common laser sources used for PBM applications are diode lasers emitting between 630 and 980 nm, but other lasers have been used: **(I)** gas lasers (e.g., argon [variable visible λ of 454.6–528.7 nm], krypton [variable λ of 521, 530, 568, and 647 nm], He–Ne [632.8 nm], and CO_2 [10600 nm]); **(II)** semiconductor lasers (e.g., GaAlInP [variable λ of 635, 650–660, 670, and 760 nm], GaAlAs [variable λ of 785, 808, and 1064 nm], GaAs [904 nm], InGaAs [980 nm]); **(III)** solid state lasers (e.g., Ruby [694 nm], Nd:YLF [1047/1053 nm], Nd:Glass [1060 nm], Nd:YAG [1064 nm], Nd:YAP [1080 nm]), Ho:YAG [2100 nm], Er:YAG [2940 nm]); and **(IV)** frequency-doubled Nd:YAG laser, also called a potassium-titanyl-phosphate (KTP) laser [532 nm] [5].

Laser light has many characteristics including monochromaticity, coherence, and polarization. The unidirectional beam produced by laser devices allows delivery of relatively high levels of concentrated energy into the circumscribed area. Besides, lasers can easily be coupled into fiber optics, and by modifying the beam width, the desired area of tissue can be irradiated. Lasers can also be operated in either continuous wave (CW) or pulsed wave (PW) modes. On the other hand, some laser devices are not portable. Also, lasers are capable of delivering photons at a high power density, which could result in painful localized skin warming and even potentially induce tissue damage (burns) because of rapid heat production [6].

LEDs are semiconductor light sources based on the electroluminescence principle. In 1962, the first practical GaAs infrared (IR) LEDs were developed, however, it took 4 years for the patent to be issued by U.S. patent office due to the interference with patents submitted by others [7]. In the early days of PBM technology, it was supposed that the coherence and monochromatic nature of laser light were essential to achieve the therapeutic benefits, but by the 1980s, this claim was challenged by successful application of LED sources to treat various medical conditions [8, 9]. Nowadays, LEDs can be produced from a variety of inorganic semiconductor materials, and therefore they can produce many different wavelengths over a very wide spectral range from the ultraviolet to the IR. Typically, LEDs emit quasi-monochromatic light generally with a 30 nm bandwidth (full width half maximum, FWHM). The area of each individual LED surface is often smaller than 1 mm^2 [10]. LEDs offer an effective and safe alternative to lasers for many reasons [11]: **(I)** LED arrays are compact, portable, and can easily be used in medical clinics or at home with minimal training required and no safety concerns; **(II)** the cost per mW of output power for LEDs is $\sim$100 times lower than that for lasers; **(III)** LEDs have received Food and Drug Administration (FDA) clearance for human applications; **(IV)** LEDs produce insignificant amounts of tissue heating at the target region which decreases the risk of thermal damage; **(V)** LEDs can be mounted on a support to form evenly spaced large arrays, which is useful when the target tissue has a large surface area; **(VI)** using integrated optical components, the illumination pattern of LEDs can also be shaped to fit the tissue surface; **(VII)** organic LEDs (OLEDs) have a narrower bandwidth than typical LEDs and can be fabricated on a flexible substrate with a large area (>1 cm^2). With higher light generation efficiency, OLEDs may emerge as a promising technology for next-generation PBM.

There is some controversy in the literature regarding as to whether LEDs are as effective as lasers for PBM purposes [12, 13]. The main focus of this debate is on the differences between light parameters for lasers and LEDs (e.g., coherence, monochromaticity, and penetration depth), which may result in an observable and measurable difference in treatment outcomes [11].

Finally, it should also be added that in addition to lasers and LEDs, other artificial light sources including lamps that are filtered by band-pass filters are occasionally used in PBM treatments [11, 14]. However, these sources can potentially cause tissue thermal damage because they consist of broad wavelength spectrum light. Furthermore, for each specific wavelength contained in a broadband source, the estimation of the actual delivered light dose received by the tissue is an important concern that should be considered [15].

2.2 Dosimetry and Treatment Parameters Involved in Photobiomodulation

Historically, laboratory studies have identified PBM parameters using cell cultures and small animal models. These studies generally applied low levels of irradiance and fluence which worked well for superficial tissues [2]. But, when researchers began to translate these findings into clinical use for the treatment of deeper structures, they used the same parameters with ineffective or negative outcomes. In those early days, some negative reports were published, and it was thought that there was insufficient evidence to propose PBM for clinical application. Nowadays, it has been realized that these negative studies were due to the lack of knowledge of the optimum dosimetry/treatment parameters for transcutaneous irradiation of deeper structures. The wavelength of light and the optical properties of tissue are the main issues that influence the depth of light penetration. It is essential for the clinician to consider that one wavelength and one set of dosimetry/ treatment parameters will not be effective for the treatment of all medical conditions [16].

In contrast to pharmacology, in which the drug dose is generally the main determinant of the effect, PBM is dependent on several interrelated parameters. Since PBM is a light-driven modality, its results depend on various device and dosimetry-related factors [17, 18]. Over the years, photobiologists and physicists have attempted to establish the range of dosimetry/treatment parameters that lead to successful PBM outcomes. A substantial body of evidence has concluded that a wide range of dosimetric variables can actually affect the chances of success, such as wavelength, irradiance, fluence, total delivered energy, pulsing mode, and beam spot size [19, 20]. Apparently, in order to achieve effective results with PBM for any specific application, each of the aforementioned parameters must be adjusted within a certain range of values. Moreover, other less important factors such as coherence, irradiation time, delivery method (headpiece probe or fiber optics), application method (contact, scanning, projection, or pressure), and treatment repetition regimen have been shown to partly contribute to treatment outcomes [21, 22]. This considerable level of complexity derived from multiple variables has stymied the development of definitive and consensus PBM protocols for each medical condition. Besides, in many of the historical PBM studies, the main device and dosage parameters were omitted or incorrectly reported, which made it difficult to reproduce the treatment protocols by following literature studies [23, 24]. Nevertheless, nowadays, besides the dosage guidelines recommended by the World Association of Laser Therapy (WALT) (https://waltza.co.za/docume ntation-links/recommendations/dosage-recommendations/), a recent comprehensive study has made effort to address the best PBM factors that produce positive results in diverse circumstances [25].

In the following sub-sections, we will explain the terminology used in PBM in terms of optical dosimetry (e.g., coherence, wavelength, power, irradiance, fluence, photon deposition, dose, pulsing mode, beam shape, and spot size) and treatment parameters (e.g., irradiation site, irradiation time, and treatment repetition regimen). Furthermore, their impact on the performance of PBM will be discussed.

2.2.1 Coherence

Coherence is an important difference between laser light and LED light. During the stimulated emission process, the phase of the emitted laser photons has a relation to each other, which is interpreted as the coherence property. Coherent light is described by its spatial and temporal coherence synchronization. Spatial coherence describes a relationship between the phase of a laser beam at two different geometrical points, whereas temporal coherence describes a relationship between the phase of a laser beam at a specific point but at different times. The coherence property allows lasers to have a bandwidth of 1 nm or less [26]. Laser devices produce light with different coherence lengths depending on the exact type of lasing medium. Laser coherence lengths can vary from several meters (for He–Ne laser) to a few millimeters (for diode lasers) [11]. Given this, it has been suggested that because the coherence lengths are comparatively longer for laser light than non-coherent light sources (e.g., LEDs and lamps), a larger volume of tissues/organs may be irradiated using monochromatic laser devices [27].

Upon the interaction between coherent laser photons with biological tissues, small geometrical imperfections in the subcellular organelles could cause the individual wavefronts to interfere with each other resulting in the mutual interference pattern known as laser speckle. The size of each laser speckle is linked to the light wavelength, so that for example, for light in the visible spectrum the diameters of the speckles are less than 1 μm, which is almost in the size range of individual cellular mitochondria [11]. Given this, some studies have suggested that laser speckles induced by coherent light play a significant role in PBM interaction with cells/organelles and have some advantages in the stimulation of mitochondria in comparison to non-coherent LED light at the same wavelengths and intensity [28, 29]. Nonetheless, most evidence has suggested that the coherence property is not needed for the photobiological effects in PBM [30–32]. As a support for the latter claim, a recent comparative review study on lasers/LEDs for PBM concluded that PBM is not dependent on coherence [11].

2.2.2 Wavelength

Wavelength refers to the distance between two points/particles in the same phase of two consecutive electromagnetic waves. The light wavelength is typically denoted by λ, which has units of [nm]. The λ is equal to the speed of a wave (V) divided by its frequency (ν), so:

$$\lambda \, (nm) \, = \, V \, [m/s]/\nu \, [Hz] \tag{2.1}$$

Optical electromagnetic radiation is divided into three spectral regions, namely, ultraviolet (100–400 nm), visible (400–750 nm), and IR (750 nm−1 mm). The visible spectrum is subdivided into five wavebands, blue (400–495 nm), green (495–570 nm), yellow (570–590 nm), orange (590–620 nm), and red light (620–750 nm). The IR spectrum is also subdivided into sub-sections, including NIR (750–1400 nm), mid-IR (1400–3000 nm), and far-IR (3000 nm−1 mm). Of note, the aforementioned wavelength values and ranges depend on various standards set down and can vary from one organization to another.

Monochromaticity refers to the spectral purity of the light beam, which is linked to the variation of light intensity over the bandwidth of the emission peak. In the context of monochromatic laser light, the maximum light intensity is obtained at a specific wavelength at which almost all the photons are emitted. Laser devices can produce very narrow monochromatic light with FWHM of 10^{-3} nm. In contrast, LEDs emit quasi-monochromatic light with a broader bandwidth (FWHM of 4–30 nm), so the spectral bandwidth and center wavelength should be specified. Polychromatic broadband light sources such as white lamps (400–700 nm) and water-filtered IR-A (760–1400 nm) have also been applied in PBM studies [11, 33]. In fact, it is believed that monochromaticity is not a very critical factor in PBM, however, it has been suggested that it can potentially allow for higher specificity and molecular-targeted PBM [34, 35]. It has been proposed that if one wavelength is stimulatory and another wavelength is inhibitory, then using them together may cancel the beneficial effects. Nevertheless, with the rapid progress of science and technology, multi-wavelength LED devices have been developed to meet the needs of various combinations of PBM. These devices can emit photons at different wavelengths independently or jointly depending on clinical purposes [10].

There are a couple of points to note here about the role of wavelength in PBM. Although PBM mostly employs wavelengths in so-called "therapeutic optical window" that lies between 600 and 1300 nm, blue and green wavelengths have gained attention because of their potential to stimulate other cellular photoacceptors instead of just CCO [36, 37]. NIR wavelengths (800–1100 nm) are preferable for deep tissue PBM, due to lower absorption and scattering by tissue chromophores. The wavelength range between 700 and 780 nm has been found to have limited therapeutic benefits so these wavelengths are not much used in PBM [38].

2.2.3 Power

Power or radiant flux (Φ) is described as the amount of light energy (Q), which is transferred per unit of time (t). Power has unit of Watts [W] and is calculated as:

$$\Phi\,(\text{W}) = Q\,[\text{J}]/t\,[\text{s}] \tag{2.2}$$

The power sometimes is called total output power. However, if the light beam is pulsed rather than continuous, the "peak power" is the energy delivered in each pulse divided by the duration of the pulse. The "average power" is usually the effective power measured by a power meter, and is calculated as follows:

$$\text{Average power [W]} = \text{peak power [W]} \times \text{pulse width [s]} \times \text{pulse frequency [Hz]} \tag{2.3}$$

In the PBM literature, the output or average power is usually reported in manufacturers' information or is directly measured by photodiodes or power meters. For PBM purposes, the typically applied output power can vary between 1 and 500 mW in order not to allow thermal damage [39]. However, recently, high-power Class IV lasers with a total output power of >5 W have been tested in clinical PBM studies [40, 41]. Nowadays, LEDs with a power of 3 W per each diode are also available.

2.2.4 Irradiance

The irradiance (E) is described as the output power (Φ) received per unit of area (A). The irradiance has unit of [W/cm^2] and is calculated as:

$$\text{Irradiance}\left[\text{W/cm}^2\right] = \text{output power [W]/beam spot size}\left[\text{cm}^2\right] \tag{2.4}$$

The irradiance is often called power density or intensity in the PBM literature, which technically is incorrect terminology. For PBM purposes, the applied irradiance can vary between 5 mW/cm^2 and 5 W/cm^2 at the culture surface or target tissue [42]. However, if a high irradiance is used (>1 W/cm^2), it is common practice to continuously move the spot around to avoid burning the tissue. In fact, irradiance values outside of a defined range do not have prominent photobiomodulatory effects. Values below the lower threshold of ∼0.5 mW/cm^2 would be no different from average room lighting (∼1 mW/cm^2), and much less than solar radiation on the earth's surface On the other hand, the irradiance values over than $\sim$100 mW/cm^2 at 400–500 nm, $\sim$300 mW/cm^2 at 600–700 nm, or $\sim$750 mW/cm^2 at 800–900 nm, may produce unacceptable tissue heating and induce possible photothermal effects [25]. For instance, Class IV lasers often produce irradiance >750 mW/cm^2, and active surface cooling units or constant movement techniques are recommended for the clinical applications of this type of laser [40]. However, LED devices have an irradiance in the range of 10–50 mW/cm^2, which is quite tolerable for irradiated tissue [10]. In the context of LEDs, it should be noted that manufacturers' quoted irradiance values are typically measured at the aperture, however, because beams from LEDs are highly divergent, the quoted irradiance could not be used for in vitro studies where the light beam may be projected to the cell culture and should be measured with a power meter [17].

It has been postulated by some authors that depending on the size of the area illuminated by a specific light power, the irradiance and the subsequent induced cellular effects might be different [25]. For instance, 100 mW delivered through an optical fiber with 300 μm diameter will produce a tiny spot with an irradiance of 141,000 mW/cm^2, while the same 100 mW delivered through a handpiece probe with a 1 cm diameter will produce an irradiance of 127 mW/cm^2 which is more than three orders of magnitude lower than previous one. Mitochondria are one of the most important sources of reactive oxygen species (ROS) in the mammalian cells. Indeed, in physiological condition, produced superoxide inside mitochondria is dismutated to hydrogen peroxide (H_2O_2) by manganese-dependent superoxide dismutase (MnSOD). But, if the superoxide levels exceed the capacity of MnSOD to detoxify it, accumulated superoxide anion (O_2^-) within the mitochondria matrix can cause mitochondrial/cell damage. Therefore, it could be claimed that the ability of MnSOD to detoxify superoxide may depend on the rate at which O_2^- is generated, which in turn may depend on the rate at which photons are delivered, that is, the irradiance [43]. This claim may be supported by a number of reports, which have demonstrated that the positive biological effects of PBM may depend on the irradiance of the light, and not on the fluence [44–46].

2.2.5 Fluence (Radiant Exposure)

The radiant exposure or more commonly termed fluence (H) is described as the radiant energy (Q) received by a unit of area (A). The fluence has a unit of [J/cm^2] and is calculated as:

$$\text{Fluence} \left[\text{J/cm}^2\right] = \text{energy [J]/beam spot size} \left[\text{cm}^2\right] \tag{2.5}$$

The fluence is often called energy density or dose in the PBM literature, which technically is incorrect terminology. The applied fluence is commonly adjusted according to the severity of the pathology, depth of the target tissue, and the desired treatment outcomes. For PBM purposes, fluences can vary between 0.03 and 100 J/cm^2 at the culture surface or target tissue [47].

As can be seen from Eq. (2.5), the fluence can be varied by changing either the irradiance or the irradiation time. According to the Bunsen-Roscoe law of reciprocity (called the second law of photobiology), the photobiological effects are independent of irradiance and irradiation time, with the responses being directly relative to the fluence (or total number of photons) delivered. In other words, the therapeutic effects should be the same for all combinations of irradiance and irradiation time that deliver the same number of photons [48]. However, this law has been refuted many times by studies that have shown that at the same fluence, variations in either the irradiance or irradiation time may lead to different PBM effects (e.g., 1 min exposure at 100 mW/cm^2 does not demonstrate the same biological effects compared with 10 min exposure at 10 mW/cm^2) [35, 49–51]. For example, it has been shown that PBM using a very low irradiance for a very long time, or a high irradiance for too short irradiation time may not result in any positive effects [39]. Moreover, applying a very high irradiance or using an excessively long delivery time may cause thermal or even sometimes bioinhibitory effects, rather than photobiomodulatory effects [52, 53]. Recent advances have been made in understanding the mechanisms of PBM dose–response at a cellular level [51]. Many PBM scientists have focused on fluence as the key metric of dose, while others favor irradiance or irradiation time. Further insights into the dose–response issue will be discussed in the next chapter, in Sect. 3.3. Taken together, it could be suggested that PBM may only act as an effective therapeutic strategy, within a certain window of dose parameters that should be defined for different cell types and different medical conditions. An overview of a large number of experimental studies suggests that the delivered fluence is a more important factor in determining the success or failure of the in vitro studies, rather than the irradiance applied. However, for in vivo studies, both light fluence and irradiance could be critical in determining the treatment outcomes [25].

It should also be noted that the three-dimensional energy deposition with a unit of [J/cm^3], which is defined as the total amount of energy delivered per unit volume, has been used in some optical simulation studies for showing the deposited flux into each voxel of tissue [54].

2.2.6 Radiant Energy (Dose)

The radiant energy, more commonly termed radiation dose or, simply, dose (Q), is calculated for a single photon as the product of Planck's constant (h) and the photon frequency (ν). Energy has a unit of Joules [J]. In terms of biophotonics applications, the dose is calculated as:

$$\text{Radiant energy [J]} = \text{power [W]} \times \text{time [s]} \qquad (2.6)$$

The radiant energy is often assumed to be the key PBM parameter, which is sufficient alone to determine treatment efficacy, assuming reciprocity (an inverse relationship between power and time). Thus, an infinite combination of values of power and irradiation time could result in the same radiant energy. Nevertheless, it has been suggested that amongst the many dosimetric parameters of PBM, the total radiant energy is the most useful measure of dose and can be as large as several thousand Joules when fairly large surfaces are irradiated [55].

2.2.7   Pulse Mode of Operation

There are two basic types of electromagnetic radiation, namely, continuous (CW) and pulsed (PW) modes. A CW mode is a wave of constant amplitude and frequency. In the context of CW lasers, the source must be excited constantly leading to the emission of a continuous stream of photons. This requires a constant supply of energy from the laser pump source. CW operating lasers generally have average power values lower than the peak power of PW mode lasers. As example, gas lasers (e.g., He–Ne and CO_2), crystal lasers (e.g., Nd:YAG), and semiconductor lasers can continuously emit light.

PW operating lasers refer to any laser not classified as CW, which emits light in the form of pulses. There are a number of pulse parameters that are specified for PW light sources, including pulse width (or pulse duration or ON time), pulse interval (or OFF time), pulse frequency (or pulse repetition rate), and duty cycle (fraction of time ON). The pulse width and pulse interval are measured in seconds, while pulse frequency has a unit of Hz. Also, the duty cycle is a unitless fractional number [%]. These parameters are related by the following equation:

$$\text{Duty cycle } [\%] = \text{pulse frequency } [\text{Hz}] \times \text{pulse width } [\text{s}] \qquad (2.7)$$

As mentioned earlier, for PW mode light, peak power measures the light intensity during the ON time and is related to the average power and duty cycle by:

$$\text{Peak power } [\text{W}] = \text{average power } [\text{W}]/\text{duty cycle } [\%] \qquad (2.8)$$

As can be seen from Eqs. (2.7) and (2.8), if the irradiation is pulsed, then the average power is only a fraction of the peak power because of the existence of the on/off periods. Therefore, only a fraction of the fluence would be delivered by PW compared with CW operation mode at the same irradiance and irradiation time, and this may affect the efficacy of PBM therapy. If the pulse duration is short (nanoseconds), the peak irradiance can be very high (tens of W) compared to only mW of average power. Two out of three parameters (pulse width, pulse frequency, and duty cycle) must be provided in order to precisely define the dosimetric parameters of the treatment protocol.

Besides the naturally pulsed laser sources, there is another way to generate PW light using pulsed power supply generated by a laser driver to "chop" the beam. The generated wave is described as "chopped" or "gated" beam. Although this technique seems to be effective for producing the chopped PW, its biological effects may not equal to the effects of naturally pulsed lasers [56]. This technology can also be applied to produce pulsed LEDs [57].

In terms of comparison between both PW and CW operation modes, it is believed that PW delivery may decrease thermal effects in the biological tissues because the OFF interval allows the heat to disperse within the tissue. Some evidence also suggests that PW mode is more effective than CW in terms of biological effects. In addition, another common claim is that PW light penetrates deeper than CW light [20]. Besides the light wavelength and tissue optical properties, the light penetration depth into tissue is known to be directly related to the irradiance at the surface of the tissue. For example, a super-pulsed GaAs laser (904–905 nm) can emit light with a typical average power of 60 mW and a peak power of about 20 W, which may increase the penetration depth [58]. However, it should be paid noted that when using super-pulsed lasers, an actual fluence that penetrates to a depth of tissues will be only a fraction since the pulses are only "on" in a small portion of the time; for instance, this time is around 100–200 ns for GaAs laser. So, this should be taken into account in calculating the dosimetry of PBM therapy using super-pulsed lasers.

2.2.8 Beam Area and Beam Profile

Beam area is usually called the "spot size" in the PBM literature. Beam area is measured in square centimeters (cm^2). Divergence or dispersion describes the angular spread of the laser beam as the distance from the source increases. Divergence is measured in units of milliradians (mrad) or degrees of angle. The geometrical shape of the laser medium and the cavity influence the divergence of the beam. Transverse electromagnetic (TEM) modes are very important characteristics for laser beam divergence. Laser beam divergence differs from one mode structure to another. The TEM mode patterns are defined by a combination of a Gaussian spatial beam profile with a Laguerre polynomial. Beam profile or aperture can be parallel, convergent, or divergent. Ideally, the Gaussian mode (TEM$_{00}$) is considered to be optimum for the laser beam because of the predictable variation in irradiance over the output area, however, laser beams generally exhibit fluctuations in the beam structure. The beam profile can influence the efficiency and tissue penetration depth in PBM [59]. Nevertheless, in most of the PBM studies, the actual spatial distribution across the active beam area is not reported [17]. Depending on the type of light source, the beam profile can be non-uniform or uniform. Lasers can emit non-circular elliptical or circular Gaussian beams. The output irradiance of most PBM lasers is Gaussian in spatial profile. Thus, cells located in the center of the incident beam receive more light than those on the margin of the irradiation spot, which has been suggested as an explanation for some observed negative effects of higher fluence PBM using Gaussian beam handpieces [60]. For the Gaussian beam profile, the beam area is defined as the diameter at which the intensity is $1/e^2$ of the maximum intensity. At this radial distance from the center of the spot, the irradiance is lower by about 13% relative to the central peak irradiance. For instance, if cells at the center of the beam received a fluence of 4 J/cm^2, those at the periphery would receive 0.26 J/cm^2. In contrast to a heterogeneous Gaussian profile, a homogeneous flat-top beam is better for spatially uniform irradiation, especially for in vitro applications. It should be noted that the beam uniformity is less important for in vivo and clinical applications due to the fact that light scattering by tissue diffuses the photons, and this can make heterogeneous light sources less critical for clinical efficacy.

2.2.9 Irradiation Time

In early PBM procedures, very low-power He–Ne laser devices were used, so that irradiation times to deliver therapeutic fluencies could be very long. Nowadays, using more powerful light sources, the treatment time can be reduced to reasonable levels. PBM irradiation times for human transcranial applications are typically about 5–20 min [61], but for intranasal applications, as well as intravascular and acupuncture procedures, the treatment times can be in the range of 20–30 min [62].

2.2.10 Treatment Repetition Regimen

The treatment repetition regimen refers to the number of treatments and the treatment intervals, which are very important parameters when designing PBM protocols [63]. In some studies using transcranial PBM, a single treatment has been shown not to have long-lasting effects [64]. On the other hand, too many or too frequent irradiation sessions have also been reported to result in ineffective or negative outcomes in animal models [65]. Many of the earlier animal studies carried out for stroke or TBI used a single treatment session, delivered within 24 h of the traumatic event. The promising results obtained by this regimen encouraged human clinical trials for acute stroke, again using a single session of PBM [66]. In this respect, although the NEST-1 and NEST-2 trials showed success with a single treatment session, the large NEST-3 trial did not. Would NEST-3 have been successful if it had used several PBM sessions repeated in the days following onset of the stroke? Even though the trial of PBM for depression and anxiety only used a single session, and positive outcomes were obtained at examination 2 weeks later, the benefit had largely disappeared by 4 weeks. Given these, it is unreasonable to expect a single session of PBM in humans to have the best long-lasting effects, and treatments should probably be repeated for several days in case of acute events such as TBI or stroke, and for more extended periods in the case of chronic diseases such as AD, PD, depression, and post-traumatic stress disorder (PTSD). Nevertheless, it is noteworthy that it might be possible to overdo the number of sessions of treatment, since a study in a mouse TBI model found that 14 daily sessions of PBM were indeed too much, while 7 daily sessions were adequate [65].

2.3 Safety Concerns

Under normal environmental condition, the human eye does not easily perceive photons outside the visible band at wavelengths <400 nm and >700 nm [67]. Thus, this increases the risk of unintentionally exposing the naked eyes to excessive doses of ultraviolet or NIR radiation. In general, visible and invisible light from the lasers and LEDs used in PBM therapy should not be directly aimed into the naked eyes. However, it should be noted that directly delivering red/NIR light into the eyes has been shown to have benefits for some ophthalmological diseases, such as dry age-related macular degeneration. The patient, therapist, and any other operators should generally wear safety goggles with a suitable optical density for the power and wavelength applied, before and during PBM procedures. In addition, a consent form including details about the safety issues should be signed by patients or the family at the beginning of the procedure.

It should be noted that since, in contrast to other non-drug therapies (e.g., transcranial magnetic stimulation [TMS], deep brain stimulation [DBS], transcranial direct current stimulation [tDCS], and electroconvulsive therapy [ECT]), which are usually administered in a clinical facility, transcranial LED-based PBM procedures can also be applied at home by the patient themselves or by a caregiver. In this case, after proper training of the patient or the family in terms of the light dosage and treatment regimen, it is essential that they also must be informed by the therapist about all safety issues and risks.

Given the lack of evidence of safety and any risks for the embryo and fetus, PBM therapy should not be used during pregnancy. PBM therapy is also not recommended for women who are lactating. There are some other precautions that should be considered including not exposing the heart region in patients with cardiac disease, and avoiding hemorrhagic regions and the gonads. PBM therapy should be applied with caution in cancer patients because it may possibly stimulate tumor growth.

2.3.1 Laser Classes

According to the Federal Laser Product Performance Standard, all laser systems are classified into four classes, depending on the potential hazard (mostly the risk of eye damage) associated with the device operation [67]. Class I lasers do not emit hazardous levels of radiation and are safe under all conditions of normal application, thus, exempt from user control measures. Class II lasers are divided into two sub-classes, namely, Class IIa and Class IIb. The safety concerns for Class IIa lasers are the same as Class I for all wavelengths and exposure durations except for durations greater than 10^3 s in the visible spectrum between 400 and 700 nm. Class IIb lasers have an output power level less than 1 mW and are capable of producing eye damage through chronic irradiation. Class III lasers are also divided into two sub-classes. Class IIIa lasers have an output power range of 1–5 mW in the visible spectrum. Class IIIa lasers are generally not hazardous when viewed quickly with the naked eye, however, they can potentially cause eye damage when viewed via an optical device. The output power of Class IIIb lasers is typically 5–500 mW. Lasers in this class can cause eye damage by direct observation of the beam but can cause no skin damage. Class IV lasers consist of lasers with output power values greater than 500 mW. Observation of the Class IV laser beam or its reflection from optical instruments and surfaces can cause damage to both the eyes and the skin. The use of Class IV laser-focused spots frequently requires active skin cooling or constant movement to avoid overheating of the skin. The beams derived from this class of lasers can potentially start fires when they are focused on flammable substances and materials. Nowadays, Class IIIa lasers are seldom used, but Class IIIb and IV lasers are often used in the PBM field.

2.3.2 Thermal Effects and Concerns

PBM therapy is photobiological rather than a thermal-based intervention. In other words, PBM therapy is the application of low levels of light energy to achieve a therapeutic effect in a non-thermal and non-invasive manner. However, given that photon absorption by biological tissues can lead to heat generation, the safety of PBM delivered to the head in order to affect the brain is a great concern in clinical settings [68].

Generally, depending on the two factors of light-tissue interaction time and light irradiance (at the surface level or in the interaction volume), the following three forms of light-tissue interaction have been defined: **(I)** photochemical effect (10^1–10^3 s; 10^{-3}–1 W/cm^2); **(II)** photothermal effect (10^{-3}–10^2 s; 1–10^6 W/cm^2); and **(III)** photomechanical effect (10^{-11}–10^{-7} s; 10^8–10^{12} W/cm^2) [69]. In the context of PBM, the red/NIR photon energy is not high enough to break chemical bonds, so that photochemical damage does not typically occur. However, photomechanical damage could result from super-pulsed laser irradiation with short pulses at high average power, which can cause tissue ablation/vaporization, cavitation, and fragmentation of biological tissues. Lastly, photothermal effects resulting in tissue heating and burns are expected when a high amount of photon energy is rapidly delivered to the target point. Indeed, the level of photothermal damage is dependent on various factors such as the area of the tissue irradiated, irradiation time, and particularly on the light irradiance. The processes involved in photothermal interaction are highly complex. The transition to photothermally induced effects starts with the irradiance increasing above a certain threshold. The tissue temperature distribution depends on the absorbed photon energy by the tissue volume, the thermal properties of the tissue (e.g., the heat conductivity and capacity), and the cooling capacity of blood circulation and external heat loss by air flow. The high water content of biological tissues leads to a substantial absorption of NIR wavelengths, mostly beyond 1100 nm, and thus results in higher energy transfer and tissue heating [69]. In an interesting study conducted by Souza-Barros et al. [22], the effect of skin color and light dose on the changes in temperature associated with 635 and 808 nm lasers PBM was investigated. They observed that dark skin heated more than light skin, and the difference between light and dark skin heating was slightly higher for 635 nm compared to 808 nm. In addition, laser irradiation using light doses of 6, 9, and 12 J ($\sim$0.2 W/cm^2; corresponding to $\sim$33, 50, and 67 J/cm^2), resulted in non-painful skin temperature increases of 0.7, 1.2, and 1.6 °C, respectively, across both skin color and wavelength groups [22]. In another study, Joensen et al. [70] also found greater heating for dark skin, but in contrast, they showed skin temperature rises of up to 22 °C for 6–12 J doses of 810 nm laser. It should be noted that laser irradiance may likely justify the painful skin heating in their study, so that they used skin irradiance of 6.37 W/cm^2, which was more than 30-fold greater than irradiance applied in Souza-Barros et al.'s study [22]. Of note, according to the American National Standards Institute (ANSI) guidelines [71], a skin irradiance value of 6.37 W/cm^2 at 810 nm wavelength is higher than the maximum permissible exposure (MPE) for safe laser irradiation to human. It has also been reported that CO_2 laser irradiation ($\sim$0.4 W/cm^2; beam area of 0.126 cm^2) can increase human skin temperature up to 4.3 °C for 65.5 J/cm^2 and to 5.2 °C for 196.5 J/cm^2 [72]. Taken together, the risk of thermal damage from the typical irradiance/dose used in PBM is considered minimal and limited to the skin.

In the context of brain PBM, several animal studies have shown a safety profile of transcranial PBM therapy in terms of thermal effects [73–77]. In a rabbit stroke model, transcranial 808 nm laser irradiation (25 mW/cm^2) increased scalp surface temperature right below the laser probe up to 3 °C. Also, the focal brain temperature directly beneath the probe was elevated by 0.8–1.8 °C during the 10 min irradiation [74]. It has also been shown that CW mode transcranial 808 nm laser (55 mW/cm^2) did not cause any cortical heating in rabbits during the 2 min irradiation, whereas irradiance of 111 mW/cm^2 increased the cortical temperature by 0.5 °C [75]. Transcranial 810 nm laser irradiation (1 W) of the rat head increased temperature of the scalp by 3.6 °C and skull by 2.9 °C during the 4 min irradiation [76]. In another study on rats, both 652 and 830 nm laser irradiation (4.8 W/cm^2) increased brain temperature by $\sim$4.5 °C, when the probe was directly applied to the parietal cortex through the arachnoid membrane. There were no measurable ablation, coagulation or large bleeding in the arachnoid for both tested lasers [78]. The rise in temperature of the mouse brain tissue directly irradiated with a CW mode 808 nm laser was found to be 0.8 °C for 1.6 W/cm^2 and 3.8 °C for 3.2 W/cm^2 irradiance values [73]. Of note, in another study on mice, transcranial 808 nm laser irradiation with 0.56 W/cm^2 (40 mW) and 2.83 W/cm^2 (200 mW) was safe, whereas irradiance of 5.66 W/cm^2 (400 mW) induced scalp drying and scaling due to the high thermal load. Nevertheless, after modification of the irradiation protocol with application of ice in direct contact with the shaved scalp during the 1 min irradiation interval, no further scalp damage was observed [79].

The concern about the risk of tissue heating is even greater when high-power light sources are applied. According to an investigation on ex vivo lamb head tissue [80], transcranial high-power LED irradiation (6 W) at 670/970 nm caused no significant temperature change at the scalp or brain tissue surfaces (<0.3 °C). On the other hand, after applying CW mode high-power lasers (10–15 W) at 810 and/or 980 nm, the temperature changes at the scalp surface ranged from 0.2 to 3.3 °C. Moreover, temperature change at the interior surface of the brain was less than 1 °C for all lasers, except when using CW mode 15 W laser at 980 nm, where the temperature increase was 2.9 °C. A comparison of data between CW and PW modes also showed less temperature rise using a PW laser regardless of wavelength and power at both scalp and brain tissue surfaces [80]. In addition, a 1064 nm laser applied to the cat cortical surface increased the brain temperature at a depth of 5 mm by 4 °C for 5 W and 7 °C for 10 W, after 8 s of irradiation [77]. Transcranial 1064 nm laser irradiation (3.4 W) to the human forehead gradually increased skin temperature from 32 to 41 °C during 4 min of irradiation, and the temperature remained constant at 41 °C for the following 4 min [81]. Taken together, when using Class IV lasers for human transcranial PBM therapy, the application of a thermal imaging camera is recommended to allow real-time thermal monitoring of the scalp surface. Besides, active cooling units could be used when the irradiation time is prolonged enough to cause possible thermal damage.

Whole-brain PBM using light helmets has recently been shown to be an effective strategy for treating neurodegenerative and traumatic brain disorders [55, 82]. Typically, these types of devices consist of a large number of LED modules, which consume a large amount of energy and thus generate considerable heat within the device. In order to manage the large amount of heat generated in the helmet, an integral thermal cooling solution has been proposed [83]. In this respect, one study showed that during 100 min of irradiation using a helmet with 430 LEDs, the maximum temperature increase inside the helmet without or with a water-cooling system was 37 °C and 28 °C, respectively [83]. Continuous monitoring of scalp temperature by sensitive sensors located inside the surface of the helmets is recommended for this type of transcranial brain PBM. Furthermore, it should be noted that if helmets are powered by external electrical sources rather than internal batteries, an electrical shock could conceivably be caused by unsafe devices.

2.3.3 Safety Profiles in Pre-clinical and Clinical Studies

In order to translate PBM technology from the basic *bench* laboratory research to the patient's *bedside*, a full evaluation of PBM safety concerns is warranted [68, 84].

So far, several pre-clinical studies have shown the short-term [53, 85, 86] and long-term [87–90] safety profile of brain PBM therapy. The safety concerning the effect of transcranial PBM on the intact brain has been studied in rats [53]. An 808 nm laser with irradiances of 7.5, 75, and 750 mW/cm^2 were irradiated at two modes of CW and PW and behavioral and histochemical assessments were evaluated. No long-term difference in neurological deficits, as well as no significant tissue damage, was found at the two lower doses, up to 70 days post-treatment. However, CW mode laser irradiation at 750 mW/cm^2 resulted in a significant behavioral change and histological damage, which was most likely due to adverse hyperthermal effects. The study concluded that applied scalp irradiances below 75 mW/cm^2 are safe and well-tolerated, at least in animals with a 1 mm skull thickness [53]. In addition, the long-term safety of transcranial PBM (70 mW, 2230 mW/cm^2 at the scalp) based on clinical hematologic and histopathologic assessments was investigated in naïve rats. No abnormalities in blood rheology or brain histopathology were found after daily treatment for 1 year [87]. PBM applied intracranially in rodents [91] and primates [92] also caused no detectable behavioral abnormalities or tissue necrosis on the surrounding brain parenchyma or on the neighboring dopaminergic cells.

In almost all clinical studies, brain PBM was well-tolerated and showed no major adverse side effects at the optimal doses [68, 93–96]. The safety of a single-session treatment with 808 nm transcranial laser PBM was evaluated in three large randomized clinical trials on acute stroke, called "NeuroThera Effectiveness and Safety Trials" (NEST-1 [97], NEST-2 [98], and NEST-3 [99]), which showed no adverse effects. A number of small-scale clinical studies have also reported no negative side effects for the application of LED PBM on patients with Alzheimer's disease [100] and major depression [94], as well as on healthy older adults [101]. Larger randomized clinical trials are essential to evaluate the safety of laser/LED PBM therapy in the aforementioned conditions. It is worth noting that no serious adverse events have also been reported in patients who received PBM in combination with antidepressant and pro-cognitive medications or psychotherapy [102, 103]. Nevertheless, some mild and transient side-effects consisting of a single episode of epileptic fits [104], a slight headache, difficulty sleeping and insomnia, and mild itching on the scalp [55] have been reported in a few clinical brain PBM therapy studies. For instance, in a study on patients with MDD, LED PBM therapy resulted in more adverse effects and more early-onset events compared to the sham group [105]. Nevertheless, all adverse events were either mild or moderate, except in two cases who reported severe restless sleep, headaches, and taste and visual illusions [105]. Likewise, the emergence of headaches associated with the combination of transcranial and intranasal LED PBM therapy has been reported in one MDD patient with anxiety disorder [106]. Most recently, in a double-blind clinical trial, Cassano et al. [107] evaluated the incidence of side effects (e.g., weight and blood pressure changes) during repeated sessions of transcranial LED PBM in MDD patients. Although individual abovementioned side effects did not reach significance, some transient side effects such as headache, poor coordination or unsteadiness, and delayed or absent orgasm have been reported by patients [107]. In a study on patients with depression comorbidity along with traumatic brain injury (TBI), 15% of patients reported headache, and 28% of patients described feeling tired or fatigued after the initial one to three sessions of NIR laser PBM therapy, however, this resolved with further treatment sessions [6]. Laser acupuncture PBM therapy has also been shown to induce some mild and transient adverse effects in depressed patients including fatigue, insomnia, dry mouth, and headache, however, the effects resolved in less than 24 h post-treatment [108].

References

1. Wan, S., et al. 1981. Transmittance of nonionizing radiation in human tissues. *Photochemistry and Photobiology* 34 (6): 679–681.
2. Hamblin, M.R., and Y. Huang. 2013. *Handbook of photomedicine*. Taylor & Francis.
3. Mathewson, I. 2015. Did human hairlessness allow natural photobiomodulation 2 million years ago and enable photobiomodulation therapy today? This can explain the rapid expansion of our genus's brain. *Medical Hypotheses* 84 (5): 421–428.
4. Roelandts, R. 2005. A new light on Niels Finsen, a century after his Nobel Prize. *Photodermatology, Photoimmunology & Photomedicine* 21 (3): 115–117.
5. Vo-Dinh, T. 2003. Basic instrumentation in photonics. In *Biomedical photonics handbook*, 193–222. CRC Press.
6. Henderson, T.A., and L.D. Morries. 2017. Multi-watt near-infrared phototherapy for the treatment of comorbid depression: An open-label single-arm study. *Frontiers in Psychiatry* 8: 187.
7. Biard, J.R., and G.E. Pittman. 1966. Semiconductor radiant diode. Google Patents.
8. Mashiko, S., et al. 1984. Effect of near-infrared light irradiation on wound healing using high power light emitting diodes. *Nippon Laser Igakkaishi* 4 (1): 187–188.
9. Pratesi, R. 1985. Potential use of incoherent and coherent light-emitting-diodes (LEDs) in photomedicine. In *Laser photobiology and photomedicine*, 293–308. Springer.
10. Gu, Y., et al. 2019. Light-emitting diodes for healthcare and well-being. In *Light-emitting diodes*, 485–511. Springer.
11. Heiskanen, V., and M.R. Hamblin. 2018. Photobiomodulation: Lasers vs. light emitting diodes? *Photochemical & Photobiological Sciences* 17 (8): 1003–1017.
12. Hipskind, S.G. 2019. Near infrared light-emitting diodes do more than you think (re: DOI: https://doi.org/10.1089/photob.2018.4603). *Photobiomodulation, Photomedicine, and Laser Surgery* 37 (2): 126–127.
13. Henderson, T.A., and L.D. Morries. 2019. Infrared light cannot be doing what you think it is doing (re: DOI: https://doi.org/10.1089/photob.2018.4489). *Photobiomodulation, Photomedicine, and Laser Surgery* 37 (2): 124–125.
14. Feehan, J., et al. 2019. Polarized light therapy: Shining a light on the mechanism underlying its immunomodulatory effects. *Journal of Biophotonics*.
15. Makropoulou, M., et al. 2019. Non-ionizing, laser radiation in Theranostics: The need for dosimetry and the role of Medical Physics. *Physica Medica* 63: 7–18.
16. Hamblin, M.R., T. Agrawal, and M. de Sousa. 2016. *Handbook of low-level laser therapy*. CRC Press.
17. Hadis, M.A., et al. 2016. The dark art of light measurement: Accurate radiometry for low-level light therapy. *Lasers in Medical Science* 31 (4): 789–809.

18. Sliney, D.H. 2007. Radiometric quantities and units used in photobiology and photochemistry: Recommendations of the Commission Internationale de l'Eclairage (International Commission on Illumination). *Photochemistry and Photobiology* 83 (2): 425–432.
19. Chung, H., et al. 2012. The nuts and bolts of low-level laser (light) therapy. *Annals of Biomedical Engineering* 40 (2): 516–533.
20. Hashmi, J.T., et al. 2010. Effect of pulsing in low-level light therapy. *Lasers in Surgery and Medicine* 42 (6): 450–466.
21. Gavish, L., and N.N. Houreld. 2019. Therapeutic efficacy of home-use photobiomodulation devices: A systematic literature review. *Photobiomodulation, Photomedicine, and Laser Surgery* 37 (1): 4–16.
22. Souza-Barros, L., et al. 2018. Skin color and tissue thickness effects on transmittance, reflectance, and skin temperature when using 635 and 808 nm lasers in low intensity therapeutics. *Lasers in Surgery and Medicine* 50 (4): 291–301.
23. Tunér, J., and P.A. Jenkins. 2016. *Parameter reproducibility in photobiomodulation.* New Rochelle, NY, USA: Mary Ann Liebert, Inc.
24. Jenkins, P.A., and J.D. Carroll. 2011. How to report low-level laser therapy (LLLT)/photomedicine dose and beam parameters in clinical and laboratory studies. *Photomedicine and Laser Surgery* 29 (12): 785–787.
25. Zein, R., W. Selting, and M.R. Hamblin. 2018. Review of light parameters and photobiomodulation efficacy: Dive into complexity. *Journal of Biomedical Optics* 23 (12): 120901.
26. Khalid, M.Z. 2016. Mechanism of laser/light beam interaction at cellular and tissue level and study of the influential factors for the application of low level laser therapy. arXiv preprint arXiv:1606.04800.
27. Karu, T.I. 2003. Low-power laser therapy. *Biomedical Photonics Handbook* 48: 1–20.
28. Zalevsky, Z., and M. Belkin. 2011. *Coherence and speckle in photomedicine and photobiology.* New Rochelle, NY, USA: Mary Ann Liebert, Inc.
29. Hode, L. 2005. The importance of the coherency. *Photomedicine and Laser Therapy* 23 (4): 431–434.
30. Enwemeka, C.S. 2006. The place of coherence in light induced tissue repair and pain modulation. *Photomedicine and Laser Therapy* 24 (4): 457–457.
31. Laakso, L., C. Richardson, and T. Cramond. 1993. Quality of light—Is laser necessary for effective photobiostimulation? *Australian Journal of Physiotherapy* 39 (2): 87–92.
32. Vacca, R., et al. 1996. Increase in cytosolic and mitochondrial protein synthesis in rat hepatocytes irradiated in vitro by He–Ne laser. *Journal of Photochemistry and Photobiology B: Biology* 34 (2–3): 197–202.
33. Dimitrios, S., and L. Stasinopoulos. 2017. Treatment of carpal tunnel syndrome in pregnancy with polarized polychromatic non-coherent light (Bioptron light): A preliminary, prospective, open clinical trial. *Laser Therapy* 26 (4): 289–295.
34. Rojas, J.C., and F. Gonzalez-Lima. 2011. Low-level light therapy of the eye and brain. *Eye and Brain* 3: 49.
35. Karu, T.I., and S. Kolyakov. 2005. Exact action spectra for cellular responses relevant to phototherapy. *Photomedicine and Laser Therapy* 23 (4): 355–361.
36. Serrage, H., et al. 2019. Under the spotlight: Mechanisms of photobiomodulation concentrating on blue and green light. *Photochemical & Photobiological Sciences.*
37. Gu, Q., et al. 2012. Stimulation of TRPV1 by green laser light. *Evidence-Based Complementary and Alternative Medicine* 2012.
38. Karu, T.I. 2010. Multiple roles of cytochrome c oxidase in mammalian cells under action of red and IR-A radiation. *IUBMB Life* 62 (8): 607–610.

39. Huang, Y.-Y., et al. 2009. Biphasic dose response in low level light therapy. *Dose-Response* 7 (4).
40. Morries, L.D., P. Cassano, and T.A. Henderson. 2015. Treatments for traumatic brain injury with emphasis on transcranial near-infrared laser phototherapy. *Neuropsychiatric Disease and Treatment* 11: 2159.
41. Cassano, P., et al. 2015. Near-infrared transcranial radiation for major depressive disorder: Proof of concept study. *Psychiatry Journal* 2015.
42. de Freitas, L.F., and M.R. Hamblin. 2016. Proposed mechanisms of photobiomodulation or low-level light therapy. *IEEE Journal of Selected Topics in Quantum Electronics* 22 (3): 348–364.
43. Hamblin, M.R. 2018. Mechanisms and mitochondrial redox signaling in photobiomodulation. *Photochemistry and Photobiology* 94 (2): 199–212.
44. Lanzafame, R.J., et al. 2007. Reciprocity of exposure time and irradiance on energy density during photoradiation on wound healing in a murine pressure ulcer model. *Lasers in Surgery and Medicine: The Official Journal of the American Society for Laser Medicine and Surgery* 39 (6): 534–542.
45. Vasilenko, T., et al. 2010. The effect of equal daily dose achieved by different power densities of low-level laser therapy at 635 and 670 nm on wound tensile strength in rats: A short report. *Photomedicine and Laser Surgery* 28 (2): 281–283.
46. Oron, U., et al. 2001. Attenuation of infarct size in rats and dogs after myocardial infarction by low-energy laser irradiation. *Lasers in Surgery and Medicine: The Official Journal of the American Society for Laser Medicine and Surgery* 28 (3): 204–211.
47. Salehpour, F., et al. 2018. Brain photobiomodulation therapy: A narrative review. *Molecular Neurobiology* 55 (8): 6601–6636.
48. Brindley, G. 1952. The Bunsen-Roscoe law for the human eye at very short durations. *The Journal of Physiology* 118 (1): 135.
49. Lubart, R., et al. 2006. Photochemistry and photobiology of light absorption by living cells. *Photomedicine and Laser Therapy* 24 (2): 179–185.
50. Anders, J., et al. 2007. Light interaction with human central nervous system progenitor cells. In *NAALT conference proceedings*.
51. Huang, Y.-Y., et al. 2011. Biphasic dose response in low level light therapy—An update. *Dose-Response* 9 (4).
52. Mohammed, H.S. 2016. Transcranial low-level infrared laser irradiation ameliorates depression induced by reserpine in rats. *Lasers in Medical Science* 31 (8): 1651–1656.
53. Ilic, S., et al. 2006. Effects of power densities, continuous and pulse frequencies, and number of sessions of low-level laser therapy on intact rat brain. *Photomedicine and Laser Therapy* 24 (4): 458–466.
54. Cassano, P., et al. 2019. Selective photobiomodulation for emotion regulation: Model-based dosimetry study. *Neurophotonics* 6 (1): 015004.
55. Hamblin, M.R. 2019. Photobiomodulation for Alzheimer's disease: Has the light dawned? In *Photonics*. Multidisciplinary Digital Publishing Institute.
56. Al-Watban, F.A., and X. Zhang. 2004. The comparison of effects between pulsed and CW lasers on wound healing. *Journal of Clinical Laser Medicine & Surgery* 22 (1): 15–18.
57. Valchinov, E.S., and N.E. Pallikarakis. 2005. Design and testing of low intensity laser biostimulator. *Biomedical Engineering Online* 4 (1): 5.
58. Joensen, J., et al. 2012. Skin penetration time-profiles for continuous 810 nm and superpulsed 904 nm lasers in a rat model. *Photomedicine and Laser Surgery* 30 (12): 688–694.

59. Li, T., et al. 2017. Photon penetration depth in human brain for light stimulation and treatment: A realistic Monte Carlo simulation study. *Journal of Innovative Optical Health Sciences* 10 (05): 1743002.
60. Amaroli, A., et al. 2016. An 808-nm diode laser with a flat-top handpiece positively photo-biomodulates mitochondria activities. *Photomedicine and Laser Surgery* 34 (11): 564–571.
61. Salehpour, F., et al. 2019. Transcranial photobiomodulation improves cognitive performance in young healthy adults: A systematic review and meta-analysis. *Photobiomodulation, Photomedicine, and Laser Surgery* 37 (10): 635–643.
62. Meneguzzo, D.T., et al. 2016. Intravascular laser irradiation of blood. In *Handbook of low-level laser therapy*, 969–988. Pan Stanford.
63. Rojas, J.C., and F. Gonzalez-Lima. 2013. Neurological and psychological applications of transcranial lasers and LEDs. *Biochemical Pharmacology* 86 (4): 447–457.
64. Fink, L., et al. 2018. F77. Effect of transcranial infrared laser stimulation to left prefrontal cortex on verbal cognition. *Biological Psychiatry* 83 (9): S267.
65. Xuan, W., L. Huang, and M.R. Hamblin. 2016. Repeated transcranial low-level laser therapy for traumatic brain injury in mice: Biphasic dose response and long-term treatment outcome. *Journal of Biophotonics* 9 (11–12): 1263–1272.
66. Stemer, A.B., B.N. Huisa, and J.A. Zivin. 2010. The evolution of transcranial laser therapy for acute ischemic stroke, including a pooled analysis of NEST-1 and NEST-2. *Current Cardiology Reports* 12 (1): 29–33.
67. Tuchin, V.V. 2010. *Handbook of photonics for biomedical science*. CRC Press.
68. Wang, E.B., et al. 2019. Safety and penetration of light into the brain. In *Photobiomodulation in the brain*, 49–66. Elsevier.
69. Knappe, V., F. Frank, and E. Rohde. 2004. Principles of lasers and biophotonic effects. *Photomedicine and Laser Surgery* 22 (5): 411–417.
70. Joensen, J., et al. 2011. The thermal effects of therapeutic lasers with 810 and 904 nm wavelengths on human skin. *Photomedicine and Laser Surgery* 29 (3): 145–153.
71. Institute, A.N.S. 2007. *American national standard for safe use of lasers*. Laser Institute of America.
72. Shen, D., et al. 2019. Besides photothermal effects, low-level CO_2 laser irradiation can potentiate skin microcirculation through photobiomodulation mechanisms. *Photobiomodulation, Photomedicine, and Laser Surgery* 37 (3): 151–158.
73. Uozumi, Y., et al. 2010. Targeted increase in cerebral blood flow by transcranial near-infrared laser irradiation. *Lasers in Surgery and Medicine* 42 (6): 566–576.
74. Lapchak, P.A., J. Wei, and J.A. Zivin. 2004. Transcranial infrared laser therapy improves clinical rating scores after embolic strokes in rabbits. *Stroke* 35 (8): 1985–1988.
75. Chen, Y., et al. 2013. Thermal effects of transcranial near-infrared laser irradiation on rabbit cortex. *Neuroscience Letters* 553: 99–103.
76. Lychagov, V.V., et al. 2006. Experimental study of NIR transmittance of the human skull. In *Complex dynamics and fluctuations in biomedical photonics III*. International Society for Optics and Photonics.
77. Wharen, R.E., et al. 1984. The Nd: YAG laser in neurosurgery: Part 1. Laboratory investigations: Dose-related biological response of neural tissue. *Journal of Neurosurgery* 60 (3): 531–539.
78. Mochizuki-Oda, N., et al. 2002. Effects of near-infra-red laser irradiation on adenosine triphosphate and adenosine diphosphate contents of rat brain tissue. *Neuroscience Letters* 323 (3): 207–210.

79. De Taboada, L., et al. 2011. Transcranial laser therapy attenuates amyloid-β peptide neuropathology in amyloid-β protein precursor transgenic mice. *Journal of Alzheimer's Disease* 23 (3): 521–535.

80. Henderson, T.A., and L.D. Morries. 2015. Near-infrared photonic energy penetration: Can infrared phototherapy effectively reach the human brain? *Neuropsychiatric Disease and Treatment* 11: 2191.

81. Wang, X., et al. 2017. Impact of heat on metabolic and hemodynamic changes in transcranial infrared laser stimulation measured by broadband near-infrared spectroscopy. *Neurophotonics* 5 (1): 011004.

82. Poiani, G.D.C.R., et al. 2018. Photobiomodulation using low-level laser therapy (LLLT) for patients with chronic traumatic brain injury: A randomized controlled trial study protocol. *Trials* 19 (1): 17.

83. Wang, P., et al. 2017. Cooling-controlled and reliable driving module for low-level light therapy LED helmet. *Microelectronics Reliability* 78: 370–373.

84. Wolbarsht, M., D. Sliney, and J. Mellerio. 1980. *Safety with lasers and other optical sources: A comprehensive handbook.* New York: Plenum Press.

85. Lapchak, P.A., et al. 2008. Safety profile of transcranial near-infrared laser therapy administered in combination with thrombolytic therapy to embolized rabbits. *Stroke* 39 (11): 3073–3078.

86. DeTaboada, L., et al. 2006. Transcranial application of low-energy laser irradiation improves neurological deficits in rats following acute stroke. *Lasers in Surgery and Medicine: The Official Journal of the American Society for Laser Medicine and Surgery* 38 (1): 70–73.

87. McCarthy, T.J., et al. 2010. Long-term safety of single and multiple infrared transcranial laser treatments in Sprague-Dawley rats. *Photomedicine and Laser Surgery* 28 (5): 663–667.

88. Choi, D.-H., et al. 2012. Effect of 710-nm visible light irradiation on neuroprotection and immune function after stroke. *NeuroImmunoModulation* 19 (5): 267–276.

89. El Massri, N., et al. 2017. Photobiomodulation-induced changes in a monkey model of Parkinson's disease: Changes in tyrosine hydroxylase cells and GDNF expression in the striatum. *Experimental Brain Research* 235 (6): 1861–1874.

90. Moro, C., et al. 2016. Effects of a higher dose of near-infrared light on clinical signs and neuroprotection in a monkey model of Parkinson's disease. *Brain Research* 1648: 19–26.

91. Moro, C., et al. 2014. Photobiomodulation inside the brain: A novel method of applying near-infrared light intracranially and its impact on dopaminergic cell survival in MPTP-treated mice. *Journal of Neurosurgery* 120 (3): 670–683.

92. Moro, C., et al. 2017. No evidence for toxicity after long-term photobiomodulation in normal non-human primates. *Experimental Brain Research* 235 (10): 3081–3092.

93. Caldieraro, M.A., and P. Cassano. 2018. Transcranial and systemic photobiomodulation for major depressive disorder: A systematic review of efficacy, tolerability and biological mechanisms. *Journal of Affective Disorders.*

94. Schiffer, F., et al. 2009. Psychological benefits 2 and 4 weeks after a single treatment with near infrared light to the forehead: A pilot study of 10 patients with major depression and anxiety. *Behavioral and Brain Functions* 5 (1): 46.

95. Lapchak, P.A. 2019. The challenge of effectively translating transcranial near-infrared laser therapy to treat acute ischemic stroke. In *Photobiomodulation in the brain*, 289–297. Elsevier.

96. Zomorrodi, R., et al. 2019. Pulsed near infrared transcranial and intranasal photobiomodulation significantly modulates neural oscillations: A pilot exploratory study. *Scientific Reports* 9 (1): 6309.

97. Lampl, Y., et al. 2007. Infrared laser therapy for ischemic stroke: A new treatment strategy: Results of the NeuroThera Effectiveness and Safety Trial-1 (NEST-1). *Stroke* 38 (6): 1843–1849.

98. Zivin, J.A., et al. 2009. Effectiveness and safety of transcranial laser therapy for acute ischemic stroke. *Stroke* 40 (4): 1359–1364.

99. Hacke, W., et al. 2014. Transcranial laser therapy in acute stroke treatment: Results of neurothera effectiveness and safety trial 3, a phase III clinical end point device trial. *Stroke* 45 (11): 3187–3193.

100. Saltmarche, A.E., et al. 2017. Significant improvement in cognition in mild to moderately severe dementia cases treated with transcranial plus intranasal photobiomodulation: Case series report. *Photomedicine and Laser Surgery* 35 (8): 432–441.

101. Chan, A.S., et al. 2019. Photobiomodulation improves the frontal cognitive function of older adults. *International Journal of Geriatric Psychiatry* 34 (2): 369–377.

102. Caldieraro, M.A., and P. Cassano. 2019. Photobiomodulation. In *The Massachusetts General Hospital guide to depression*, 233–246. Springer.

103. Salehpour, F., M.R. Hamblin, and J.O. DiDuro. 2019. Rapid reversal of cognitive decline, olfactory dysfunction, and quality of life using multi-modality photobiomodulation therapy: Case report. *Photobiomodulation, Photomedicine, and Laser Surgery* 37 (3): 159–167.

104. Hesse, S., C. Werner, and M. Byhahn. 2015. Transcranial low-level laser therapy may improve alertness and awareness in traumatic brain injured subjects with severe disorders of consciousness: A case series. *International Archives of Medicine* 8.

105. Cassano, P., et al. 2018. Transcranial photobiomodulation for the treatment of major depressive disorder. The ELATED-2 pilot trial. *Photomedicine and Laser Surgery* 36 (12): 634–646.

106. Caldieraro, M.A., et al. 2018. Long-term near-infrared photobiomodulation for anxious depression complicated by Takotsubo cardiomyopathy. *Journal of Clinical Psychopharmacology* 38 (3): 268–270.

107. Cassano, P., et al. 2019. Reported side effects, weight and blood pressure, after repeated sessions of transcranial photobiomodulation. *Photobiomodulation, Photomedicine, and Laser Surgery* 37 (10): 651–656.

108. Quah-Smith, J.I., W.M. Tang, and J. Russell. 2005. Laser acupuncture for mild to moderate depression in a primary care setting—A randomised controlled trial. *Acupuncture in Medicine* 23 (3): 103–111.

Light Penetration into Brain

3

3.1   Introduction

The determination of the ideal irradiation parameters for photobiomodulation (PBM) delivered to the brain is a major challenge that still needs to be fully addressed. The interaction between light and tissue is very complex and depends on the light parameters and the optical characteristics of the tissue. The fundamental properties of light are the wavelength, irradiance, total power, pulse frequency, and duration, coherence, polarization, and spot size. On the other hand, reflection, absorption, scattering, and light transport (anisotropy) are fundamental properties of tissue [1]. A critical problem influencing light delivery and therapy efficacy is the propagation of light within brain tissue. That is to say, researchers and clinicians interested in this area of biomedical photonics should be familiar with quantitative assessment of light tissue penetration, in order to optimize PBM. Generally, light irradiance attenuates by a constant factor depending on the tissue depth as the light scatters and is absorbed by biomolecules. The Beer-Lambert law describes how the propagation of light in biological tissue is attenuated via absorption. The attenuation of transmittance in tissue is proportionate to the depth and to the tissue absorption coefficient. However, this law is sometimes used erroneously to calculate irradiance loss in tissue. The Radiation Transfer Equation (RTE) is a numerical method, which provides a more appropriate mathematical approach with the only exception being for very simple tissue geometry [2]. The penetration depth is conventionally defined as the distance in which the light intensity falls to 36.8% (1/e) of the initial power.

The treatment efficacy and patient compliance for brain PBM depend strongly on the amount and the rate of light energy arriving at the target neurons. In this regard, biophotonics investigations have provided values for the optical properties of various head tissues, which can be sued to calculate PBM dose parameters to achieve maximum light

© The Author(s), under exclusive license to Springer Nature Switzerland AG 2023
F. Salehpour et al., *Photobiomodulation for the Brain*, Synthesis Lectures
on Biomedical Engineering, https://doi.org/10.1007/978-3-031-36231-6_3

penetration [3]. Nonetheless, the tissue optical characteristics are considerably heterogeneous in the current literature, making it difficult to construct accurate light penetration models, especially for complex tissue structures. In addition, comparisons are also difficult considering the diverse study designs used in the literature for transmission measurements of PBM dose parameters. Numerous different study designs comprising ex vivo, in vivo, or simulation methods have been carried out to investigate the light penetration through brain tissue in several animal species, and also in humans at distinct wavelengths [4–6].

3.2   Absorption Features of Biological Tissue Components

A variety of chromophores are found in biological tissues, including water molecules, oxyhemoglobin (Hb), deoxyhemoglobin (Hb), oxymyoglobin (MbO$_2$), reduced myoglobin (MbR), lipids, and many pigments like melanin, bilirubin, and cytochromes.

The single most important component of all tissues is water, which is relatively transparent to visible and NIR light, with a minimum absorption at about 420 nm, but exponentially increasing up to ~11,000 nm with discrete peaks within this range [7, 8]. For instance, water absorption is roughly 190 times higher at 970–980 nm compared to 600 nm, and about 5,500,000 times higher at the absorption maximum of 2,950 nm. As another example, at 810 nm, water absorption is just 9 times higher than that at 600 nm [9]. In the absorption histogram, HbO$_2$ shows peaks at 414, 542, and 576 nm in the visible region and at 924 nm in the NIR spectrum. The lowest absorption of HbO$_2$ is at 688 nm over the whole visible and NIR wavelength ranges. Likewise, Hb has two absorption peaks in the visible (at 434 and 556 nm) and also in the NIR spectrum (at 758 and 914 nm) with a minimum measured at 856 nm [10]. Also, both MbO$_2$ and MbR show quite similar absorption spectra in the red/NIR region. MbO$_2$ mostly absorbs between 660 and 1000 nm in the red to NIR region and rarely absorbs at 680 nm. Reduced myoglobin absorbs light between 670 and 1000 nm [11]. Lipids mostly absorb wavelengths ranging from 430 to 1100 nm with peak absorption from 930 to 1050 nm and small shoulders at ~750 and 830 nm [11].

HbO$_2$, Hb, water, and fat absorb less light at wavelengths between 400 and 1200 nm when compared to melanin, the principle skin pigment [12]. For bilirubin, the absorption peak in the visible spectrum shows a peak at 460 nm and reaches 540 nm [11]. Flavins also show absorption up to 520 nm in the visible spectrum with a maximum at 450 nm [13].

Apart from the chromophores mentioned above, cytochrome c oxidase (CCO) also absorbs red/NIR light and shows changes in its structure and function. CCO contains two copper centers (Cu$_A$ and Cu$_B$) and two heme centers, heme(a) and heme(a3) and functions as the terminal enzyme in the respiratory electron transport chain in the mitochondria. The absorption spectrum of CCO shows peaks in the red (heme a, 605 nm; reduced Cu$_A$, 620 nm; heme a3/Cu$_B$, 655 nm), far-red (Cu$_B$ oxidized, 680 nm), and NIR regions (Cu$_B$ reduced, 760 nm; Cu$_A$ oxidized, 825 nm) [14, 15]. Cu$_A$, heme a, and heme a$_3$/

Cu_B-binaries were suggested to respectively account for 77%, 18%, and 5% of NIR light absorption at 815 nm [16]. It must be highlighted that due to the low relative abundance of CCO in tissues, despite its relatively high molar excitation coefficient compared to HbO_2 and Hb in the red/NIR spectrum, its overall importance is unclear [17].

3.3   Light Penetration Profiles of Scalp and Skull Tissues

3.3.1   Data from Animal Studies

A study on male Sprague-Dawley rats showed that an 810 nm laser penetrated through the skull to a level of 51%, and 40% through the scalp and skull combined in the prefrontal region [18]. Another study on the same species/gender showed, however, that the cerebral cortical layer only received 7.1% of 808 nm laser light when the laser probe was positioned at 1.8 mm anterior to the bregma and 2.5 mm lateral to the midline [19]. Studies using red light reported that for wavelengths of 660 nm (light-emitting diodes [LEDs]) [20] and 630 nm (laser) [18], only 5.8% and 23% respectively, of light could penetrate both the shaved scalp and the skull to reach the cortical surface.

The penetration of 800 nm laser light into the wet skulls of different animal species was compared by Lapchak et al. [21]. For *Oryctolagus cuniculus* (rabbit), it was 18%, for Sprague-Dawley rats, it was 39%, and for male C57BL/6 mice, it was 63%. Similarly, 70% of 808 nm laser light could penetrate the skull of a male C57BL/6 J mouse to reach the parietal region [22].

On the other hand, other studies using BALB/c mice have reported comparatively much lower penetration values through the skull [23–26]. In this regard, 15% of 810 nm laser could penetrate the scalp, while 6% could penetrate the combined scalp and skull, both independent of pulse frequency as expected [23]. In addition, laser light of 670 nm and 810 nm wavelengths penetrated the intact skull by a factor of 1.20% and 1.75%, respectively [24]. Furthermore, the light intensity in the mouse cranium was reduced by a factor of some 90% for 670 nm LED light [25]. Similarly, when the light was positioned on the bregma of 18-month-old mice, about 16% of 660 nm laser light was transmitted through the combined skull and scalp [26].

In one study using an adult pig skull (parietal region), the highest light penetration was found at wavelengths between 700 and 850 nm with a small shoulder at ~750 nm, which is believed to be related to residual deoxygenated hemoglobin (Hb) [27]. In addition, laboratory measurements at 808 nm showed that about 0.11% of the laser light was transmitted through the combined scalp and skull of a pig [28].

Overall, the results of these animal studies demonstrate substantial variations in reported light penetration levels through the scalp and skull of various species, perhaps owing to different water content and protein composition, and mainly due to the variable tissue thickness between species. Also, some variation in penetration depth may probably be explained by heterogeneous PBM parameters used in the literature.

3.3.2 Data from Human Studies

3.3.2.1 Laboratory Data

In an adult human head, the total thickness (t) of tissues above the brain surface adds up to 14–17 mm, comprising the scalp (t = 5–6 mm) [29], the skull (t = 6–7 mm), and additional tissues and spaces, including periosteum, dura mater, and sub-arachnoid space (t = 3–4 mm) [3]. Laboratory measurements gathered from adult human cadaver heads reported a mean scalp and skull thickness of 8 ± 2.0 mm and 10.7 ± 5.7 mm in occipital region, 7.2 ± 2.2 mm and 8 ± 2.0 mm in parietal region, 6.7 ± 0.7 mm and 8 ± 2.7 mm in temporal region, and 6.8 ± 1.1 mm and 7.2 ± 2.2 mm in frontal region, all respectively [30]. Despite the comparatively thicker measurement of the scalp/skull in the occipital region as stated above, this difference was not reported to be statistically significant [30].

The relatively high optical scattering and absorption of the skull is because the calvarial tissue is composed of mineral substances (58%), proteins (24.6%), carbohydrates (5.2%), and water (12.2%) [31]. Also, the morphology of the tissues must be taken into account when considering the scattering characteristics. The light transmittance through the temporal bone increased from 0.2 to 2% in the wavelength interval of 600–670 nm, and another rise from 6 to 8% in the 800–814 nm interval [12]. For the bregma zone and the parietal region of the human skull/calvaria, the measurements by Lapchak et al. [21] have shown that 808 nm laser light shows transmittance of 4.18% and 4.24%, respectively, when samples are dehydrated, and 4.63% and 4.74%, respectively, when hydrated. A significant negative correlation was reported between 800 nm light penetration and thickness at bregma and parietal skull as expected, yet no correlation was found for skull density [21].

According to a report by Henderson and Morries [32], there was no measured photonic energy transmission through 1.9 mm of ex vivo human skin for LED devices with 0.05 W of 810 nm and 0.2 W of 650/880 nm. However, at the same thickness of human skin, there was a penetration of 11.5% and 16.7% for 10 W of 810/980 nm and 15 W of 810 nm lasers, respectively [32]. This report of poor transmission using lower-power LED devices contradicts previous results which favored transcranial LED light with an output power of 0.012 W for 870 nm and 0.001 W for 633 nm in order to reach the cortical surface and produce therapeutic benefit [33, 34]. It seems most probable that the light detector used in their study was not sensitive enough to measure the transmitted energy at lower input power, so further studies are needed to investigate the exact light transmission through the human skin.

Furthermore, in a comparison of red and NIR light from LED devices, 633 nm and 830 nm light penetration through the parietal bone was reported to be 3.4% and 9%, and for frontal bone was reported to be 0.9% and 7.7%, both respectively [35]. In addition to that, at 10 mm depth (an approximate cadaver thickness with intact soft tissue) using the same wavelengths of light, the penetration was measured to be 0.0 and 0.9% for the temporal region, 0.5 and 2.1% for the frontal region, and 0.7 and 11.7%

for the occipital region, all respectively [35]. These results highlight the importance of the anatomical location for the light penetration depth, since the skull bone is not of uniform thickness and also has distinct tissue compositions in different head regions [36]. Nonetheless, measurements on adult human cadaver heads have indicated that 850 nm LED light penetrates various cranial regions (occipital, parietal, temporal, and frontal) with a statistically non-significant difference in attenuation; yet, for the temporal region, this measurement was the least [30]. Unlike the conclusions of [35], they suggested that modifications in the positioning of the LED probe from one cranial location to another have no significant effect on light penetration/distribution in the brain [30].

In another study, the transmission of a 810 nm laser through the skull plus scalp tissue was measured to be 5–0.5% with the respective sample thickness varying between 0.7 and 1.9 cm [37]. These results are in good agreement with those obtained by computational simulation modeling which indicated that only 0.2% of 810 nm light can penetrate from the scalp through the skull and cerebrospinal fluid (CSF) to the gray matter across a distance of more than 2 cm, when the point source was located at Cz on the scalp surface [6]. Similarly, a study on the post-mortem human frontal skull revealed that 1064 nm NIR light penetrated the supraorbital frontal bone (Fp2 point in electroencephalogram [EEG]) with a fraction of 2% [38]. For the human cranial bone over a spectral range of 800–2000 nm, the maximal absorption bands were measured to be at 978, 1192, 1464, 1745, and 1930 nm, which was attributed to the higher absorption coefficients for water and lipids [39]. Using a high-power laser (5 W), 808 nm light with flat-top beam profile and a normalized scalp irradiance of 1 mW/cm^2, penetration gradients varying over 4 orders of magnitude up to 4 cm depth in a cadaver brain were reported regardless of location of the laser source on the scalp, or the laser operation mode (continuous or pulsed mode) [40]. In human head transcranial irradiation study using 850 nm LEDs on occipital, parietal, temporal, and frontal regions, it was reported that the photon flux was attenuated by 3–4 orders of magnitude at 3–6 cm in depth in brain tissue [30].

3.3.2.2 Simulation Data

The modeling of light propagation in biological tissues can be carried out using different mathematical methods [30, 41–43]. For scientists concerned with diagnostic and therapeutic applications, the main question is whether actual human brains with normal anatomy and accurate optical characteristics can be modeled for each tissue type [44]. While scattering and absorption are the most important interaction of light with tissue [45], the complete optical characteristics of tissue include four parameters measured at each wavelength: absorption coefficient (μ_a), scattering coefficient (μ_s), scattering anisotropy factor (g), and refractive index (n) [46]. Since it is impossible to assess light penetration in a living human head by direct detection, light propagation by means of voxelized homogeneous media provides an adequate estimate of light propagation in the brain using anatomically correct human head models and computer simulations (e.g., Monte Carlo, Finite Element modeling, etc.) [47, 48].

Recently, a team of Chinese researchers used the Monte Carlo 3D model and visible Chinese human head model in three distinct studies to estimate the photon fluence distribution in brain tissue [47, 49, 50]. The study included eight types of tissue in the visible Chinese human head model: scalp, skull, CSF, muscle, arterial and venous blood vessels, gray and white mater, and simulated the beam sources applied to the middle of the forehead surface. The first two studies reported the different penetration into brain tissue at three different wavelengths (810 nm, 660 nm, and 980 nm) [47, 50]. Simulation data showed a normalized irradiance of 10^{-5} W/cm^2 at depths, below the scalp, of 4.13, 3.45, and 1.45 cm in the head tissue, for 810, 660, and 980 nm wavelengths, respectively. They also suggested that 810 nm light may be spread across a broader area of the human brain in comparison with wavelengths of 660 and 980 nm. Although the light penetration of 980 nm was not as high as 660 or 810 nm, it was suggested that this wavelength was more suitable for acupuncture therapy as it was less scattered in the head tissue and provided a more concentrated spot. Furthermore, a smaller beam size (e.g., 2 cm) could penetrate more widely and deeply in the brain compared to bigger spots (e.g., 4 or 6 cm), but there was a non-significant difference between them. The Gaussian beam profile penetrated slightly deeper than a top-hat beam profile, but the difference was quite small [47]. Nevertheless, in another study [50], the top-hat beam profile had 3% deeper penetration than the Gaussian beam profile for all combinations of beam sizes and wavelengths. The most important factors governing the depth of penetration into the brain were the wavelength, beam size, and beam profile [47]. In the third study [49], they compared the differences in light fluence, penetration depth, and light absorption of cerebral tissue for wavelengths of 660, 810, 980, and 1064 nm. Their simulation findings suggested that in terms of deeper or wider penetration of light into the brain tissue, 660 and 810 nm were better than 980 and 1064 nm. Furthermore, the fluence distribution of photons was somewhat better for 660 nm compared to 810 nm, whereas 1064 nm was better than 980 nm. The results revealed a normalized incident irradiance of 10^{-5} W/cm^2 at depths of 6.0, 5.8, 5.0, and 4.7 cm for 660, 810, 1064, and 980 nm, respectively. Gray and white matter absorption was approximately 3–4 times higher for 660 and 810 nm compared to 980 and 1064 nm [49]. Their two studies [47, 50] showed some inconsistencies when comparing penetration data (in particular for 660 and 810 nm). It is noteworthy that in their third study [49], the authors used rather different optical properties to create the phantoms for 660, 810, and 980 nm light compared with their prior studies [47, 50] which yielded different findings, as anticipated. It is worth mentioning that Paolo Cassano's team lately questioned the above findings by introducing a more clinically-relevant energy deposition simulation, rather than just using penetration measurement [48]. In their simulation, the penetration of 670 nm was higher than 1064 nm when the energy deposition was simulated in deeper brain regions; both wavelengths produced a quite similar deposition for more superficial brain regions, yet sometimes, 1064 nm was superior. In addition, their findings indicated that nearly 2% of 810 nm LED light could penetrate to the dorsolateral prefrontal cortex (dlPFC) when the light source was placed at F3 or

F4, EEG points [48]. Additional laboratory experiments are required to evaluate the exact light distribution for these wavelengths in human post-mortem heads.

In another relevant study [51], using modification of the visible Chinese head model, three head models were created with differing volumes and amounts of hemorrhage in the cerebral subdural region of the forehead (Level I, 0.96 cm^3; Level II, 9.58 cm^3; Level III, 29.4 cm^3). Using Monte Carlo modeling, the propagation of 808 nm light with varying beam size and beam type was simulated. The findings indicated a greater light penetration and a higher total lesion fluence using a Gaussian beam profile compared to a top-hat beam profile, when the hemorrhagic region was not larger than the beam size. On the other hand, if the hemorrhagic region was much larger than the beam size, the top-hat beam penetrated more deeply. Considering these results, they proposed that a Gaussian beam was more suitable for the recovery from the stroke, where the beam size is intended to cover a hemorrhagic lesion. It has to be mentioned that a hemorrhagic stroke is often considered to be a contraindication to transcranial PBM, owing to the theoretical hazard of further bleeding, and additional heating due to localized absorption of light by blood. Moreover, their report showed that with increasing hemorrhage size, the maximum penetration depth was sharply decreased so that penetration depths of 3.99 cm were obtained (for intact brain), for example, for a Gaussian beam with a 1 cm radius, of 3.34 cm (for level I), 2.47 cm (level II) and 1.54 cm (level III) [51].

In two separate studies [30, 42], Lan Yue et al. utilized different modeling techniques to evaluate the effectiveness of various LED array emitters compared to a single point source for transcranial light penetration into the human brain. Using the finite element method and the COMSOL Multi-physics package, they found roughly 4–5 orders of magnitude reduction for penetration of 850 nm light at 3 cm deep in the frontal area. Furthermore, as photons propagate deeper into the brain (depth > 4 cm), multi-source irradiation exhibited a more gradual decline than that found with a single source. This can be explained by the idea that an overall increase in intracranial photon flux can result from the superposition of the laterally scattered photons generated by multiple sources [30]. Another simulation study was carried out using the Monte Carlo method with the Colin27 brain template as an anatomical human head model to evaluate the light propagation of single-source and multi-source LEDs (690 and 850 nm) [42]. It should be noted that, due to its reported minor effect on the photon flux attenuation rate, blood flow was not taken into account in this simulation. Single light sources had the same attenuation profiles in the five distinct areas (frontal, occipital, temporal, parietal, and crown areas) with ~4 orders of magnitude for 850 nm and ~5 orders of magnitude for 690 nm after penetration to 3 cm depth. Nevertheless, there was a moderately decreased light penetration of 850 nm at 6 cm depth for simulations with single sources in the occipital and temporal areas. A multi-source simulation with 277 separate sources uniformly placed on the scalp resulted in an increased intracranial flux so that this flux increase started at a 2 cm depth for both 690, and 850 nm, increased to ~10× for 690 nm and ~5× for 850 nm at 4 cm depth, and reached ~15× for 690 nm and ~10× for 850 nm at a depth of 6 cm. The enhanced increase in photon flux recorded at 690 nm in comparison to 850 nm, appears to result from the higher tissue scattering coefficients for shorter wavelengths. Furthermore, multiple simulations using 13, 53, 105, 181, 229, and 277 separate sources indicated that the optimum source density of 181 point sources on the scalp could increase photon flux and uniform distribution of photons within the brain [42]. However, this finding may only apply to the particular sources studied, and may not be applicable to all settings. The clinical significance of photon flux amplification seen after a reduction of 4–5 orders of magnitude is also debatable because the quantity of light left is probably negligible and cannot be therapeutic.

A recent study conducted by Cassano et al. [48] was carried out to examine intranasal and transcranial irradiation routes to deliver red/NIR light to the dlPFC and ventrome-dial prefrontal cortex (vmPFC). Using a hardware-accelerated Monte Carlo model and standard adult brain atlas, they investigated the LED light penetration profiles of various wavelengths of 670, 810, 850, 980, and, 1064 nm through the skull and scalp. They found that transcranial irradiation at the F3 and F4 EEG sites provided better light delivery to the dlPFC, while a LED source positioned in close proximity to the cribriform plate was well-suited for reaching the vmPFC. Both transcranial irradiation at the Fp1–Fpz–Fp2 location and intranasal irradiation via a source located in the mid-nose region were shown to provide a good flux to the vmPFC. Among the intranasal approaches, positioning of a 810 nm light source at the cribriform plate led to 46-fold and 658-fold higher energy deposition on the vmPFC compared to mid-nose and nostril positions, respectively. They also suggested that among the various wavelengths, 810 nm would provide the overall highest photon flux to the targeted cortical regions.

Although the effect of hair on light scattering/absorption has not been taken into account in the simulation designs of all the above-mentioned studies, a recent experimental measurement of laser light penetration through 2 mm of hair samples showed strong light absorption and scattering by a factor of 98%, suggesting hair acts as a serious barrier for transcranial PBM procedures [52]. In fact, hair and hair follicles can be separately measured to have varying optical properties and are differently distributed on the head from individual to individual [53]. Future simulation studies should explore the exact impact of hair on the penetration depth in transcranial PBM.

3.4   Light Penetration Profiles of Brain Tissues

3.4.1   Data from Animal Studies

In a study on living brains of female New Zealand White rabbits, it was found that only 0.1, 0.4, and 2.5% of laser light, with wavelengths of 635, 671, and 808 nm, respectively, could penetrate through 10 mm of cerebral tissue [54].

In the Sprague-Dawley rat brain, the penetration of 808 nm laser light into 1.8 cm of cortical surface was measured to be roughly 0.08% [55]. Concerning the lower penetration depth of 808 nm light compared to 830 nm, this finding agrees with another study in Sprague–Dawley rats, which showed that about 0.22% of 830 nm laser light penetrated 1.8 cm into the cortex [56]. In Sprague-Dawley female rats, about 12% of 808 nm laser light could reach the substantia nigra (SNc) at ~7 mm distance from the cortex when the laser probe was placed transcranially on the shaved scalp [57].

In a notable study measuring the light transmission in rodent brain tissue, a 65% decrease in light intensity at 670 nm was observed for every one millimeter of depth into the brain [58]. This study revealed that the laser intensity was reduced by a factor of <0.001% at 10 mm depth of brain parenchyma in Sprague-Dawley rats, and the LED intensity was reduced by a factor of <1% at 5 mm depth of brain in male BALB/c mice [58]. Other studies using male BALB/c mice [26, 59] demonstrated that for wavelengths of 670 and 810 nm the intensity of light remaining at 5 mm of depth, i.e., the distance from the skull surface to SNc region, reduced to ~2.5% and 3%, respectively [59]. It was also demonstrated that about 10% of 660 nm laser light was transmitted through 1 mm of brain tissue [26].

A study on an ex vivo pig head revealed that the intensity of 808 nm laser light was reduced by a factor of 91% for each 5 mm of brain tissue after the penetrating the scalp and skull [28].

In another original study using an ex vivo lamb head, it was found that 2.9% of 810 nm and 1.2% of 980 nm light from high-power (15-W) laser devices could penetrate 3 cm of scalp, skull, and brain tissues. Although a 10-W 810/980 nm laser device showed only a 0.35% penetration at 3 cm depth in the brain from the skin surface, 0.05-W 810 nm and 0.2-W 650/880 nm LED devices did not show any detectable light power at that depth [32].

3.4.2   Data from Human Studies

3.4.2.1   Laboratory Data

One study was carried out on the intact brains of a 2-month-old infant and a 67-year-old female adult obtained at autopsy. Penetration depths of 5.4 mm and 1.2 mm were measured for 660 nm, while 8.8 mm and 3.2 mm were measured for 1064 nm light, respectively. Overall, the penetration depths were reportedly 2–3 times deeper in the neonatal brain compared to the fully myelinated adult brain for wavelengths between 488 and 1064 nm, which suggests an association between the degree of myelination and the penetration depth [60].

Tedford et al. [40] carried out an analysis of the findings of penetration depths of various wavelengths using eight intact cadaver adult brains and found less fluence rate attenuation in the intraparenchymal brain tissue for 808 nm compared to 940 nm (with an approximate one order-of-magnitude difference) and compared to 660 nm (with about 1.9 orders of magnitude), where the distance between the laser probe and the detector was more than 2 cm. Moreover, penetration depths of 0.92 mm, 1.38 mm, 2.17 mm, and 2.52 mm were recorded with human postmortem intact brain tissue, respectively, with 632.8 nm, 675 nm, 780 nm, and 835 nm lasers [61].

For the effective and adequate irradiation of the substantia nigra pars compacta (SNpc) using transsphenoidal light delivery, Pitzschke et al. [62] carried out an interesting experiment measuring light delivery and dosimetry in an intact cadaver head. An optical fiber-based light diffuser was coupled to 671 or 808 nm laser diodes, and then positioned within the sphenoidal sinus, through a nostril, under the guidance of endoscopy in order to direct the light toward the SNpc. The findings indicated the penetration of 0.03% and 0.36% to the SNpc for wavelengths 671 and 808 nm, respectively. In addition, their Monte Carlo simulation modeling revealed that the delivered light fluence to the SNpc was 133-fold and 26-fold higher for 671 nm and 808 nm, respectively, when placing the light source near the third ventricle in the proximity of the pons and thalamus, compared to irradiation through the sphenoidal sinus. It is worth noting that the transmission value was still lower at 3.99% ($133 \times 0.03\%$) than 9.36% ($26 \times 0.36\%$). Furthermore, the percentage of transmitted light was reduced by 60-fold and 18-fold for red or NIR wavelengths, respectively, in comparison with the transsphenoidal approach when light sources were placed in the oral cavity pointing toward the SNpc [62].

3.4.2.2 Simulation Data

Within the 360–1100 nm spectral range, the optical properties of intact brain tissue including white matter, gray matter, cerebellum, and brainstem (thalamus and pons) were studied in vitro using integrating-sphere measurements combined with an inverse Monte Carlo method [43]. In the brain white matter, there was more light penetration for 630 nm (μ_a = 0.08) and 670 nm (μ_a = 0.07) than for 850 and 1064 nm (both μ_a = 0.1). The gray matter data also demonstrated more light penetration for 630 and 670 nm (both μ_a = 0.02), than 850 nm (μ_a = 0.03) and 1064 nm (μ_a = 0.05). In general, the depth of light penetration for white matter was significantly less than that for gray matter. For wavelengths of 670 nm (μ_a = 0.06) and 630, 850, and 1064 nm (all μ_a = 0.07), the cerebellum had almost the same penetration depth. Pons and thalamus data showed almost the same light penetration for 630 nm (μ_a = 0.05–0.06), 670 nm (μ_a = 0.04–0.05), and 1064 nm (μ_a = 0.1), except for 810 nm (μ_a = 0.07 for pons, and μ_a = 0.05 for thalamus). Overall, it could be concluded that the thalamus showed better penetration than white matter, but less penetration compared with gray matter [43].

3.5   Concluding Remarks Regarding Penetration Depth

All the head tissues (scalp, skull, dura mater, blood, and CSF) attenuate light penetration into the brain. If the light energy attenuation by these tissues could be quantitatively evaluated, optimized PBM parameters might be obtained, resulting in improvements to brain PBM therapy. Over the course of the past two decades, researchers have used mathematical optical models, sophisticated simulation systems, and experimental measurements in ex vivo animal and human tissues, in order to discover and test optimized irradiation parameters for brain PBM. Earlier trials suggested that cortical light fluences within the range of 1–20 J/cm^2 were ideal for activating neurobiological processes involved in neuroprotection and neural repair. A recent review reported that the wavelength of 810 nm was the most prevalent (55.4%) among a total of 464 studies, followed by 660 nm (14.4%), 1064 nm (14.0%), 630 nm (9.3%), and 980 nm (6.9%) [49]. Of all these reviewed wavelengths, 810 nm penetrates the most deeply through the head and is also well absorbed by CCO, which makes it the first choice for deep brain PBM. Two factors should be considered to select the most effective wavelength for brain PBM therapy; the suitable wavelength for maximal penetration, and the absorption/action spectrum of the chromophore needed to trigger the biological response.

Infrared radiation comprises wavelengths between 700 and 1 mm. The IR band is generally subdivided into smaller ranges including NIR (700–1100 nm), short-wavelength IR (1100–3000 nm), medium IR (3000–5000 nm), and far-IR (5000 nm–1 mm) regions. Wavelengths in the "third optical window" (1550–1870 nm) are ideal for light penetration into the brain, followed by additional windows II (1000–1350 nm), IV (2100–2300 nm), and I (700–1000 nm) [63–65]. Although penetration values for wavelengths in window III are deeper than wavelengths used in conventional PBM (600–1100 nm), and light in the window III might be thought a good candidate for deep brain PBM, the photobiological responses to these wavelengths are probably not efficient and not well understood. Recently, the idea that nanoparticle engineering might be combined with biophotonic excitation has been suggested as a way to overcome the penetration limitations in deep brain PBM using conventional wavelengths. Up to now, a number of high-performance lanthanide-doped upconversion nanoparticles (UCNPs) have been effectively synthesized for use in several medical applications, including bioimaging, optogenetics, drug delivery, photoactivation, and photodynamic therapy [66]. It has been proposed that UCNPs could be used as an effective photon transducer using light in the optical window III to activate biological photoacceptors, by transforming it into light at conventional PBM wavelengths. The hypothesis is that the delivery of UCNPs to the deeper structures of the brain (e.g., limbic system, brainstem, etc.), followed by irradiation using wavelengths in the third optical window could be a promising approach to maximize light transmission. The ability of UCNPs to cross the brain–blood barrier (BBB) and their low toxicity make them highly promising candidate for applications in the brain [66].

The debate about whether continuous wave (CW) or pulsed wave (PW) light is better for increased depth of penetration in PBM is not settled. According to evidence in the literature, pulsed light appears to have a slightly greater penetration depth in living tissues. For example, light in PW mode yielded somewhat better penetration through an ex vivo lamb skull and brain tissue, but the difference was not statistically significant compared to CW mode [32]. As a hypothetical explanation for this observation, it could be that at an identical average output power between CW mode and PW mode, the PW mode may potentially emit more photons deeper into the head tissue at the pulse peaks. Nonetheless, in fixed human cadaver heads, there was no significant difference between the penetration of PW or CW mode laser light [35]. Additional experimental and simulation studies are needed to understand the advantages and disadvantages of PW and CW mode irradiation. Another very common claim is that laser light can penetrate deeper than LED light. Although some researchers have proposed that light from LEDs may penetrate deeper with longer irradiation times, a laboratory measurement indicated that NIR light from low-power 0.2-W LED cannot penetrate the thickness of the lamb scalp [32]. It is proposed that a collimated laser beam is more likely to be forward scattered in living tissue than a divergent LED beam [67]. There is another assumption that an increase in light output power may lead to greater penetration in the tissue. For instance, NIR light produced by laser or LED devices with <1 W, 6 W, 10 W, and 15 W output powers penetrated 3 cm deep into brain tissue with factors of 0.0%, 0.005%, 0.14%, and 1.26%, respectively [32]. Moreover, the penetration of NIR laser light into porcine brain tissue from a 1.5-W emitter was deeper than that from a 0.5-W emitter [28].

An apparent concern that experimental scientists have raised is that, in most studies using animals, light penetration is significantly higher than similar measurements in humans. Two experimental studies appear to be especially useful as they are in accordance with light penetration measurements in the human head: (1) the 808 nm laser light penetration through the scalp plus skull in a pig head which was 0.11%; (2) the 810 nm laser light penetration through the intact skull in BALB/C mice which was 1.75%. Based on these penetration values, these two models could recapitulate the results for transcranial PBM in humans. However, the application of red light for PBM in animal models would not be appropriate for humans, because although the penetration of red light in animals is sufficient to trigger photochemical reactions in the brain, this would not occur with humans who have much larger heads [4]. Beside the two mentioned examples above, an in vitro brain organoid study could be considered as a third promising model. Although this model may not be useful in evaluating light penetration alone, it would allow to investigate the effects of very low power light on organized brain tissue, instead of only using flat neuronal cell cultures [68]. Both models of computerized mesh modeling of the human head and the human cadaver have shown that light penetrates 3–4 cm into the brain, albeit at a very low level.

The fundamental theory of transcranial PBM suggests that photochemical reactions cannot be triggered by very low levels of light, which will therefore have no biological effects. Studies on the organoid brain would better investigate whether transcranial light irradiation could lead to biological effects deep within brain tissue. This sort of study is also important for the larger transcranial PBM field because the available light penetration studies prompt the question of whether some parameters could provide adequate energy to trigger any photochemical reactions at all within the brain. Data from skin penetration studies using a 0.2 W, 880 nm, LED source have shown only minimal penetration through approximately 2 mm of tissue. However, preliminary, uncontrolled clinical studies have demonstrated a therapeutic effect using the same light power. Photochemical processes such as induction of electromagnetic fields caused by very low energy PBM could possibly be responsible for the presumed disconnect [4].

Besides the obvious role of light power in the penetration depth, the composition, morphology, and inter-individual differences of human head tissues are a less predictable issue, which might hamper the development of widespread therapeutic application of transcranial PBM. The larger thickness of the intervening tissue between scalp surface and brain, and tissue dehydration, both typically increased with advancing age, will reduce light penetration. It is then possible that dosimetric protocols and target effects on brain function in young healthy subjects might underestimate the effective dosimetry of transcranial PBM for the middle-aged and elderly populations. This presumed disconnect is particularly problematic given the overall aging of our population, and given that clinical trials to test important applications in neurodegenerative diseases are often undertaken without prior dose-ranging studies of transcranial PBM [4].

References

1. Jacques, S.L. 2013. Optical properties of biological tissues: A review. *Physics in Medicine & Biology* 58 (11): R37.
2. Tuchin, V.V. 2015. Tissue optics and photonics: Light-tissue interaction. *Journal of Biomedical Photonics & Engineering* 1 (2).
3. Chung, H., et al. 2012. The nuts and bolts of low-level laser (light) therapy. *Annals of Biomedical Engineering* 40 (2): 516–533.
4. Salehpour, F., et al. 2019. Penetration profiles of visible and near-infrared lasers and light-emitting diode light through the head tissues in animal and human species: A review of literature. *Photobiomodulation, photomedicine, and laser surgery* 37 (10): 581–595.
5. Morse, P.T. et al. 2020. *Cytochrome c oxidase-modulatory near-infrared light penetration into the human brain: Implications for the noninvasive treatment of ischemia/reperfusion injury.* IUBMB life.
6. Bhattacharya, M., and A. Dutta. 2019. Computational modeling of the photon transport, tissue heating, and cytochrome C oxidase absorption during transcranial near-infrared stimulation. *Brain Sciences* 9 (8): 179.
7. Pope, R.M., and E.S. Fry. 1997. Absorption spectrum (380–700 nm) of pure water II. Integrating cavity measurements. *Applied Optics* 36 (33): 8710–8723.
8. Sogandares, F.M., and E.S. Fry. 1997. Absorption spectrum (340–640 nm) of pure water I. Photothermal measurements. *Applied Optics* 36 (33): 8699–8709.

9. Hale, G.M., and M.R. Querry. 1973. Optical constants of water in the 200-nm to 200-μm wavelength region. *Applied optics* 12 (3): 555–563.
10. Prahl, S. 1999. *Optical absorption of hemoglobin*. http://omlc.ogi.edu/spectra/hemoglobin.
11. Yao, J., and L.V. Wang. 2014. Sensitivity of photoacoustic microscopy. *Photoacoustics* 2 (2): 87–101.
12. Wan, S., et al. 1981. Transmittance of nonionizing radiation in human tissues. *Photochemistry and Photobiology* 34 (6): 679–681.
13. Yu, X., et al. 2012. Flavin as a photo-active acceptor for efficient energy and charge transfer in a model donor–acceptor system. *Physical Chemistry Chemical Physics* 14 (19): 6749–6754.
14. Karu, T.I., and S. Kolyakov. 2005. Exact action spectra for cellular responses relevant to phototherapy. *Photomedicine and Laser Therapy* 23 (4): 355–361.
15. Ball, K.A., P.R. Castello, and R.O. Poyton. 2011. Low intensity light stimulates nitrite-dependent nitric oxide synthesis but not oxygen consumption by cytochrome c oxidase: Implications for phototherapy. *Journal of Photochemistry and Photobiology B: Biology* 102 (3): 182–191.
16. Szundi, I., G.-L. Liao, and Ó. Einarsdóttir. 2001. Near-infrared time-resolved optical absorption studies of the reaction of fully reduced cytochrome c oxidase with dioxygen. *Biochemistry* 40 (8): 2332–2339.
17. Gupta, P.K. 2013. *Light–tissue interactions*. Handbook of photomedicine, 25.
18. Salehpour, F., et al. 2016. Therapeutic effects of 10-HzPulsed wave lasers in rat depression model: A comparison between near-infrared and red wavelengths. *Lasers in Surgery and Medicine* 48 (7): 695–705.
19. Yang, L., et al. 2018. Photobiomodulation therapy promotes neurogenesis by improving post-stroke local microenvironment and stimulating neuroprogenitor cells. *Experimental Neurology* 299: 86–96.
20. Rojas, J.C., A.K. Bruchey, and F. Gonzalez-Lima. 2012. Low-level light therapy improves cortical metabolic capacity and memory retention. *Journal of Alzheimer's Disease* 32 (3): 741–752.
21. Lapchak, P.A., et al. 2015. Transcranial near-infrared laser transmission (NILT) profiles (800 nm): Systematic comparison in four common research species. *PLoS ONE* 10 (6): e0127580.
22. Uozumi, Y., et al. 2010. Targeted increase in cerebral blood flow by transcranial near-infrared laser irradiation. *Lasers in Surgery and Medicine* 42 (6): 566–576.
23. Ando, T., et al. 2011. Comparison of therapeutic effects between pulsed and continuous wave 810-nm wavelength laser irradiation for traumatic brain injury in mice. *PLoS ONE* 6 (10): e26212.
24. Zhang, Y., et al. 2015. Quantitative evaluation of SOCS-induced optical clearing efficiency of skull. *Quantitative Imaging in Medicine and Surgery* 5 (1): 136.
25. Shaw, V.E., et al. 2010. Neuroprotection of midbrain dopaminergic cells in MPTP-treated mice after near-infrared light treatment. *Journal of Comparative Neurology* 518 (1): 25–40.
26. Salehpour, F., et al. 2018. A protocol for transcranial photobiomodulation therapy in mice. *JoVE (Journal of Visualized Experiments)* 141: e59076.
27. Firbank, M., et al. 1993. Measurement of the optical properties of the skull in the wavelength range 650–950 nm. *Physics in Medicine & Biology* 38 (4): 503.
28. Aulakh, K. et al. 2016. Transcranial light-tissue interaction analysis. In *Optical interactions with tissue and cells XXVII*. International Society for Optics and Photonics.
29. Garn, S.M., S. Selby, and R. Young. 1954. Scalp thickness and the fat-loss theory of balding. *AMA Archives of Dermatology and Syphilology* 70 (5): 601–608.
30. Yue, L. et al. 2015. Simulation and measurement of transcranial near infrared light penetration. In *Optical interactions with tissue and cells XXVI*. International Society for Optics and Photonics

31. White, D. et al. 1991. The composition of body tissues.(II) Fetus to young adult. *The British Journal of Radiology* 64 (758): 149–159.
32. Henderson, T.A., and L.D. Morries. 2015. Near-infrared photonic energy penetration: Can infrared phototherapy effectively reach the human brain? *Neuropsychiatric Disease and Treatment* 11: 2191.
33. Naeser, M.A., et al. 2011. Improved cognitive function after transcranial, light-emitting diode treatments in chronic, traumatic brain injury: Two case reports. *Photomedicine and Laser Surgery* 29 (5): 351–358.
34. Naeser, M.A., et al. 2014. Significant improvements in cognitive performance post-transcranial, red/near-infrared light-emitting diode treatments in chronic, mild traumatic brain injury: Open-protocol study. *Journal of Neurotrauma* 31 (11): 1008–1017.
35. Jagdeo, J.R., et al. 2012. Transcranial red and near infrared light transmission in a cadaveric model. *PLoS ONE* 7 (10): e47460.
36. Hamdy, O., and H.S. Mohammed. 2020 Investigating the transmission profiles of 808 nm laser through different regions of the rat's head. *Lasers in Medical Science*, 1–8.
37. Lychagov, V.V. et al. 2006. Experimental study of NIR transmittance of the human skull. In *Complex dynamics and fluctuations in biomedical photonics III*. International Society for Optics and Photonics.
38. Barrett, D.W., and F. Gonzalez-Lima. 2013. Transcranial infrared laser stimulation produces beneficial cognitive and emotional effects in humans. *Neuroscience* 230: 13–23.
39. Bashkatov, A.N. et al. 2006. Optical properties of human cranial bone in the spectral range from 800 to 2000 nm. In *Saratov fall meeting 2005: Optical technologies in biophysics and medicine VII*. International Society for Optics and Photonics.
40. Tedford, C.E., et al. 2015. Quantitative analysis of transcranial and intraparenchymal light penetration in human cadaver brain tissue. *Lasers in Surgery and Medicine* 47 (4): 312–322.
41. Fang, Q., and D.A. Boas. 2009. Monte Carlo simulation of photon migration in 3D turbid media accelerated by graphics processing units. *Optics Express* 17 (22): 20178–20190.
42. Yue, L., and M.S. Humayun. 2015. Monte Carlo analysis of the enhanced transcranial penetration using distributed near-infrared emitter array. *Journal of Biomedical Optics* 20 (8): 088001.
43. Yaroslavsky, A., et al. 2002. Optical properties of selected native and coagulated human brain tissues in vitro in the visible and near infrared spectral range. *Physics in Medicine & Biology* 47 (12): 2059.
44. Okada, E., et al. 1997. Theoretical and experimental investigation of near-infrared light propagation in a model of the adult head. *Applied Optics* 36 (1): 21–31.
45. Jacques, S.L. 1996. Origins of tissue optical properties in the UVA, visible, and NIR regions. *OSA TOPS on Advances in Optical Imaging and Photon Migration* 2: 364–369.
46. Simpson, C.R., et al. 1998. Near-infrared optical properties of ex vivo human skin and subcutaneous tissues measured using the Monte Carlo inversion technique. *Physics in Medicine & Biology* 43 (9): 2465.
47. Li, T., et al. 2017. Photon penetration depth in human brain for light stimulation and treatment: A realistic Monte Carlo simulation study. *Journal of Innovative Optical Health Sciences* 10 (05): 1743002.
48. Cassano, P., et al. 2019. Selective photobiomodulation for emotion regulation: Model-based dosimetry study. *Neurophotonics* 6 (1): 015004.
49. Wang, P., and T. Li. 2019. Which wavelength is optimal for transcranial low-level laser stimulation? *Journal of Biophotonics* 12 (2): e201800173.

50. Li, T., et al. 2015. Effects of wavelength, beam type and size on cerebral low-level laser therapy by a Monte Carlo study on visible Chinese human. *Journal of Innovative Optical Health Sciences* 8 (01): 1540002.
51. Li, T., et al. 2018. Optimize illumination parameter of low-level laser therapy for hemorrhagic stroke by Monte Carlo simulation on visible human dataset. *IEEE Photonics Journal* 10 (3): 1–9.
52. Henderson, T.A., and L.D. Morries. 2019. Near-infrared photonic energy penetration—principles and practice. In *Photobiomodulation in the brain*, 67–88. Elsevier.
53. Strangman, G.E., Q. Zhang, and Z. Li. 2014. Scalp and skull influence on near infrared photon propagation in the Colin27 brain template. *NeuroImage* 85: 136–149.
54. Pitzschke, A., et al. 2015. Optical properties of rabbit brain in the red and near-infrared: Changes observed under in vivo, postmortem, frozen, and formalin-fixated conditions. *Journal of Biomedical Optics* 20 (2): 025006.
55. DeTaboada, L., et al. 2006. Transcranial application of low-energy laser irradiation improves neurological deficits in rats following acute stroke. *Lasers in Surgery and Medicine: The Official Journal of the American Society for Laser Medicine and Surgery* 38 (1): 70–73.
56. Abdo, A., A. Ersen, and M. Sahin. 2013. Near-infrared light penetration profile in the rodent brain. *Journal of Biomedical Optics* 18 (7): 075001.
57. Oueslati, A., et al. 2015. Photobiomodulation suppresses alpha-synuclein-induced toxicity in an AAV-based rat genetic model of Parkinson's disease. *PLoS ONE* 10 (10): e0140880.
58. Moro, C., et al. 2014. Photobiomodulation inside the brain: A novel method of applying near-infrared light intracranially and its impact on dopaminergic cell survival in MPTP-treated mice. *Journal of Neurosurgery* 120 (3): 670–683.
59. Reinhart, F., et al. 2017. The behavioural and neuroprotective outcomes when 670 nm and 810 nm near infrared light are applied together in MPTP-treated mice. *Neuroscience Research* 117: 42–47.
60. Svaasand, L.O., and R. Ellingsen. 1983. Optical properties of human brain. *Photochemistry and Photobiology* 38 (3): 293–299.
61. Stolik, S., et al. 2000. Measurement of the penetration depths of red and near infrared light in human "ex vivo" tissues. *Journal of Photochemistry and Photobiology B: Biology* 57 (2–3): 90–93.
62. Pitzschke, A., et al. 2015. Red and NIR light dosimetry in the human deep brain. *Physics in Medicine & Biology* 60 (7): 2921.
63. Sordillo, L.A., et al. 2014. Deep optical imaging of tissue using the second and third near-infrared spectral windows. *Journal of Biomedical Optics* 19 (5): 056004.
64. Sordillo, L.A. et al. 2014. Third therapeutic spectral window for deep tissue imaging. In *Optical biopsy XII*. International Society for Optics and Photonics.
65. Golovynskyi, S., et al. 2018. Optical windows for head tissues in near-infrared and short-wave infrared regions: Approaching transcranial light applications. *Journal of Biophotonics* 11 (12): e201800141.
66. Meynaghizadeh-Zargar, R., et al. 2019. Potential Application of Upconverting Nanoparticles for Brain Photobiomodulation. *Photobiomodulation, Photomedicine, and Laser Surgery* 37 (10): 596–605.
67. Heiskanen, V., and M.R. Hamblin. 2018. Photobiomodulation: Lasers vs. light emitting diodes? *Photochemical & Photobiological Sciences* 17(8): 1003–1017.
68. Fitzgerald, M., et al. 2013. Red/near-infrared irradiation therapy for treatment of central nervous system injuries and disorders. *Reviews in the Neurosciences* 24 (2): 205–226.

Action Mechanisms of Photobiomodulation in Neuronal Cells and the Brain

4

4.1 Molecular and Cellular Mechanisms of Photobiomodulation

Wavelengths in the electromagnetic spectrum ranging from 400 to 1100 nm have energy levels of 3.09–1.12 eV/photon (according to Planck's equation: E (eV) = 1,240/λ [nm]). These energies are sufficient for electronic excitation of organic molecules and the initiation of photochemical reactions. The mechanisms underlying the interaction between light with different wavelengths and tissues are very complex due to the various photoacceptors present inside the cells. Biological tissues contain a wide range of photoacceptors, including water molecules, oxyhemoglobin, deoxyhemoglobin, myoglobin, melanin, endogenous porphyrins, mitochondrial and membrane bound cytochromes, flavoproteins such as the plasma membrane NADPH oxidase system, which contains flavoproteins and cytochrome b [1]. Each of these photoacceptors is sensitive to a different specific light spectrum.

One of the most established mechanisms for light-cell interaction was proposed by Russian photobiologist, Karu [2]. Her early work revealed a close similarity between the spectral characteristics in the cellular proliferation action spectrum and the cellular absorption spectrum with red/near-infrared (NIR) light irradiation [3]. She postulated that light absorption by some cellular photoacceptors could result in photon-induced modulation of the cellular proliferation process. It was thought that the majority of the cellular photoacceptors, such as flavins, iron-sulfur centers, or heme are located in the mitochondria, hence light-cell interactions could be due to light-absorption by the mitochondria [4, 5]. Further studies with red/NIR light between 600 and 850 nm proved the above-mentioned hypothesis, and one of the mitochondrial respiratory enzymes, cytochrome c oxidase (CCO), was suggested to be the dominant photoacceptor responsible for light absorption in the cells. Mitochondria play a central role in cellular metabolism through the process of oxidative phosphorylation taking place within their inner membrane. The mitochondrial electron transport chain is composed of five separate complexes: complex I (NADH dehydrogenase), complex II (succinate dehydrogenase), complex III (cytochrome bc_1 complex), complex IV (CCO), and complex V (ATP synthase). The CCO enzyme consists of 13 protein subunits containing two copper centers (Cu_A and Cu_B) and two heme centers (heme a and a_3). CCO transfers electrons from cytochrome c to oxygen via Cu_A and from heme a to the heme a_3/Cu_B metal center. This enzyme has absorption peaks in the violet-blue (at 400 nm), the red to far-red (Heme a, 605 nm; Cu_A reduced, 620 nm; heme a_3/Cu_B, 655 nm; Cu_B oxidized, 680 nm) and the NIR regions (Cu_B reduced, 760 nm; heme a_3, 784 nm; Cu_A oxidized, 825 nm) [6, 7].

Based on four decades of accumulating evidence derived from both in vitro and in vivo studies, the following cellular response model has been proposed to be the main action mechanism for PBM at 600–850 nm wavelengths [8]. Upon the absorption of photons by CCO, the energy is absorbed by the metal centers of CCO resulting in excitation of electrons [9]. Along with this photoexcitation, nitric oxide (NO) could be photodissociated from the binuclear center in CCO (a_3/Cu_B) leading to an increase in the mitochondrial membrane potential (MMP or [$\Delta\Psi_m$]). This, in turn, induces an increase in ATP production and changes in the concentrations of some signaling molecules such as Ca^{2+} and reactive oxygen species (ROS). The ROS include the superoxide radical anion ($O_{2\bullet}^-$) and its stable product hydrogen peroxide (H_2O_2). In a secondary cascade of events, the aforementioned primary responses alter the intracellular redox potential, the intracellular pH, cyclic adenosine monophosphate (cAMP) levels, and expression of redox-sensitive factors (e.g., nuclear factor kappa B [NF-κB]). These phototransduction cascades originating from the electron transport chain occur at times after light irradiation. Retrograde mitochondrial signaling is a pathway for communication inside cells occurring from mitochondria to the nucleus, which affects several cellular processes under various conditions [10]. The transcription factor NF-κB is considered to be an important mediator in oxidative stress-induced activation of retrograde mitochondrial signaling. Considering this pathway, signal transduction processes induced by PBM lead to activation of other transcription factors, gene expression, and ultimately to promotion of several beneficial biological processes (e.g., cell metabolism, cell viability, proliferation, and differentiation) [11–13]. For example, it has been shown that 660 nm laser PBM can upregulate genes coding for subunits of complex I (NDUFA11 and NDUFS7 genes), complex IV (COX6B2 and COX6C genes), and ATP synthase (ATP5F1 gene) in wounded, diabetic wounded, and ischemic fibroblast cells ni vitro [14]. Another upregulated gene following NF-κB activation is the mitochondrial enzyme superoxide dismutase (SOD), which is a strong antioxidant [15]. It was also found that 825 nm NIR light at fluences less than 10 J/cm^2 can stimulate neuronal differentiation of adipose stem cells (ASCs) in vitro through modulation of cellular metabolism and redox status [13].

Mitochondrial respiratory complexes I-III have absorption bands in the visible region ranging 400–550 nm (violet-blue to green) [16]. A 532 nm green laser has been shown to increase ATP levels and cellular proliferation in vitro, most likely via activation of the mitochondrial complex III (cytochromes b, c1, and c) [17]. Another report has stated that only 420 nm blue light, not 477, 511, and 544 nm wavelengths, can increase ATP production, likely through the regulation of the mitochondrial complex I (NADH-dehydrogenase) [10]. Besides the mitochondrial chromophores, it has been suggested that green light can also be absorbed by neuronal opsin photoreceptors (OPN2-5) and activate transient receptor potential channels (TRPCs) leading to nonselective permeabilization to Ca^{2+}, Na^+, and Mg^{2+} [18]. TRP channels are categorized into seven subfamilies, among which TRP vanilloid subfamily member 1 (TRPV1) has been found to be activated by 532 nm green light [19]. In addition, 479 nm blue light can be absorbed by melanopsin (OPN4) leading to a rise in intracellular Ca^{2+} $[Ca^{2+}]_i$ [20]. The cryptochromes have also been proposed as blue-light sensitive flavoproteins, which can activate and transduce cellular signals via the optic nerve to the suprachiasmatic nucleus in the brain [21].

In addition to CCO, other photoacceptors for PBM were found in a study by Wang et al. [22] who showed that heat/light sensitive calcium channels were sensitive to 980 nm NIR light. It was suggested that NIR light at wavelengths ranging 900–1100 nm could be absorbed by nanostructured water clusters in these ion channels. The increase in the vibrational energy of water clusters could result in perturbation of the protein conformation and opening the channel, which ultimately allows modulation of $[Ca^{2+}]_i$ levels. Furthermore, modulation of inducible NO synthase (iNOS) [23] and heat-shock proteins (HSP) [24] have been proposed as alternative mechanisms for 1072 nm NIR light. The quantum-dot laser diodes in the range of 1265–1270 nm have been recently gained attention in PBM research. Oxygen molecule has strong optical absorption band in the 1265–1270 nm wavelengths, thus it has been suggested that these wavelengths at low fluencies (typically <30 J/cm^2) could directly generate singlet oxygen from ground state triplet oxygen, leading to changes in the cytosolic Ca^{2+} concentration, thereby resulting in PBM effects [25–27]. It has also been postulated that light in the range of 1600–1900 nm can be absorbed by ATP synthase, leading to a change in its conformation and subsequent release of more ATP molecules [28]. Beyond these wavelengths, there is the mid-infrared region in which CO_2 laser (10,600 nm) is the most commonly used medical laser in this spectral region. The proposed photoacceptor for this region is almost certainly water [29]. It has been proposed that during CO_2 laser irradiation, the heat generated by water absorption promotes ATP synthesis to release more cellular ATP [30]. This hypothesis requires further experimental explorations in future work.

Recently, Andrei Sommer from Ulm, Germany, has suggested an alternative molecular mechanism for PBM based on the nanoscopic interfacial water layers (IWL) inside cells. According to his hypothesis, if the IWL were inside the cellular mitochondria, then the lowering of viscosity as a result of the energy absorption during red/NIR PBM could allow the mitochondrial rotary motor (ATP synthase) to rotate faster and generate more ATP molecules. On the other hand, if the IWL were localized within the cellular plasma membrane, photon absorption could enhance the uptake of nutrients accounting for increased cellular proliferation [31, 32].

There are therefore many issues that are important to take into consideration: (I) PBM can also increase the activity of complexes I-III and succinate dehydrogenase in mitochondrial electron transfer chain, however, CCO is generally considered to be the primary photoacceptor [33]; (II) CCO has two distinct maximum absorption peaks at 670 and 830 nm wavelengths, while 728 nm does not match with the absorption spectrum of CCO [34]; (III) although several studies indicate that NIR light consistently activates CCO [7, 35], NIR light at specific wavelengths of 750 and 950 nm has been shown to inhibit CCO activity and cause a reduction in MMP and mitochondrial respiration in vitro. On the other hand, 810 nm has been proposed as the most effective wavelength for activation of CCO [36]; (IV) evidence suggests that ~810 nm NIR light has almost exclusive biostimulatory impact on the mitochondrial complex IV [37], nevertheless, 1064 nm NIR light (30 J/cm^2) has also been shown to be able to interact and modulate the activity of the respiratory complexes I, III, and IV [38]; (V) PBM with 660 nm light (2.4–3.2 J/cm^2) has been shown to increase mitochondrial metabolism and stimulate proliferation in cells lacking CCO, indicating that CCO might be not required for PBM and alternative targets appear more likely [39]; (VI) NO could be released directly by a photodissociation mechanism from CCO during irradiation or shortly thereafter, nevertheless, NO may also be generated enzymatically following an increase in the activity of NOS long after irradiation, possibly by increasing $[Ca^{2+}]_i$ levels [40]; NO can also be released by photodissociation from nitrosothiols [41] as well as nitrosylated hemoglobin and myoglobin [42]; and (VII) because damaged tissues and hypoxic cells are more likely to have inhibitory levels of NO compared to healthy tissue. It is believed that PBM has considerable effects on diseased cells and tissues but does not dramatically affect healthy cells [1].

4.2 Neurobiological Mechanisms of Photobiomodulation

4.2.1 Neuronal Mitochondrial and Metabolic Functions

The presence of mitochondrial dysfunction has been long considered a key etiological factor in the development of numerous neurologic and psychiatric diseases [43, 44]. Under various pathological conditions, mitochondria may experience significance variation in functional aspects (i.e., a decrease in respiratory chain complex activity, a decrease in ATP production, loss of MMP, and ROS overproduction) [45]. The advantages of PBM for improving mitochondrial function and metabolic homeostasis have been examined in different cell types [46]. Because of the high mitochondrial content of neuronal tissues, their light exposure would result in activation of CCO leading to improved energy metabolism via increase in oxygen consumption and subsequent ATP formation [47]. It is widely accepted that red/NIR light absorption by neuronal CCO is the first stage in the brain PBM process [48].

4.2.1.1 Cytochrome C Oxidase Levels/Activity

Studies have shown that 780 nm light is absorbed up to 50% by CCO [49]. The Cu_A, heme a, and heme a_3/Cu_B centers are also, respectively, responsible for 77%, 18%, and 5% of light absorption at 815 nm [50]. It has been shown by Wong-Riley and colleagues that 670 nm light-emitting diodes (LEDs) can not only increase CCO activity in intact cultured neurons but also can significantly reverse the down-regulation of CCO activity induced by tetrodotoxin [51]. Their subsequent work in rat cultured visual cortical neurons also revealed a higher CCO up-regulation by LED irradiation at 670 and 830 nm wavelengths compared to 770 and 880 nm [7]. Pre-treatment with 808 nm LED can considerably restore CCO activity to control level in cobalt chloride ($CoCl_2$)-treated cultures of cortical neurons [52]. In another research, mitochondrial suspensions from adult rat brains were irradiated with 660 nm laser at low and high fluences, and the results showed that only the activity of mitochondrial complex IV (but not complexes I or II) was increased at both 5 and 60 min post-irradiation in all PBM groups (10, 30, and 60 J/cm^2) [53].

In addition, it has been demonstrated in vivo that transcranial 660 nm LED PBM can up-regulate CCO activity by 13.6% in the prefrontal cortex (PFC) in naïve rats [54]. Also, transcranial 633 nm LED PBM increased the whole brain CCO activity by 56.7% in a rat model of rotenone-induced neurotoxicity [55]. Whole-body 670 nm LED PBM of intact rats for 30 days also resulted in more than three-fold increase in CCO levels in brain tissue [56]. A fascinating series of studies conducted by Quanguang Zhang and co-workers showed that transcranial 808 nm laser PBM could remarkably increase CCO activity in the PFC of mice in a chronic stress model [57], in the hippocampus of a rat Alzheimer's disease (AD) model [58], in the hippocampal CA1-region of rat global cerebral ischemia model [59], and in the cortical peri-infarct region of a rat ischemic stroke model [60]. Transcranial NIR laser PBM applied to the healthy human forehead could also markedly up-regulate cerebral oxidized CCO levels [61, 62] and synchronize brain activity [63] via non-thermal mechanisms.

4.2.1.2 ATP Levels

Since the membrane depolarization/repolarization process in mitochondria consumes a large amount of energy, neural tissue is highly dependent on mitochondrial ATP. Accumulating evidence suggests that optimum amounts of visible to NIR light (regardless of whether laser or LED sources are used) can influence neuronal ATP synthesis [21, 64]. In vitro studies showed that red (670 nm) and NIR (770, 830, and 880 nm) LED PBM (4 J/cm^2, culture surface) significantly restored the ATP content to control levels in potassium cyanide (KCN)-treated neurons [7]. Despite the enhancement in the ATP levels of 1-methyl-4-phenylpyridinium (MPP^+)-treated neuronal cells following one session of 670 nm LED PBM (4 J/cm^2) [65], a single-session laser PBM resulted in no improvement in ATP content in Aβ-treated PC12 cells (670 nm laser, 1 J/cm^2) [66] or Parkinson's disease (PD) cybrid cell lines (810 nm laser, 2 J/cm^2) [67]. In intact neuronal cells, a 808 nm laser using a much lower fluence of 0.05 J/cm^2, compared to the above-mentioned studies, increased ATP production in normal human neural progenitor cells [68]. In contrast, 532 nm green laser PBM with a much higher fluence of 3042 J/cm^2 resulted in increment of ATP synthesis in human-derived glioblastoma cells (A-172) [17]. In addition to these, a 970 nm NIR laser PBM (6 J/cm^2) resulted in somewhat significantly increased ATP production in an in vitro model of SH-SY5Y cells, while 445 nm blue (3 J/cm^2) did not [69]. These contradictory in vitro observations could be explained by the differences between damaged or healthy cells, where various optimum levels of light energy (depending on the applied wavelengths) are required for improving neuronal metabolic capacity. More studies addressing the peak response of cellular ATP formation following low-level light irradiation could be helpful for better PBM treatment planning [70]. In this respect, a study in cultured human neuronal cells suggested that the ATP production becomes maximum at 10 min after laser PBM [68]. Besides these, several in vivo studies have also shown that red/NIR PBM promotes cerebral ATP generation in various neuropathological models [58, 71–73].

4.2.1.3 Mitochondrial Membrane Potential

The mitochondrial respiratory chain-generated ATP is produced by an electrochemical gradient, which is called the MMP. Accumulating evidence indicates that PBM with visible to NIR light can raise the MMP and stimulate ATP production, and at the same time produce a burst of ROS leading to activation of NF-κB [1, 2]. In an early effort to explore the influence of light on the mitochondrial inner membrane in isolated rat liver mitochondria, Passarella et al. showed that in the presence of mitochondrial inhibitors such as rotenone (inhibitor of complex I), antimycin A (inhibitor of complex III), or oligomycin (inhibitor of ATP synthase), 632.8 nm laser PBM (5 J/cm^2, at culture level) caused an increase in MMP and the proton gradient [74]. PBM on the human melanoma cell line (A2058) at increasing fluences starting at 0.5 J/cm^2 showed an increase in MMP with a marked effect at 2 J/cm^2 [75]. In a remarkable study, Vos et al. assessed the response of MMP in mitochondria from *pink1* mutant *Drosophila* fruitflies to PBM (2.5 J/cm^2) at several wavelengths between 635 and 900 nm and found that red (652 and 690 nm) and NIR (808, 830, and 865 nm) wavelengths were the most effective ones in terms of MMP restoration [76]. Alexandratou et al. also evaluated the response of MMP in single cells following 647 nm laser PBM (1.5 mJ/cm^2). Their results showed that a maximum rise in MMP by 30% of its basal level occurred at 2 min post-irradiation, and then the MMP reduced gradually back to the basal value at 4 min post-irradiation [77].

Giuliani and colleagues were the first to use laser-confocal microscopy for live monitoring of MMP changes after laser irradiation in PC12 cells. They showed that the drop in MPP after H$_2$O$_2$ exposure was prevented by both short-term (20 s; 0.11 and 0.22 J/cm^2) and long-term (15 min; 5.06 and 10.12 J/cm^2) PBM at 670 nm [78]. During the years 2011–2014, studies from Hamblin's laboratory provided deep insights into the MMP alterations induced by PBM in primary cortical neurons in vitro [79–81]. First, they irradiated intact neurons with 810 nm laser at a wide range of fluences between 0.03 and 30 J/cm^2 (at culture level) and found a biphasic dose–response governed the behavior. According to their results, MMP was significantly increased at 0.3 J/cm^2 reaching a peak at twice basal value at 3 J/cm^2, and then, it decreased at 10 J/cm^2, and significant outright depolarization of the MMP was observed at 30 J/cm^2 [79]. They showed, in their following studies, that 810 nm laser PBM (3 J/cm^2, culture level) could restore the MMP reduction induced by different sources of oxidative stress, such as $CoCl_2$, rotenone, or H_2O_2 [80]. Moreover, similar results were found with excitotoxicity in primary cortical neurons caused by exposure to glutamate, NMDA, or kainate [81]. Taken together, it was suggested that PBM at a fluence range of 0.1–10 J/cm^2, with a peak at ~3 J/cm^2 [79, 82], would be optimal for the restoration of MMP disruption in neuronal cultures. Transcranial ~810 nm laser PBM (3 J/cm^2 [83] and 8 J/cm^2 [84], at cortical level) has also been shown to restore cerebral MMP in various animal models. In addition to NIR light, transcranial 670 nm LEDs PBM (4 J/cm^2, at scalp level) also increased synaptic MMP in both wild-type and Tg2576 AD mouse models [85].

4.2.1.4   Mitochondrial Biogenesis

The definition of "mitochondrial biogenesis" refers to the process by which cells increase the mass and number of their individual mitochondrial units. Genes regulating mitochondrial biogenesis (e.g., PGC-1α, NRF1, TFAM) have been shown to be decreased during aging and in neuropathological conditions [86, 87]. SIRT1 is involved in the mitochondrial biogenesis process where it directly interacts with PGC-1α and increases its activity/ expression. PGC-1α, is a master regulator of mitochondrial biogenesis, which regulates mitochondrial mass/content via transcription factors like NRF1 and NRF2, which in turn up-regulate mitochondrial gene expression, such as TFAM, in nuclear DNA [88]. Stimulation of neuronal mitochondrial biogenesis, and the level of mitochondrial activity, is therapeutically relevant since it could improve brain bioenergetics in neuropathological conditions. In the case of brain tissue, it has been shown that transcranial 810 nm laser PBM can stimulate neuronal mitochondrial biogenesis in a cerebral ischemic aging model through the activation of the SIRT1/PGC-1α/NRF1/TFAM signaling pathway. In fact, this activity has been suggested to be a key mechanism for PBM to increase cerebral mitochondrial content [71]. Normal respiration and ATP production in the brain are linked with the levels of functional mitochondria in neurons. Besides the above study [71], some other works have demonstrated that short [89] and long-term [71, 73] exposure to 660 or 810 nm laser PBM (8 J/cm^2, at cortical level) can increase mitochondrial number and function in the brain tissue. Transcranial 670 nm LED PBM for 4 weeks has also been shown to increase the synaptic mitochondrial number in Tg2576 AD mice, most likely through a reduction in the synaptic binding of Aβ oligomers [85]. Of note, PBM did not change the mitochondrial content in the synapses in wild-type animals suggesting that under normal physiological conditions, the neuroprotective effects of PBM may not be significant [85].

4.2.2 Cerebral Blood Flow and Oxygen Consumption

Because red/NIR light stimulates CCO activity and accelerates electron transfer, neuronal mitochondrial oxidative phosphorylation, and oxygen consumption are increased by PBM. Impaired cerebral vascular perfusion has been widely recognized as an early manifestation of many brain disorders [90, 91]. NO is a powerful vasodilator, which is released upon light absorption by CCO accompanied by restoration of oxygen consumption. As preclinical observations suggest, PBM can enhance the neuronal NO concentration and cerebral blood flow (CBF), resulting from activation of endothelial nitric oxidase synthase [92] and also increase the blood vessel diameter [93]. It is suggested that PBM of specific brain areas could affect the regional CBF, most likely mediated by NO and the neurotransmitter, glutamate [92].

Investigation of mouse brain mitochondria in vitro indicated that among the 590, 627, and 660 nm LED ($10 \, W/m^2$), PBM, only 590 nm wavelength could significantly increase NO bioavailability through direct NO synthesis from nitrite catalyzed by CCO [94]. None of the above-mentioned wavelengths showed a significant biostimulatory effect on CCO activity on oxidizing cytochrome c [94]. As stated by Uozumi et al. [92], the transient increase in CBF by PBM is dependent on both the NOS activity and NO content. According to their in vivo findings on naïve mice, transcranial 808 nm laser PBM ($10.6 \, W/cm^2$) increased cortical NO content (by 50%) immediately after turning on the laser, and gradually enhanced CBF in the laser-exposed mice (by 30%) and the opposite hemisphere (by 19%) at 45 min after starting the irradiation [92]. It is interesting to note that to date, only one in vivo study has shown CBF changes after PBM using ^{18}F-FDG positron emission tomography (PET) scans. The researchers created bilateral common carotid artery stenosis in mice, and then the animals underwent transcranial 810 nm LEDs PBM for 4 weeks. Neuroimaging results showed that CBF was significantly restored to normal levels in animals with chronic cerebral hypoperfusion [95]. These findings indicate that PBM could be a promising preventive approach in subjects with chronic brain hypoperfusion who are at increased risk of dementia [96, 97].

So far, some clinical researches have evaluated the impact of PBM on the CBF and cerebral oxygenation in healthy subjects [61, 98, 99] as well as in patients with different brain conditions [100, 101]. In a study in healthy adults, a 810 nm laser PBM delivered transcutaneously onto the bilateral frontal points F3 and F4 (on the electroencephalography (EEG) electrode placement system) resulted in an increase in regional CBF as measured by blood-oxygen-level-dependent (BOLD) functional magnetic resonance imaging (fMRI). The changes were most pronounced in the dorsolateral PFC (dlPFC) just beneath the tip of the fiber and were also seen in widespread cerebral regions such as the ipsilateral parietal cortex [102]. Given the duration of the laser irradiation and duration of fMRI data acquisition, the authors claimed that the alteration in blood flow was most likely due to the increased neuronal activation of the fronto-parietal network rather than laser-induced local NO release [102]. Two studies carried out in Hanli Liu's laboratory also reported positive effects [61, 99]. Right forehead exposure of healthy volunteers to transcranial 1064 nm laser irradiation rapidly enhanced the CBF, blood volume, and cerebral oxygenation during the 8 min irradiation. The observed effects were still seen 5 min after the laser was turned off [61]. In a similar study using the same laser (1064 nm), PBM to the center and right side of the forehead led to an increase in cerebral oxygenation in the right and left PFC during the 10 min irradiation; the effects persisted during a 6 min recovery period following the stimulation [99]. Besides, whole-brain PBM of healthy individuals using two different types of transcranial helmet significantly increased the regional cerebral oxygen saturation (rSO$_2$) not only during the 15 min irradiation but also for 20 min after the completion of stimulation [103, 104]. The reported sustained hemodynamic effects of 5–20 min by PBM [61, 99, 103, 104] are much longer than the 1 min, which has been reported for transcranial magnetic stimulation (TMS) [105]. The difference might be due to the fact that TMS rapidly and directly affect the excitability of neurons, while PBM acts by stimulating cerebral hemodynamics which is actually slower than neuronal excitability, and is connected to the intracellular metabolic cascade [99].

It should be mentioned that, in contrast to the transient metabolic stimulatory effects of PBM suggested by above-mentioned clinical studies, a study in dogs revealed that only repeated transcranial 808 nm laser PBM for longer than 2 weeks could enhance cerebral bioenergetics, while significant changes were not detectable immediately after a single irradiation session [106]. Schiffer et al. [101] have demonstrated that during transcranial 810 nm LED irradiation bilaterally to the forehead, the prefrontal CBF was non-significantly increased in patients with major depressive disorder (MDD).

The development of sustainable devices for brain stimulation has been a global aim for many years. The duration of PBM effects on human cerebral perfusion has also been addressed by some researchers. Salgado et al. [98] showed that twice/week for 4 weeks of transcranial 627 nm LED PBM (applied to 4 points in the frontal and parietal regions) significantly improved the blood flow velocity in the left middle cerebral (by 30%) and the basilar (by 25%) arteries of healthy elderly women. Besides, twice/day for 73 days of bilateral transcranial 805 nm LED PBM to the forehead of a patient in a vegetative state enhanced CBF (by 20%) in the left anterior frontal lobe [100].

4.2.3   Neuroprotection and Neuronal Survival

One of the most prominent and potentially important effects of PBM in the brain and in neurons is its ability to protect cells against a variety of toxic insults. Accumulating in vitro evidence has demonstrated that red/NIR PBM has protective effects on neuronal damage induced by a wide range of mitochondrial inhibitors (e.g., rotenone [65, 80, 107], MPP$^+$ [65, 107, 108], KCN [7, 107, 109], and NaN$_3$ [7]), excitotoxins (e.g., glutamate, NMDA, and kainate) [81], and oxidative stressors (e.g., CoCl$_2$ [52, 80] and H$_2$O$_2$ [78, 80]), as well as tetrodotoxin (a sodium channel blocker) [7, 51] and sodium nitroprusside

(SNP) [110]. In a couple of studies, the neuroprotective benefits of PBM at a culture fluence of 4 J/cm^2 have been addressed [7, 51, 65]. In this regard, irradiation with 670 and 830 nm LED light for 5 days significantly reversed the detrimental effects of tetrodotoxin (0.4 μM) on neuronal mitochondrial function [7, 51]. In fact, the toxins KCN and NaN_3 irreversibly inhibit CCO (complex IV), while the neurotoxin MPP^+ and the pesticide rotenone are both selective mitochondrial complex I inhibitors. Application of 670 nm LED PBM twice a day for each of the 5 days of exposure to KCN (10–100 μM) or NaN_3 (10 μM, 100 μM, and 1 mM) significantly prevented cell death in primary visual cortical neurons [7]. LED treatment twice a day for 2 days during MPP^+ (250 μM) or rotenone (200 nM) exposure, partially but significantly, rescued cortical and striatal neuronal death [65]. In addition, a 810 nm laser PBM at a culture fluence of 3 J/cm^2 has been shown to produce a modest but significant increase in neuronal survival when treated with glutamate (30 μM), NMDA (100 μM) or kainate (50 μM)-mediated excitotoxicity [81]. Furthermore, a 810 nm laser (3 J/cm^2, culture surface) significantly protected cultured murine cortical neurons from the cytotoxic effects of $CoCl_2$ (0.2, 0.5, 1, and 2 mM) and H_2O_2 (10 and 20 μM) [80]. The PC12 cell death induced by an even higher concentrations of H_2O_2 was also reduced by PBM. According to the study [78], a 20 s 670 nm laser irradiation (0.11 or 0.22 J/cm^2) at the beginning of the 15 min H_2O_2 exposure (300 μM) slightly but significantly protected cell culture viability. Taken together, this evidence indicates that the antitoxin effects of PBM can be attributed to its activation of CCO, and the most effective action spectrum of neuroprotective PBM would correspond to the red (~660–670 nm) and NIR (~810–830 nm) absorption peaks of CCO. Indeed, in damaged cells where the MMP is lower than normal due to existing excitotoxicity or inhibition of mitochondrial electron transport, free radicals are generated from the dysfunctional mitochondria. In this case, red/NIR photon absorption by electron transport chain results in an increase in the MMP and thereby a decrease in oxidative stress. Besides, it has been suggested that the neuroprotective effects of PBM could be mediated by a cascade of events leading to altered gene expression in neurons, especially those that are involved with neuronal activity [111].

Moreover, a study by Yu et al. [112] demonstrated that 810 nm laser PBM (3 J/cm^2, culture surface) could protect primary cortical neurons against oxygen-glucose deprivation (OGD)-induced neurotoxicity via multiple mechanisms. They suggested that an improvement in key survival signaling (measured by Akt phosphorylation and Bcl-2) and decreased death signaling (measured by Bax and BAD) pathways, as well as downregulation of NO generation, possibly via inhibiting neuronal NOS (nNOS) activity, may be at least part explain the neuroprotective mechanisms of NIR light. They also tested the direct effects of PBM on NO-induced neurotoxicity and found that 810 nm laser significantly attenuated neuronal death induced by a 50 μM concentration of S-nitroso-N-acetyl-DL-penicillamine (SNAP), a NO donor [112]. PBM with 632.8 nm laser irradiation has also been shown to increase the time required for loss of excitability in hippocampal brain slices after a transient exposure to OGD and markedly improve recovery from ischemic injury [113].

Several other researchers have investigated the neuroprotective effects of PBM against toxicity induced by accumulation of two abnormally folded proteins, amyloid-beta (Aβ) [114–116] and tau [85]. Three studies performed in the Da Xing laboratory reported novel findings [114, 115, 117]. In the first study [114], they suggested that activated Akt induced by 632.8 nm laser PBM (2 J/cm^2, culture surface) interacted with and then inactivated GSK3β which had been induced by treatment with 25 μM Aβ$_{25\text{-}35}$. Following this process, as a consequence of inhibition of GSK3β, β-catenin accumulated in the cytoplasm and subsequently translocated into the nucleus, where it acted as a transcriptional cofactor to promote neuronal survival [114]. In another study [115], they used the same laser and found three results. First, a 2 J/cm^2 fluence of PBM could protect SH-SY5Y cells against Aβ$_{25\text{-}35}$ (25 μM) neurotoxicity only at 24 and 48 h post-irradiation. Second, after exposure to Aβ$_{25\text{-}35}$, the cell viability of both SH-SY5Y cells and hippocampal neurons was increased by PBM in a dose-dependent manner, while a significant increase was only observed at fluences of 2 and 4 J/cm^2. Third, a 2 J/cm^2 fluence of PBM was enough to protect hippocampal neurons against Aβ$_{1\text{-}42}$ (25 μM) neurotoxicity [115]. More recently, Da Xing and co-workers suggested that by increasing the mitochondrial CCO activity and subsequently increasing the levels of cAMP and ATP, 632.8 nm laser PBM (2 J/cm^2, culture surface) was able to activate the PKA/SIRT1 signaling pathway in SH-SY5Y-APPswe cells resulting in reduced Aβ levels [117]. Moreover, transcranial 670 nm LED PBM (4 J/cm^2, at scalp level) for 4 weeks was shown to provide neuroprotection against the synaptic accumulation of toxic tau oligomers in two Tg mouse models of human tauopathies (hTau and 3xTgAD) [85, 118]. A study by Duggett and Chazot also demonstrated the neuroprotective effects of a relatively long-wavelength LED, 1068 nm at 4.5 J/cm^2, which was able to protect CAD neuroblastoma cells from cell death induced by Aβ$_{1\text{-}42}$ at 3.5–25 μM concentrations [116].

Age-related attenuation of HSP expression can result in abnormal polypeptide folding, leading to toxic accumulation of aggregates in the brain, which can trigger the extrinsic and intrinsic apoptotic pathways, and contribute to the increased risk of neurodegenerative diseases such as AD. Another study by the Chazot group [24] reported that PBM could up-regulate a panel of stress-response (heat-shock) proteins in the brain, known to inhibit both protein aggregation and apoptosis. Chronic whole-body 1072 nm LED PBM for 5 months significantly increased HSP60, HSP70, HSP105, and phosphorylated-HSP27 (p-HSP27) protein levels. They reported that PBM induced the following effects: 1) HSP60 may contribute to the decreased Aβ deposition by aiding protein folding; 2) HSP70 may prevent neuronal cell death through the c-Jun N-terminal kinases (JNK)-BID-mediated mitochondrial apoptotic pathway and apoptosis-inducing factor (AIF); 3) HSP105 could inhibit the mitogen-activated protein kinase-38 (MAPK38) and JNK stress-pathways resulting in a decrease in pro-apoptotic proteins [24].

4.2.4   Neuronal Oxidative Stress

It is well established that mitochondria are the main intracellular source of ROS and free radicals such as H_2O_2 and superoxide radical anion. The increased mitochondrial Ca^{2+} levels also increase free radical generation. Oxidative stress causes a variety of irreversible alterations to biological molecules such as DNA, proteins, and lipids, and this can potentially affect neurons in part by damaging their mitochondrial function [119]. A growing body of literature has shown a significant role of oxidative stress in various brain conditions, such as AD, stroke, traumatic brain injury (TBI), and psychiatric disorders [120–122]. Furthermore, studies have shown correlations between the cognitive impairment involved in normal aging and vulnerability to oxidative stress [123]. Indeed, the beneficial or inhibitory/harmful effects of PBM are in part associated with mitochondrial ROS and NO production. On the one hand, it is believed that PBM at the optimum fluence of <10 J/cm^2 produces a brief burst of mitochondrial ROS accompanied by a rise in MMP and this, in turn, mediates cellular signaling pathways involved in cell survival and proliferation. On the other hand, much higher fluencies of PBM (accompanied by a drop in MMP) can generate a massive amount of ROS [79]. These observations may explain the biphasic dose-response nature of PBM, which will be discussed in detail in the next chapter. PBM-generated modest levels of ROS allow mutual communication between mitochondria and the cytosol and/or the nucleus and could alter gene expression via inducing redox-sensitive transcription factors [124]. These signaling cascades may regulate the expression of genes related to antioxidant activity and improve the cellular ability to scavenge free radicals [125, 126].

The neuroprotective effect of PBM against ROS-induced neurotoxicity has been reported in vitro against damage caused by Aβ [127], CoCl$_2$ [72, 80], H_2O_2 [80, 128], KCN and MPP$^+$ [107, 109], glutamate, NMDA, and kainate [81], as well as rotenone [80]. In various animal models, PBM has also been shown to improve cerebral antioxidant defense system [58, 83, 84, 89]. A study in a mouse model of sleep deprivation revealed that acute PBM with 810 nm laser could enhance hippocampal total antioxidant capacity and antioxidant enzyme activity, such as SOD and GPx, as well as decrease levels of ROS and lipid peroxidation (MDA) [89]. The same results were observed in a mouse model of depression where 810 nm laser PBM significantly decreased levels of SOD, GPx, MDA and increased total antioxidant capacity (TAC) in the hippocampus and PFC, as well as reduced serum glutathione (GSH) levels [84].

Blue laser acupuncture (405 nm) also significantly elevated SOD and catalase and inhibited acetylcholinesterase activity in the hippocampus in a rat AD model [129]. In another series of studies in AD rats [58] and hypoxia–ischemia models [83], 808 nm laser PBM significantly ameliorated increased G6PDH and NADPH activity, as well as the increasing the levels of NADPH. Furthermore, PBM-treated animals showed a significant decrease in mitochondrial and cytoplasmic superoxide anion production in the hippocampal CA1 region. PBM also mitigated oxidative damage in CA1 neurons as measured by oxidative stress markers for lipid peroxidation, peroxynitrite production, DNA double-strand breaks, and oxidative DNA damage [58, 83].

NO is an essential messenger molecule, that at physiological levels is associated with several beneficial biological functions, such as immunomodulation, neurotransmission, and inhibition of platelet aggregation [130]. However, at high levels, NO can react with superoxide anion (O_2^-) and produce peroxynitrite $(ONOO^-)$, which can stimulate cellular apoptosis by acting as an oxidant or inducing free radicals. Prolonged and high levels of NO have neurotoxic effects and potentially contribute to neuronal damage. It has been shown in sodium nitroprusside (SNP)-treated SH-SY5Y cells, that 635 nm laser PBM protected against neuronal oxidative damage via blocking the mitochondrial apoptotic pathway induced by $ONOO^-$ synthesis and ROS production, most likely through promoting the scavenging of superoxide anion (O_2^-) [110]. In addition, suppression of NOS isoform activity (namely, endothelial NOS [eNOS], nNOS, and iNOS) after transcranial 660 nm laser treatment has been proposed as a likely mechanism responsible for antioxidant functions of PBM against oxidative stress in vivo [131].

SIRT1 has been shown to directly control PGC-1α activity through phosphorylation and deacetylation [132]. PGC-1α is a transcriptional coactivator that up-regulates the activity of antioxidant enzymes such as GPx and SOD [133]. Increasing the PGC-1α expression levels has been recommended as a neuroprotective strategy against ROS-mediated neuronal death [134]. Besides the aforementioned mechanisms for the antioxidant action of PBM, a recent study in a rat global ischemia model showed that 810 nm laser PBM alone or in combination with CoQ_{10} could abate ROS-induced neuronal damage by activation of the SIRT1/PGC-1α pathway [71].

4.2.5 Neuroinflammation

Accumulating evidence suggests that uncontrolled inflammatory responses are involved in the pathophysiology of most brain conditions. In response to several neuropathological conditions, microglia undergo a sequence of morphological and proliferative alterations leading to the excessive release of pro-inflammatory mediators, including chemokines, cytokines, ROS, and NO [135]. The increase in cerebral cytokines released by microglia has been associated with some neuropsychiatric disorders [136]. Generally, in the central nervous system, microglial cells react to inflammatory signals through the production of pro-inflammatory cytokines such as interleukin (IL)-1β, IL-2, IL-6, tumor necrosis factor-α (TNF-α), and interferon-γ (IFN-γ). The aforementioned cytokines not only stimulate inflammatory responses but also influence synaptic plasticity and memory formation, as well as neurotransmitter metabolism and mood regulation [137].

One of the important benefits of PBM is its pronounced anti-neuroinflammatory activity. It has been suggested that PBM-induced modulation of ROS, NO, cAMP, and Ca^{2+} is associated with the anti-inflammatory actions of low levels of red/NIR light [138]. PBM has been shown to attenuate pro-inflammatory cytokines through the inhibition of NF-κB signaling mediated by cAMP, resulting in a reduction of inflammatory response [139]. Evidence has suggested that the anti-neuroinflammatory activity of PBM may be partly due to its ability to regulate microglial activity, with a subsequent decrease in inflammatory mediators [60, 140]. Among the many possible cytokines, TNF-α and interleukins including IL-4, IL-6, IL-10, IL-18, and IL-1β have been the most investigated in brain/neuronal PBM [21].

White LEDs (411–777 nm) used alone or in combination with the bioflavonoid luteolin have been shown to inhibit lipopolysaccharide (LPS)-induced neuroinflammation by reducing TNF-α and IL-6 production and modulation of p38 and ERK signaling in BV2 microglial cells in vitro [141]. An 808 nm laser PBM with higher fluences of 4 and 30 J/cm^2 also induced expression of M1 polarization markers (CD86) in BV2 microglia, while markers of the M2 phenotype (CD206 and TIMP-1) were only observed at lower fluences of 0.2–10 J/cm^2. However, PBM at any light fluence had no significant effects on TNF-α and IL-1β production [142]. In another interesting study using microglia-like BV-2 cells and neuron-like neuroblastoma SH-SY5Y cells, 632.8 nm laser PBM (20 J/cm^2) inhibited LPS-activated microglia-mediated neuroinflammation and enhanced the phagocytic activity through activation of Src/PI3K/Akt/Rac1 signaling pathway [143]. It has also been reported that 632.8 nm laser pre-treatment could exert anti-neuroinflammatory activity by suppressing the Aβ-induced expression of IL-1β and iNOS in astrocytes [127].

Furthermore, numerous studies have shown anti-neuroinflammatory effects of PBM in various animal models of TBI [60, 71, 140, 147, 148], AD [149–151], PD [152], and depression [84]. In an early study on a rat model of cryogenic brain injury, laser PBM at either 660 or 780 nm significantly decreased IL-1β levels at 24 h compared to 6 h post-injury. However, this study reported no changes in brain TNF-α levels following PBM with both wavelengths at both time points [145]. Transcranial 810 nm laser PBM down-regulated the expression of some pro-inflammatory chemokines, including CXC-chemokine ligand 10 (CXCL10) and CC-chemokine ligand 2 (CCL2) in the mouse brain at both time points of 6 h and 28 days post-TBI. However, this work showed different short- and long-term neuroinflammatory responses in terms of cytokine expression. PBM reduced cerebral levels of IL-1β at 6 h, TNF-α at 28 days, and IL-6, CCL2, and CXCL10 levels at 6 h and 28 days but increased TNF-α levels at 6 h [146]. In a study using a mouse photothrombotic stroke model, transcranial 610 nm LED PBM suppressed neuroinflammatory responses, such as neutrophil infiltration and microglial activation in the ischemic cortex. PBM also decreased NLRP3 inflammasome-mediated brain damage, accompanied by down-regulation of IL-18 and IL-1β levels at 72 h post-ischemia. Apparently, the PBM-induced inhibition of NLRP3 expression and activity was achieved by reducing the activity of both intracellular NF-κB and MAPK signaling pathways, and by reducing TLR2 levels [140]. In another similar study using the same model, 610 nm LED pre-conditioning inhibited Iba-1 (a sensitive marker of microglial cells) and GFAP (a sensitive marker of astrocyte activation) positive cells in the ischemic cortex. These anti-neuroinflammatory responses were also accompanied by suppression of pro-inflammatory mediators (iNOS, COX-2, IL-1β, TNF-α, CCL2, and CXCL10) and inhibition of MAPK activation and NF-κB translocation [148]. In a rat middle cerebral artery occlusion (MCAO) stroke model, the decrease in the helper CD4$^+$ T-lymphocyte subset after MCAO was significantly reversed by transcranial 710 nm LED PBM, along with a reduction in microglial activation. While IL-4 expression was not significantly different between the ischemic and PBM-treated animals, IL-10 mRNA expression was significantly increased by PBM at 3 weeks post-stroke [147]. Furthermore, transcranial 808 nm laser PBM for 7 days suppressed cortical pro-inflammatory cytokines (TNF-α, IL-6, and IL-18) and increased anti-inflammatory cytokines (IL-4 and IL-10) in the rat at 14 days post-stroke [60]. An 810 nm laser PBM alone or in combination with CoQ$_{10}$ was shown to inhibit neuroinflammation via reduction of cerebral iNOS, TNF-α, and IL-1β levels in a mouse model of transient cerebral ischemia superimposed on a model of aging [71]. Two different studies using the 5XFAD transgenic [150] and Aβ-protein precursor (AβPP) transgenic [151] mouse models of AD, also demonstrated anti-neuroinflammatory activity of red/NIR PBM. The overall results showed that PBM could promote a reduction in the hyperactivation of microglia [150] and decrease neuroinflammation via suppression of TNF-α, IL-1β, and TGF-β [151]. In addition, another study in a mouse Aβ-induced AD model has shown that along with a reduction in the activation of astrocytes and microglia, combined red/NIR PBM significantly reduced hippocampal TNF-α, IL-1β, and IL-6 levels [149]. In a rat PD model, Salgado et al. [152] compared the effects of coherent laser (630 nm) and non-coherent LEDs (627 nm) on inflammatory markers and found that a scalp fluence of 4 J/cm^2 from both light sources was able to significantly decrease serum TNF-α and increase serum IFN-γ and IL-2 levels. Finally, in a mouse restraint stress model of depression, 810 nm laser PBM and/or CoQ$_{10}$ treatments suppressed stress-induced activation of microglia and secretion of pro-inflammatory markers (NF-κB, JNK, and p38) in the hippocampus and PFC areas, along with a decrease of serum levels of TNF-α and IL-6 [84].

4.2.6 Neuronal Apoptosis

Apoptosis is an important contributing pathophysiological mechanism in the aging brain [153] and in almost all brain injuries and disorders [154, 155]. The two well-understood mechanisms for apoptosis are the intrinsic pathway and the extrinsic pathway. The intrinsic pathway, also known as the mitochondrial pathway, is activated by intracellular signals generated when cells are under stress and depends on the release of some proteins from the mitochondrial intermembrane space. Specifically, according to this pathway, apoptosis is initiated by a decrease in MMP and release of the pro-apoptotic factor, cytochrome c, from the mitochondria into the cytoplasm, thereby activating downstream effector caspases, such as caspase-3 [156]. The pro-apoptotic and anti-apoptotic Bcl-2 family of proteins are known to be crucial regulators of apoptosis [157]. Overexpression of Bax or an elevated Bax/Bcl-2 protein expression ratio also triggers activation of the caspase cascade and results in apoptosis [158].

Duan et al. [159] reported the first in vitro evidence for the anti-apoptotic effects of PBM in neuronal cell culture and found that 640 nm LED PBM significantly prevented apoptosis induced by $A\beta_{25-35}$ toxicity in PC12 cells at 24 h post-irradiation [159]. Afterwards, it was shown that twice a day 670 nm LED PBM decreased the number of cortical and striatal neurons undergoing apoptosis caused by exposure to MPP^+ or rotenone [107]. PBM pre-treatment with 670 nm LED at culture surface fluences of 4 J/cm^2 [65, 111] or 30 J/cm^2 [109] also markedly rescued the primary neurons from apoptosis induced by various neurotoxins. A 635 nm LED PBM has been shown to inhibit the mitochondrial-dependent apoptotic pathway via the suppression of cytochrome c release, Bax protein down-regulation, and caspase-9 and caspase-3 inhibition in SNP-treated SH-SY5Y cells [110]. A 660 nm LED [160] or 810 nm laser [112] PBM also significantly diminished OGD-induced neuronal apoptosis through modulation of protein expression of apoptotic markers, namely, Bax, Bcl-2, and BAD, as well as down-regulation of caspase-3 activity. The protein kinase C (PKC) family contains serine/threonine kinases with essential roles in cellular apoptosis. The activation of PKC can affect the expression of cellular Bax and Bcl-xl, and in turn, lead to apoptosis inhibition [161, 162]. In this regard, it has been shown that 632.8 nm laser PBM at low fluences (0.15–0.62 J/cm^2) significantly inhibited apoptosis in PC12 cells by decreasing the Bax/Bcl-xl mRNA ratio through the PKC activation pathway [163]. In addition to the above-mentioned pathways [115], a 632.8 nm laser PBM has been demonstrated to inhibit the activation of glycogen synthase kinase (GSK-3b), Bax, and caspase-3 expression, resulting in the prevention of staurosporine-induced apoptosis through inactivation of the GSK-3b/Bax pathway [164]. Moreover, 632.8 nm laser exerted anti-apoptotic effects in PC12 cells by the activation of the Akt/YAP/p73 [165] and/or the Akt/GSK3b/b-catenin pathways [114].

The anti-apoptotic effects of PBM have also been reported in vivo in models of transient cerebral ischemia [83, 92, 140, 166, 167], AD [58, 149], TBI [144, 168, 169], depression [84], and normal brain aging [73]. Pre-treatment with 808 nm laser in a mouse transient cerebral ischemia model has been shown to significantly reduce the numbers of apoptotic cells in the hippocampus CA1 region and cerebral cortex, as indicated by TUNEL-positive cells [92]. Another study in a rat transient cerebral ischemia model suggested that 660 nm laser PBM could inhibit apoptosis by up-regulating Akt, pAkt, pBAD, and Bcl-2 expression, and down-regulating caspase-9 and caspase-3 expression [167]. In a mouse model of focal cerebral ischemia, a 610 nm LED PBM also attenuated neuronal apoptosis as shown by fewer TUNEL-positive cells on the ipsilateral side, mediated by inhibition of the inflammasome, caspase-1 and caspase-11 [140]. In addition, an interesting series of studies performed by Quanguang Zhang and his colleagues showed that transcranial 808 nm laser PBM could suppress cytochrome c release into the cytoplasm, and prevent neuronal apoptosis by inhibition of caspase-9 and caspase-3 activity in a rat model of AD [58], and neonatal hypoxic-ischemia injury [83, 166]. The anti-apoptotic effects of transcranial PBM via inhibiting the mitochondrial-dependent apoptotic pathway have also been shown in animal models of TBI, depression, and brain aging using various light sources, including 670 nm LEDs [168], 660 nm [73] and 810 nm lasers [73, 84, 169], and nano-pulsed 808 nm laser [144] irradiation. These significantly prevented neuronal programmed death via the down-regulation of pro-apoptotic proteins Bax, caspase-9 and caspase-3, and the up-regulation of anti-apoptotic protein Bcl-2. It should be noted that, so far, no studies have investigated the anti-apoptotic mechanisms of PBM in neuronal cells through the extrinsic pathway; thus, future studies are needed to explore this issue.

The inhibition of mitochondria-mediated apoptosis induced by PBM has been proposed as a potential therapeutic strategy to prevent neuronal death. In animal models of AD [58] and global cerebral ischemia [59], transcranial 808 nm laser PBM shifted the mitochondrial dynamics toward fusion, by balancing the mitochondrial-targeted fission proteins (Mff and Fis1) and fusion proteins (Opa1 and Mfn1). Also, by decreasing Drp1 GTPase fission protein activity and its binding to adaptor proteins (Mff and Fis1), PBM reduced mitochondrial fragmentation and protected hippocampal CA1 neurons from neuronal apoptosis.

4.2.7 Neurotrophic Factors and Neurogenesis/Synaptogenesis

Neurotrophic factors or neurotrophins are a family of proteins that support the growth, survival, and differentiation of neurons. Brain-derived neurotrophic factor (BDNF), glial cell-derived neurotrophic factor (GDNF), and neuronal growth factor (NGF) are as the main members of the neurotrophin family. BDNF can increase GAP-43 expression, a protein involved in the extension of the neurite growth cone [170]. GDNF is a member of the TGF-β family expressed by various cells such as microglia, astrocytes as well as neurons, that has a neurotrophic action on dopaminergic cells [171]. Increased expression of BDNF and NGF may account for the induction of neurogenesis and synaptogenesis [172]. Up-regulation of BDNF expression could result in reduced atrophy of cortical dendrites and hippocampus during the progression of AD [115]. Moreover, increased cerebral BDNF expression could contribute to reduced atrophy and cell death in the hippocampus and PFC regions in patients with depression and anxiety [173].

The first in vitro evidence to show that 810 nm laser PBM could significantly increase BDNF and GDNF expression (but not vascular endothelial growth factor [VEGF]), resulted in a subsequent increase in olfactory ensheathing cell proliferation [174]. A 632.8 nm laser PBM was shown to rescue dendritic atrophy and neuronal death in Aβ-treated neurons by activation of the ERK/CREB/BDNF pathway [115]. In another study using cultured dorsal root ganglion neurons [175], the Ca^{2+}/ERK/CREB cascade was suggested to be a potential signaling pathway involved in PBM-induced BDNF mRNA transcription. They found that 632.8 nm laser PBM activated intracellular IP3-sensitive receptor stores, resulting in release of $[Ca^{2+}]_I$ and subsequent activation of the ERK/ CREB pathway, which eventually improved BDNF expression [175]. In H_2O_2-treated hippocampal neurons in vitro, 660 nm LED PBM also up-regulated BDNF expression via the ERK and CREB signal transduction pathways [128]. In addition to these findings, whole-body PBM using 670 nm LED light notably increased BDNF expression in the occipital cortex [176] as well as hippocampus and serum [177] in a rat methanol-induced toxicity model. Red laser acupuncture (650 nm) to acupoints GV20 (head) and HT7 (right forepaw) also up-regulated the expression of CREB and BDNF in the hippocampus, and improved cognitive impairment in a rat ischemic model [178]. Intracranial 670 nm LED PBM has also been shown to increase GDNF expression in the brain striatum in a primate PD model [179].

Accumulating evidence has shown that hippocampal atrophy and neurogenesis deficits in the dentate gyrus play a key role in the pathophysiology of depression and AD [180, 181]. Based on the above-mentioned evidence, it is, therefore, reasonable to propose that PBM can induce a cascade of processes that lead to the induction of neurogenesis, and the migration of neurons and neuron-supporting glial cells in the damaged brain [144, 182, 183]. With respect to this assumption, in a noteworthy series of studies in a mouse TBI model, Michael Hamblin and his team showed that transcranial 810 nm laser PBM could considerably promote neurogenesis and up-regulate the migration of neural pro-genitor cells, and their differentiation into neurons. They also reported increased BDNF expression in the dentate gyrus and subventricular zone regions, along with the stim-ulation of synaptogenesis and neuroplasticity in the mouse cortex [169, 184–186]. In a photothrombotic rat model of ischemic stroke, PBM markedly enhanced cortical neuroge-nesis and maintained the survival of newly formed neurons, as evidenced by significantly higher levels of Ki67 (a marker for stem cell proliferation) and DCX (a pre-mature neu-ronal marker and differentiation marker). PBM also induced synaptogenesis between the newly formed neurons in the cortex represented by increased levels of protein markers of synaptic plasticity (spinophilin and synaptophysin) [60]. In a study conducted by Oron et al. [187] using a rat MCAO stroke model, transcranial 808 nm laser PBM significantly increased the number of proliferating cells (incorporating BrdU) in the subventricular zone, thereby stimulating neurogenesis, and induced migration of neuroprogenitor cells to the infarct area at 4 weeks post-stroke [187]. It is worth noting that in contrast to the results of Oron et al. [187], in a rat MCAO stroke model, transcranial 830 nm LED PBM applied 24 h after the onset of ischemia did not show beneficial effects in terms of cere-bral ischemic lesions and neurogenesis, as evidence by histological analysis of a neuronal marker (Fox3), glial marker (GFAP), and neurogenesis (DCX and Ki 67) at 12 weeks post-stroke. In order to explain the ineffective results, the authors postulated that since the MCAO model induces a large ischemic lesion, it is possible that this could limit the functional recovery, and possibly explain the failure of PBM in animals and humans with large stroke lesions. They also proposed that prolonged individual housing of the animals might prevent optimal recovery and negatively impact on the positive effects of PBM [188]. Finally, both acute (1 day) and chronic (10 days) transcranial NIR PBM applied to naïve rats also promoted hippocampal neurogenesis as indicated by high levels of BrdU-positive cells in the CA1 region [189].

4.2.8   Cerebral Neurotransmitter Systems

Up to date, several reports have shown the regulation of the levels of neurotransmitters in the brain following red/NIR PBM in naïve animals [190–195], animal models of epilepsy [196], and restraint-induced stress [197, 198]. In 1982, researchers from China introduced optical fibers into the caudate nucleus or frontal cortex of naïve rats and compared the effects of He-Ne (632.8 nm) and nitrogen (337.1 nm) lasers on the content of neurotransmitter aminoacids in the striatum. It was found that 632.8 nm low-level laser irradiation to the caudate nucleus of rats significantly decreased striatal levels of dopamine, serotonin, aspartic acid, and glutamic acid, and increased striatal GABA levels. However, irradiation to the frontal region decreased serotonin and its metabolites and increased aspartic acid, glutamic acid, and GABA levels. Irradiation with a nitrogen laser implanted into the frontal region resulted in a decrease of norepinephrine and serotonin and its metabolites, and an increase in aspartic acid and GABA levels. Also, irradiation to the caudate nucleus decreased dopamine and serotonin and its metabolites levels and increased only aspartic acid levels in striatal [191]. Afterward, the same group of researchers compared the effects of the above lasers on the contents of neurotransmitter aminoacids during conditioned avoidance response training [190]. Collectively, their results showed that only He-Ne laser irradiation to the caudate nucleus, promoted general movement, and facilitated the conditioned avoidance response, along with increased striatal concentrations of dopamine and norepinephrine [190]. According to a study by Ahmed and colleagues in naïve rats, low power (90 mW) 830 nm laser irradiation significantly suppressed axonal conduction of cortical tissue by decreasing cortical glutamate, aspartate, and glutamine, and increasing glycine [192]. In a subsequent study [196], Radwan et al. reported that 830 nm laser irradiation at 90 mW could reverse the neurochemical alterations in amino acid neurotransmitters induced by pilocarpine in a rat epilepsy model. They found that PBM could restore the increased cortical levels of glutamic acid, glutamine, glycine, and taurine, as well as increase the hippocampal levels of aspartate and glycine in pilocarpine-treated animals to near-control values [196]. Intracranial PBM with 840 nm PW laser into the intact rat brain has also been shown to decrease glutamate and increase dopamine levels in the striatum, suggesting the promising potential of deep brain PBM for neurological diseases by controlling the release of dopamine and glutamate to maintain the balance in the cerebrospinal fluid (CSF) [193]. A 3-month course of whole-body PBM in mice using white fluorescent light decreased dopamine and its metabolites in the striatum, while

710 nm LED PBM for the same treatment period did not produce any significant changes in striatal dopamine concentration [194]. It is clear from the above-mentioned literature that the results are fragmented and often contradictory, probably because of differences in experimental settings (e.g., animals' age, laser wavelength and dose, and irradiation area). Obviously, in this field, further research is needed to clarify the mechanisms through which PBM acts at neuronal and synaptic levels within the brain.

A growing body of literature has shown lower levels of serotonin and its metabolites in the blood, CSF, and postmortem brain tissue of depressed patients [199]. The lack of serotonin in the synapses is accompanied by a wide range of neurobehavioral abnormalities, including changes in mood and aggressive behavior, sleep disturbance, eating dysfunction, and suicide attempts [200]. Furthermore, anhedonia is a cardinal manifestation of major depression, related to a dysfunction in the reward system, in particular, dopamine system dysfunction [201]. Evidence indicates that lower levels of homovanillic acid (a dopamine metabolite) in the CSF of depressed patients is also associated with an increased risk of suicide [202]. In 1993, Cassone et al. [198] for the first time investigated the bimodulatory effects of 632.8 nm laser irradiation on brain biogenic amine levels in rats subjected to immobilization stress. They found that laser irradiation increased the cortical levels of noradrenaline in stressed animals, and decreased striatal and hippocampal levels of serotonin to control levels, while it did not significantly change dopamine levels in any of the brain areas [198]. In a second recent study, Eshaghi et al. [197] reported that transcranial 810 nm laser PBM could effectively increase serotonin in the prefrontal cortex and hippocampus of mice subjected to restraint-induced stress, along with amelioration of depression and anxiety-like behavior [197]. It could be assumed that brain PBM may stimulate neuronal mitochondria and improve ATP production, which in turn affects synaptic transmission and neurotransmitter release. Additional animal studies are needed to test the effects of PBM in the regulation of neurotransmitter systems, which are involved in mood disorders.

4.2.9 Effects on Intrinsic Brain Networks

The term "intrinsic brain networks" describes a collection of distant but integrated structures within the human brain with widespread neuronal connections. These include the default mode network (DMN), central executive network (CEN), and salience network (SN) as the main examples. These networks are not only activated upon stimulation by neural inputs, but their basal activity is detectable even in the resting state [203]. Intrinsic brain networks have been identified in animal species, including mice and monkeys [204]. These intrinsic networks can modulate the higher level emotional and cognitive functions [205]. Evidence suggests that both chronic neurodegenerative diseases and acute traumatic brain injuries lead to an imbalance in the activity of these networks [206, 207]. For instance, in TBI patients, abnormalities in higher-level cognitive performances have been shown to be associated with weak connections between the DMN, CEN, and SN nodes, resulting in impaired dynamic interactions within these networks [208, 209]. In addition, in AD patients, neuroimaging studies have demonstrated reduced functional connectivity in the DMN, using resting-state functional-connectivity MRI [210].

It was first suggested by Margaret Naeser [203, 211] that the geographical matching of transcranial light irradiation sites on the scalp, with the corresponding anatomical regions of intrinsic brain networks, could allow re-establishment of their connections. She has specifically suggested the following scalp placements to deliver light to specific cortical nodes within the intrinsic brain networks: (I) the probe location on the midline, centered over the front hairline and the upper forehead, designed to target the dorsal anterior cingulate (dACC) of the SN, and the mesial prefrontal cortex (mPFC) of the DMN; (II) the probe location on the high-parietal, midline precuneus area of DMN; (III) the probe location on the posterior to the front hairline, on a line up from the pupil of each eye, designed to target the dlPFC area of CEN [203]. Indeed, the cortical nodes within intrinsic brain networks have a high demand for energy to maintain their normal function. Thus, transcranial PBM can bring enough red/NIR light to the neuronal mitochondria within these cortical nodes, in order to enhance the impaired functional connectivity among these networks [212]. In this regard, Naeser and her colleagues [211] reported, for the first time, that transcranial LED PBM over the DMN, CEN, and SN nodes could improve cognitive function in TBI patients, likely through the increase of metabolic capacity within these intrinsic networks. Patients with post-traumatic stress disorder (PTSD) have also been found to exhibit reduced functional connectivity within the DMN, and increased connectivity within the SN [213]. It should also be noted that in the study of Naeser et al. [211] the decreased PTSD symptoms in TBI patients may have been associated with improved modulation between the DMN and SN networks [214]. Application of transcranial PBM in stroke patients with aphasia has also been shown to be beneficial through the stimulation of cortical nodes within the CEN network [215]. In addition to the above-mentioned evidence, most recently, Chao [216] applied 12 weeks of transcranial Vielight LED headset combined with an intranasal LED applicator in four patients with dementia. The Vielight headset device is fixed on the head at four specific scalp regions (according to EEG system: Fz, Pz, P3, and P4) and provides irradiation to the bilateral mPFC, precuneus/posterior cingulate cortex, and angular gyrus, which are considered to be the main hubs of the DMN. Based on their neuroimaging data, PBM increased cerebral perfusion and connectivity between the posterior cingulate cortex and lateral parietal nodes within the DMN, along with improved cognitive and behavioral functions [216]. The increased connectivity and synchronization of the brain networks in terms of integration and segregation has also been reported following a single session of transcranial LEDs PBM (810 nm, 40-Hz) in healthy elderly individuals [217].

4.2.10 Effects on Electroencephalogram Rhythms

Nowdays, several diagnostic methods for the measurement of brain response to neuro-modulation techniques have been developed, including fMRI, functional near-infrared spectroscopy (fNIRS), and EEG. The EEG, which measures the electrical fields produced by the brain at the surface of the scalp, provides neurophysiological information about brain function with a very high temporal resolution (about 1 ms). Among the above-mentioned methods, EEG is a more economical and easy-to-use method, which can be applied in many research laboratories or clinical settings [218]. Nevertheless, it has to be mentioned that the main disadvantages of EEG are its low spatial resolution (about 10 mm), and its inability to record subcortical or asynchronous neural activity [219]. Synchronized neural activity produces measurable rhythmic oscillations on the scalp surface; these signals can be grouped into five canonical frequency bands, namely, delta (1–3 Hz), theta (4–7 Hz), alpha (8–13 Hz), beta (14–30 Hz), and gamma (30–100 Hz) oscillations [220]. There are many reports demonstrating the association of delta band oscillations with cognitive function, in particular attention [221, 222], theta and alpha bands with working memory, as well as long-term and short-term memory [223, 224], beta bands with sensory and motor processing [225], and gamma bands with motor function and cognitive performance such as working memory and long-term memory [223]. Abnormal cortical oscillations and disruptions in brain wave activity have been observed in several neurodegenerative and neuropsychiatric disorders. For instance, in two different investigations, depressed [226] and AD [227] patients have shown disrupted EEG patterns, as evidenced by increases in theta and delta activity and decreases in alpha and beta activiy. During slow-wave sleep, schizophrenic patients exhibit decreased delta wave activity, while delta rhythm was demonstrated to be increased during waking periods in patients with more severe schizophrenia symptoms [228]. It is also worth mentioning that anhedonia as a non-clinical predictive factor in depression and schizophrenia is accompanied by elevated delta wave power in the anterior cingulate cortex [229]. In addition, decreased EEG delta rhythm during slow-wave sleep has been observed in PD animal models [230]. A subset of attention deficit hyperactivity disorder (ADHD) patients has been reported to show elevated theta wave activity [231].

Researchers have recently suggested that PBM can reverse disturbances in EEG patterns observed in brain disorders and can also modulate neural oscillations of the brain in healthy individuals [232–234]. In a rat model of reserpine-induced depression, 7 days of 810 nm laser PBM significantly decreased delta and increased theta, alpha, beta-1, and beta-2 frequency bands, which returned to the control values (except for theta wave) along with improvements in depression-like behavior [235]. To date, a few studies on healthy and clinical subjects have shown EEG spectral power changes in response to transcranial PBM [236]. A series of studies conducted in the laboratories of Francisco Gonzalez-Lima and Hanli Liu have revealed that transcranial CW 1064 nm laser PBM can modulate electrophysiological properties (especially neural oscillations) in the healthy human brain

[237–240]. In their first study on healthy elderly subjects (with subjective memory loss at risk for cognitive decline), each subject underwent EEG examinations before, during, and after 55 s of transcranial laser irradiation. The laser probe was focused only on a single forehead site centered at Fp2 delivering a scalp fluence of 137.5 J/cm^2. The results showed a clear improvement in alpha power, both during and after laser irradiation, with the largest peak in the occipital recording. However, only a small alpha effect was observed in the frontal recording, and no effects were seen in temporal recordings. Nonetheless, the temporal recordings showed increased gamma power (at >32-Hz) and smaller increases during the laser irradiation for beta bands (around 20-Hz). It should be noted that these alterations in EEG patterns were observed both ipsilateral (right forehead) and contralateral (left forehead) to the irradiation side, suggesting that laser PBM modulates bilateral neural networks in the resting-state [240]. In the second study, they showed that the laser-induced increase of alpha power in healthy subjects was dose-dependent, so that the most pronounced activation was found between 8 and 10 min after the irradiation start (77–96 J/cm^2 on the scalp). They proposed that the desynchronization of the alpha and beta bands, and subsequent stimulation of an ipsilateral, fronto-parieto-occipital network and a contralateral parieto-occipital network with the alpha band may in part explain the improved cognitive function following transcranial laser PBM [239]. Later, they carried out the first sham-controlled, time-resolved topographic mapping study of EEG/PBM in healthy subjects. They found that laser irradiation to the Fp2 site resulted in up to a 20% increase in spatially broad EEG alpha and beta band powers across the anterior to posterior regions, particularly at 8–11 min after the irradiation onset. The transcranial PBM had no statistically significant effects on delta and theta band powers [237, 238]. Recently, another group of researchers also investigated the effects of one session of 850 nm LED PBM on phase-amplitude coupling (PAC) between brain regions in healthy volunteers. Their results from EEG recordings showed that transcranial CW PBM to the Fp2 site could decrease brain delta-alpha coupling activity as well as cross-frequency interaction of the resting state resulting in improved cognitive function, in particular, sustained attention [241]. The effect of PW mode PBM on human neural oscillations has been explored in two clinical studies [217, 242]. A single session of transcranial LED PBM (810 nm, 40-Hz) to the DMN regions on the head significantly increased the power of alpha, beta, and gamma, and decreased the power of delta and theta frequencies in healthy elderly subjects in the resting state. As mentioned above, the reduced power in the higher frequency bands (alpha, beta, and gamma) and higher power in the lower frequency bands (delta and theta) are associated with disorders involving cognitive impairment (e.g., dementia, AD), suggesting that 40-Hz PW mode transcranial PBM might be a promising approach to ameliorate AD symptoms [217]. It has also been proposed that since the amplitude of gamma oscillations is directly associated with cerebral GABA levels, and since during memory consolidation the higher gamma oscillations could improve the cognitive function, PW mode PBM may act by stimulating GABA release, but this needs to be confirmed in future studies [217]. In the second study, Yao et al. [242] tested whether and how transcranial 810 nm LED irradiation with varying frequencies (0–Hz, 5–Hz, 10–Hz, or 20 Hz) could affect brain activity in healthy young subjects by analyzing the measured EEG signals during a single PBM session. Their results showed that with an increase in pulse frequency, the EEG signals increased more rapidly, indicating that brain activity was increased. Whereas the increase of relative energy of EEG signals was larger, suggesting that PBM using higher pulse frequencies (e.g., 10 and 20-Hz) may have better effects on improving memory function. These findings also suggested that the main site of activation induced by PBM was in the frontal region, and the activation became more significant with the increased pulse frequency [242].

Quantitative EEG (QEEG) is a useful tool involving the numerical analysis of EEG data, based on measurable parameters, such as amplitude, reaction time, and P300 event-related potential (ERP). Up to now, only a few studies have employed QEEG in a clinical setting to investigate electrophysiological changes in the brain associated with transcranial/intranasal PBM [243–246]. In a study conducted by Grover et al. [244], healthy individuals underwent QEEG event-related response tests before and after transcranial 903 nm laser PBM. Although neuronal coherence (as measured by P300) showed no meaningful changes, PBM significantly increased the amplitude in individuals initially displaying low-voltage readings, resulting in an acute change in the electrical brain state favoring faster reaction times. They suggested that since Ca^{2+} ions are involved in synaptic and extracellular processes that influence coherent neuronal firing, NIR PBM might affect calcium channel-driven neuronal activity and related cell voltage and could decrease reaction times via this mechanism [244]. In another study on healthy volunteers, one session of transcranial 850 nm LED PBM to the Fp2 site resulted in a significant decrease in both the relative and absolute power of the delta band, accompanied by a decrease in reaction time in the attentional Go/No-Go task [243]. Berman and his colleagues also reported for the first time, that transcranial 1072 nm LED PBM for 28 consecutive days could improve the QEEG amplitude and connectivity measures, shown by normalization of the absolute and relative power of the abnormal delta, as well as alpha relative power, resulting in improved executive function in subjects diagnosed with early to mid-stage dementia [245]. In another pilot study, Berman et al. [246] also reported that a single session of 810 nm intranasal LED PBM positively modified both QEEG absolute power and coherence measures in young subjects with social anxiety, bipolar disorder, or TBI. Combined transcranial plus intranasal PBM for 6 days/week for 2 weeks was also shown to increase baseline absolute power of all brain oscillations (alpha, delta, and theta) in a case study of a single AD patient [247].

References

1. Hamblin, M.R. 2018. Mechanisms and mitochondrial redox signaling in photobiomodulation. *Photochemistry and Photobiology* 94 (2): 199–212.
2. Karu, T.I. 2008. Mitochondrial signaling in mammalian cells activated by red and near-IR radiation. *Photochemistry and Photobiology* 84 (5): 1091–1099.
3. Karu, T. 1999. Primary and secondary mechanisms of action of visible to near-IR radiation on cells. *Journal of Photochemistry and Photobiology B: Biology* 49 (1): 1–17.
4. Chance, B., and B. Hess. 1959. Spectroscopic evidence of metabolic control: Rapid measurements of intracellular events afford new evidence on mechanisms for metabolic control. *Science* 129 (3350): 700–708.
5. Gonzalez-Lima, F., B.R. Barksdale, and J.C. Rojas. 2014. Mitochondrial respiration as a target for neuroprotection and cognitive enhancement. *Biochemical Pharmacology* 88 (4): 584–593.
6. Karu, T.I., and S. Kolyakov. 2005. Exact action spectra for cellular responses relevant to phototherapy. *Photomedicine and Laser Therapy* 23 (4): 355–361.
7. Wong-Riley, M.T., et al. 2005. Photobiomodulation directly benefits primary neurons functionally inactivated by toxins role of cytochrome c oxidase. *Journal of Biological Chemistry* 280 (6): 4761–4771.
8. de Freitas, L.F., and M.R. Hamblin. 2016. Proposed mechanisms of photobiomodulation or low-level light therapy. *IEEE Journal of Selected Topics in Quantum Electronics* 22 (3): 348–364.
9. Santana-Blank, L., E. Rodriguez-Santana, and K. Santana-Rodriguez. 2010. Theoretic, experimental, clinical bases of the water oscillator hypothesis in near-infrared photobiomodulation. *Photomedicine and Laser Surgery* 28 (S1): S-41–S-52.
10. Karu, T. 1988. Molecular mechanism of the therapeutic effect of low-intensity laser radiation. *Lasers Life Science* 2 (1): 53–74.
11. Karu, T.I. 2013. Cellular and molecular mechanisms of photobiomodulation (low-power laser therapy). *IEEE Journal of Selected Topics in Quantum Electronics* 20 (2): 143–148.
12. Ganeshan, V., et al. 2019. Pre-conditioning with remote photobiomodulation modulates the brain transcriptome and protects against MPTP insult in mice. *Neuroscience* 400: 85–97.
13. George, S., M.R. Hamblin, and H. Abrahamse. 2020. Photobiomodulation-induced differentiation of immortalized adipose stem cells to neuronal cells. *Lasers in Surgery and Medicine* 52 (10): 1032–1040.
14. Masha, R.T., N.N. Houreld, and H. Abrahamse. 2013. Low-intensity laser irradiation at 660 nm stimulates transcription of genes involved in the electron transport chain. *Photomedicine and Laser Surgery* 31 (2): 47–53.
15. Chen, A.C., et al. 2011. Low-level laser therapy activates NF-kB via generation of reactive oxygen species in mouse embryonic fibroblasts. *PLoS ONE* 6 (7): e22453.
16. Serrage, H., et al. 2019. Under the spotlight: mechanisms of photobiomodulation concentrating on blue and green light. *Photochemical & Photobiological Sciences*.
17. Fukuzaki, Y., et al. 2013. 532 nm low-power laser irradiation recovers γ-secretase inhibitor-mediated cell growth suppression and promotes cell proliferation via Akt signaling. *PLoS ONE* 8 (8): e70737.
18. Montell, C. 2011. The history of TRP channels, a commentary and reflection. *Pflügers Archiv-European Journal of Physiology* 461 (5): 499–506.
19. Gu, Q., et al. 2012. Stimulation of TRPV1 by green laser light. *Evidence-Based Complementary and Alternative Medicine*.

20. Oldham, M.A., and D.A. Ciraulo. 2014. Bright light therapy for depression: A review of its effects on chronobiology and the autonomic nervous system. *Chronobiology International* 31 (3): 305–319.

21. Salehpour, F., et al. 2018. Brain photobiomodulation therapy: A narrative review. *Molecular Neurobiology* 55 (8): 6601–6636.

22. Wang, Y. et al. 2017. Photobiomodulation of human adipose-derived stem cells using 810 nm and 980 nm lasers operates via different mechanisms of action. *Biochimica et Biophysica Acta (BBA)-General Subjects* 1861 (2): 441–449.

23. Bradford, A., A. Barlow, and P.L. Chazot. 2005. Probing the differential effects of infrared light sources IR1072 and IR880 on human lymphocytes: Evidence of selective cytoprotection by IR1072. *Journal of Photochemistry and Photobiology B: Biology* 81 (1): 9–14.

24. Grillo, S., et al. 2013. Non-invasive infra-red therapy (1072 nm) reduces β-amyloid protein levels in the brain of an Alzheimer's disease mouse model, TASTPM. *Journal of Photochemistry and Photobiology B: Biology* 123: 13–22.

25. Zinchenko, E., et al. 2019. Pilot study of transcranial photobiomodulation of lymphatic clearance of beta-amyloid from the mouse brain: Breakthrough strategies for non-pharmacologic therapy of Alzheimer's disease. *Biomedical Optics Express* 10 (8): 4003–4017.

26. Semyachkina-Glushkovskaya, O., et al. 2017. Laser-induced generation of singlet oxygen and its role in the cerebrovascular physiology. *Progress in Quantum Electronics* 55: 112–128.

27. Dolgova, D., et al. 2019. Anti-inflammatory and cell proliferative effect of the 1270 nm laser irradiation on the BALB/c nude mouse model involves activation of the cell antioxidant system. *Biomedical Optics Express* 10 (8): 4261–4275.

28. Drochioiu, G. 2010. Laser-induced ATP formation: Mechanism and consequences. *Photomedicine and Laser Surgery* 28 (4): 573–574.

29. Chung, H., et al. 2012. The nuts and bolts of low-level laser (light) therapy. *Annals of Biomedical Engineering* 40 (2): 516–533.

30. Shen, D., et al. 2019. Besides photothermal effects, low-level CO_2 laser irradiation can potentiate skin microcirculation through photobiomodulation mechanisms. *Photobiomodulation, Photomedicine, and Laser Surgery* 37 (3): 151–158.

31. Sommer, A.P. 2019. Revisiting the Photon/Cell Interaction Mechanism in Low-Level Light Therapy. *Photobiomodulation, Photomedicine, and Laser Surgery*.

32. Sommer, A.P., M.K. Haddad, and H.-J. Fecht. 2015. Light effect on water viscosity: Implication for ATP biosynthesis. *Scientific reports* 5: 12029.

33. Hennessy, M., and M.R. Hamblin. 2016. Photobiomodulation and the brain: A new paradigm. *Journal of Optics* 19 (1): 013003.

34. Quirk, B.J., et al. 2012. Therapeutic effect of near infrared (NIR) light on Parkinson's disease models. *Frontiers in Bioscience (Elite Edition)* 4: 818–823.

35. Karu, T. and N. Afanas' eva. 1995. Cytochrome c oxidase as the primary photoacceptor upon laser exposure of cultured cells to visible and near IR-range light. In *Doklady Akademii Nauk*.

36. Sanderson, T.H., et al. 2018. Inhibitory modulation of cytochrome c oxidase activity with specific near-infrared light wavelengths attenuates brain ischemia/reperfusion injury. *Scientific Reports* 8 (1): 3481.

37. Amaroli, A., S. Ferrando, and S. Benedicenti. 2019. Photobiomodulation affects key cellular pathways of all life-forms: Considerations on old and new laser light targets and the calcium issue. *Photochemistry and Photobiology* 95 (1): 455–459.

38. Ravera, S., et al. 2019. 1064 nm Nd: YAG laser light affect transmembrane mitochondria respiratory chain complexes. *Journal of Biophotonics* e201900101.

39. Lima, P.L., et al. 2019. Photobiomodulation enhancement of cell proliferation at 660 nm does not require cytochrome c oxidase. *Journal of Photochemistry and Photobiology B: Biology*.

40. Lubart, R., et al. 2005. Low-energy laser irradiation promotes cellular redox activity. *Photomedicine and Laser Therapy* 23 (1): 3–9.

41. Borutaite, V., A. Budriunaite, and G.C. Brown. 2000. Reversal of nitric oxide-, peroxynitrite- and S-nitrosothiol-induced inhibition of mitochondrial respiration or complex I activity by light and thiols. *Biochimica et Biophysica Acta (BBA)-Bioenergetics* 1459 (2–3): 405–412.

42. Lohr, N.L., et al. 2009. Enhancement of nitric oxide release from nitrosyl hemoglobin and nitrosyl myoglobin by red/near infrared radiation: Potential role in cardioprotection. *Journal of Molecular and Cellular Cardiology* 47 (2): 256–263.

43. Mattson, M.P., M. Gleichmann, and A. Cheng. 2008. Mitochondria in neuroplasticity and neurological disorders. *Neuron* 60 (5): 748–766.

44. Rezin, G.T., et al. 2009. Mitochondrial dysfunction and psychiatric disorders. *Neurochemical Research* 34 (6): 1021.

45. Nunnari, J., and A. Suomalainen. 2012. Mitochondria: In sickness and in health. *Cell* 148 (6): 1145–1159.

46. Passarella, S., and T. Karu. 2014. Absorption of monochromatic and narrow band radiation in the visible and near IR by both mitochondrial and non-mitochondrial photoacceptors results in photobiomodulation. *Journal of Photochemistry and Photobiology B: Biology* 140: 344–358.

47. Yang, L., et al. 2020. Mitochondria as a target for neuroprotection: Role of methylene blue and photobiomodulation. *Translational Neurodegeneration* 9: 1–22.

48. Jack, C., A. del Olmo, and S. Valles. 2019. Can mild cognitive impairment be stabilized by showering brain mitochondria with laser photons? *Neuropharmacology* 107841.

49. Beauvoit, B., T. Kitai, and B. Chance. 1994. Contribution of the mitochondrial compartment to the optical properties of the rat liver: A theoretical and practical approach. *Biophysical Journal* 67 (6): 2501–2510.

50. Szundi, I., G.-L. Liao, and Ó. Einarsdóttir. 2001. Near-infrared time-resolved optical absorption studies of the reaction of fully reduced cytochrome c oxidase with dioxygen. *Biochemistry* 40 (8): 2332–2339.

51. Wong-Riley, M.T., et al. 2001. Light-emitting diode treatment reverses the effect of TTX on cytochrome oxidase in neurons. *NeuroReport* 12 (14): 3033–3037.

52. Chen, X., et al. 2014. Effect and mechanism of 808 nm light pretreatment of hypoxic primary neurons. *International Journal of Photoenergy*.

53. Silveira, P.C.L., et al. 2019. Effects of photobiomodulation on mitochondria of brain, muscle, and C6 astroglioma cells. *Medical Engineering & Physics* 71: 108–113.

54. Rojas, J.C., A.K. Bruchey, and F. Gonzalez-Lima. 2012. Low-level light therapy improves cortical metabolic capacity and memory retention. *Journal of Alzheimer's Disease* 32 (3): 741–752.

55. Rojas, J.C., et al. 2008. Neuroprotective effects of near-infrared light in an in vivo model of mitochondrial optic neuropathy. *Journal of Neuroscience* 28 (50): 13511–13521.

56. Mathangi, D., and R. Shyamala. 2016. Effect of LED photobiomodulation on fluorescent light induced changes in cellular ATPases and Cytochrome c oxidase activity in Wistar rat. *Lasers in Medical Science* 31 (9): 1803–1809.

57. Xu, Z., et al. 2017. Low-level laser irradiation improves depression-like behaviors in mice. *Molecular Neurobiology* 54 (6): 4551–4559.

58. Lu, Y., et al. 2017. Low-level laser therapy for beta amyloid toxicity in rat hippocampus. *Neurobiology of Aging* 49: 165–182.

59. Wang, R., et al. 2019. Photobiomodulation for global cerebral ischemia: Targeting mitochondrial dynamics and functions. *Molecular Neurobiology* 56 (3): 1852–1869.

60. Yang, L., et al. 2018. Photobiomodulation therapy promotes neurogenesis by improving post-stroke local microenvironment and stimulating neuroprogenitor cells. *Experimental Neurology* 299: 86–96.

61. Wang, X., et al. 2017. Up-regulation of cerebral cytochrome-c-oxidase and hemodynamics by transcranial infrared laser stimulation: A broadband near-infrared spectroscopy study. *Journal of Cerebral Blood Flow & Metabolism* 37 (12): 3789–3802.

62. Wang, X., et al. 2017. Impact of heat on metabolic and hemodynamic changes in transcranial infrared laser stimulation measured by broadband near-infrared spectroscopy. *Neurophotonics* 5 (1): 011004.

63. Dmochowski, G.M., et al. 2020. Near-infrared light increases functional connectivity with a non-thermal mechanism. *Cerebral Cortex Communications* 1 (1): tgaa004.

64. Hamblin, M.R. and Y.-Y. Huang. 2019. *Photobiomodulation in the brain: Low-level laser (light) therapy in neurology and neuroscience.* Academic Press.

65. Ying, R., et al. 2008. Pretreatment with near-infrared light via light-emitting diode provides added benefit against rotenone-and MPP+-induced neurotoxicity. *Brain Research* 1243: 167–173.

66. Sommer, A.P., et al. 2012. 670 nm laser light and EGCG complementarily reduce amyloid-β aggregates in human neuroblastoma cells: Basis for treatment of Alzheimer's disease? *Photomedicine and Laser Surgery* 30 (1): 54–60.

67. Trimmer, P.A., et al. 2009. Reduced axonal transport in Parkinson's disease cybrid neurites is restored by light therapy. *Molecular Neurodegeneration* 4 (1): 26.

68. Oron, U., et al. 2007. Ga-As (808 nm) laser irradiation enhances ATP production in human neuronal cells in culture. *Photomedicine and Laser Surgery* 25 (3): 180–182.

69. Luisa, Z., et al. 2019. Photobiomodulation therapy at different wavelength impacts on retinoid acid–dependent SH-SY5Y differentiation. *Lasers in Medical Science* 1–6.

70. Lapchak, P.A. 2010. Taking a light approach to treating acute ischemic stroke patients: Transcranial near-infrared laser therapy translational science. *Annals of Medicine* 42 (8): 576–586.

71. Salehpour, F., et al. 2019. Photobiomodulation and coenzyme Q10 treatments attenuate cognitive impairment associated with model of transient global brain ischemia in artificially aged mice. *Frontiers in Cellular Neuroscience* 13.

72. Dong, T., et al. 2015. Low-level light in combination with metabolic modulators for effective therapy of injured brain. *Journal of Cerebral Blood Flow & Metabolism* 35 (9): 1435–1444.

73. Salehpour, F., et al. 2017. Transcranial low-level laser therapy improves brain mitochondrial function and cognitive impairment in D-galactose–induced aging mice. *Neurobiology of Aging* 58: 140–150.

74. Passarella, S., et al. 1984. Increase of proton electrochemical potential and ATP synthesis in rat liver mitochondria irradiated in vitro by helium-neon laser. *FEBS Letters* 175 (1): 95–99.

75. Hu, W.-P., et al. 2007. Helium–neon laser irradiation stimulates cell proliferation through photostimulatory effects in mitochondria. *Journal of Investigative Dermatology* 127 (8): 2048–2057.

76. Vos, M., et al. 2013. Near-infrared 808 nm light boosts complex IV-dependent respiration and rescues a Parkinson-related pink1 model. *PLoS ONE* 8 (11): e78562.

77. Alexandratou, E., et al. 2002. Human fibroblast alterations induced by low power laser irradiation at the single cell level using confocal microscopy. *Photochemical & Photobiological Sciences* 1 (8): 547–552.

78. Giuliani, A., et al. 2009. Low infra red laser light irradiation on cultured neural cells: Effects on mitochondria and cell viability after oxidative stress. *BMC Complementary and Alternative Medicine* 9 (1): 8.

79. Sharma, S.K., et al. 2011. Dose response effects of 810 nm laser light on mouse primary cortical neurons. *Lasers in Surgery and Medicine* 43 (8): 851–859.

80. Huang, Y.Y., et al. 2013. Low-level laser therapy (LLLT) reduces oxidative stress in primary cortical neurons in vitro. *Journal of Biophotonics* 6 (10): 829–838.

81. Huang, Y.Y., et al. 2014. Low-level laser therapy (810 nm) protects primary cortical neurons against excitotoxicity in vitro. *Journal of Biophotonics* 7 (8): 656–664.

82. Rhee, Y.-H., et al. 2019. Effect of photobiomodulation therapy on neuronal injuries by ouabain: The regulation of Na, K-ATPase; Src; and mitogen-activated protein kinase signaling pathway. *BMC Neuroscience* 20 (1): 19.

83. Tucker, L.D., et al. 2018. Photobiomodulation therapy attenuates hypoxic-ischemic injury in a neonatal rat model. *Journal of Molecular Neuroscience* 65 (4): 514–526.

84. Salehpour, F., et al. 2019. Near-infrared photobiomodulation combined with coenzyme Q10 for depression in a mouse model of restraint stress: Reduction in oxidative stress, neuroinflammation, and apoptosis. *Brain RESEARCH BULLETIN* 144: 213–222.

85. Comerota, M.M., B. Krishnan, and G. Taglialatela. 2017. Near infrared light decreases synaptic vulnerability to amyloid beta oligomers. *Scientific Reports* 7 (1): 15012.

86. Hagen, T.M., C.M. Wehr, and B.N. Ames. 1998. Mitochondrial decay in aging: Reversal through supplementation of acetyl-l-carnitine and N-tert-butyl-α-phenyl-nitrone a. *Annals of the New York Academy of Sciences* 854 (1): 214–223.

87. Golpich, M., et al. 2017. Mitochondrial dysfunction and biogenesis in neurodegenerative diseases: Pathogenesis and treatment. *CNS Neuroscience & Therapeutics* 23 (1): 5–22.

88. Hock, M.B., and A. Kralli. 2009. Transcriptional control of mitochondrial biogenesis and function. *Annual Review of Physiology* 71: 177–203.

89. Salehpour, F., et al. 2018. Transcranial near-infrared photobiomodulation attenuates memory impairment and hippocampal oxidative stress in sleep-deprived mice. *Brain Research* 1682: 36–43.

90. Sadigh-Eteghad, S., et al. 2015. Effect of alpha-7 nicotinic acetylcholine receptor activation on beta-amyloid induced recognition memory impairment. Possible role of neurovascular function. *Acta Cirurgica Brasileira* 30 (11): 736–742.

91. Borghammer, P., et al. 2012. Cerebral oxygen metabolism in patients with early Parkinson's disease. *Journal of the Neurological Sciences* 313 (1–2): 123–128.

92. Uozumi, Y., et al. 2010. Targeted increase in cerebral blood flow by transcranial near-infrared laser irradiation. *Lasers in Surgery and Medicine* 42 (6): 566–576.

93. Litscher, G., et al. 2015. Transcranial yellow, red, and infrared laser and LED stimulation: Changes of vascular parameters in a chick embryo model. *Integrative Medicine International* 2 (1–2): 80–89.

94. Ball, K.A., P.R. Castello, and R.O. Poyton. 2011. Low intensity light stimulates nitrite-dependent nitric oxide synthesis but not oxygen consumption by cytochrome c oxidase: Implications for phototherapy. *Journal of Photochemistry and Photobiology B: Biology* 102 (3): 182–191.

95. Lee, D.-J., et al. 2019. Photobiomodulation therapy in mice with chronic cerebral hypoperfusion using application-specific near-infrared light-emitting diode system. *Transactions on Electrical and Electronic Materials* 1–6.

96. de la Torre, J.C. 2016. Cerebral perfusion enhancing interventions: A new strategy for the prevention of Alzheimer dementia. *Brain Pathology* 26 (5): 618–631.

97. Gonzalez-Lima, F., and D.W. Barrett. 2014. Augmentation of cognitive brain functions with transcranial lasers. *Frontiers in Systems Neuroscience* 8: 36.

98. Salgado, A.S., et al. 2015. The effects of transcranial LED therapy (TCLT) on cerebral blood flow in the elderly women. *Lasers in Medical Science* 30 (1): 339–346.

99. Tian, F., et al. 2016. Transcranial laser stimulation improves human cerebral oxygenation. *Lasers in Surgery and Medicine* 48 (4): 343–349.
100. Nawashiro, H., et al. 2012. Focal increase in cerebral blood flow after treatment with near-infrared light to the forehead in a patient in a persistent vegetative state. *Photomedicine and Laser Surgery* 30 (4): 231–233.
101. Schiffer, F., et al. 2009. Psychological benefits 2 and 4 weeks after a single treatment with near infrared light to the forehead: A pilot study of 10 patients with major depression and anxiety. *Behavioral and Brain Functions* 5 (1): 46.
102. Nawashiro, H., et al. 2017. Blood-oxygen-level-dependent (BOLD) functional magnetic resonance imaging (fMRI) during transcranial near-infrared laser irradiation. *Brain Stimulation: Basic, Translational, and Clinical Research in Neuromodulation* 10 (6): 1136–1138.
103. Litscher, G. 2019. *Brain photobiomodulation—Preliminary results from regional cerebral oximetry and thermal imaging.* Multidisciplinary Digital Publishing Institute.
104. Litscher, G. 2018. *Transcranial laser stimulation research—A new helmet and first data from near infrared spectroscopy.* Multidisciplinary Digital Publishing Institute.
105. Kozel, F.A., et al. 2009. Using simultaneous repetitive transcranial magnetic stimulation/functional near infrared spectroscopy (rTMS/fNIRS) to measure brain activation and connectivity. *NeuroImage* 47 (4): 1177–1184.
106. Mintzopoulos, D., et al. 2017. Effects of near-infrared light on cerebral bioenergetics measured with phosphorus magnetic resonance spectroscopy. *Photomedicine and Laser Surgery* 35 (8): 395–400.
107. Liang, H.L., et al. 2008. Near-infrared light via light-emitting diode treatment is therapeutic against rotenone-and 1-methyl-4-phenylpyridinium ion-induced neurotoxicity. *Neuroscience* 153 (4): 963–974.
108. Gu, X., et al. 2017. Photoactivation of ERK/CREB/VMAT2 pathway attenuates MPP+-induced neuronal injury in a cellular model of Parkinson's disease. *Cellular Signalling* 37: 103–114.
109. Liang, H., et al. 2006. Photobiomodulation partially rescues visual cortical neurons from cyanide-induced apoptosis. *Neuroscience* 139 (2): 639–649.
110. Lim, W., et al. 2009. Inhibition of mitochondria-dependent apoptosis by 635-nm irradiation in sodium nitroprusside-treated SH-SY5Y cells. *Free Radical Biology and Medicine* 47 (6): 850–857.
111. Wong-Riley, M., et al. 2002. cDNA microarray analysis of the visual cortex exposed to light-emitting diode treatment in monocularly enucleated rats. *Social Neuroscience Abstract.*
112. Yu, Z., et al. 2015. Near infrared radiation protects against oxygen-glucose deprivation-induced neurotoxicity by down-regulating neuronal nitric oxide synthase (nNOS) activity in vitro. *Metabolic Brain Disease* 30 (3): 829–837.
113. Iwase, T., et al. 1996. Low power laser irradiation reduces ischemic damage in hippocampal slices in vitro. *Lasers in Surgery and Medicine: The Official Journal of the American Society for Laser Medicine and Surgery* 19 (4): 465–470.
114. Liang, J., L. Liu, and D. Xing. 2012. Photobiomodulation by low-power laser irradiation attenuates Aβ-induced cell apoptosis through the Akt/GSK3β/β-catenin pathway. *Free Radical Biology and Medicine* 53 (7): 1459–1467.
115. Meng, C., Z. He, and D. Xing. 2013. Low-level laser therapy rescues dendrite atrophy via upregulating BDNF expression: Implications for Alzheimer's disease. *Journal of Neuroscience* 33 (33): 13505–13517.
116. Duggett, N.A., and P.L. Chazot. 2014. Low-intensity light therapy (1068 nm) protects CAD neuroblastoma cells from β-amyloid-mediated cell death. *Biologie et Médecine* 1 (103): 2.

117. Zhang, Z., et al. 2019. Activation of PKA/SIRT1 signaling pathway by photobiomodulation therapy reduces Aβ levels in Alzheimer's disease models. *Aging Cell.*

118. Comerota, M.M., et al. 2019. Near infrared light treatment reduces synaptic levels of toxic tau oligomers in two transgenic mouse models of human tauopathies. *Molecular Neurobiology* 56 (5): 3341–3355.

119. Bhat, A.H., et al. 2015. Oxidative stress, mitochondrial dysfunction and neurodegenerative diseases; a mechanistic insight. *Biomedicine & Pharmacotherapy* 74: 101–110.

120. Rodriguez-Rodriguez, A., et al. 2014. Oxidative stress in traumatic brain injury. *Current Medicinal Chemistry* 21 (10): 1201–1211.

121. Manzanero, S., T. Santro, and T.V. Arumugam. 2013. Neuronal oxidative stress in acute ischemic stroke: Sources and contribution to cell injury. *Neurochemistry International* 62 (5): 712–718.

122. Maurya, P.K., et al. 2016. The role of oxidative and nitrosative stress in accelerated aging and major depressive disorder. *Progress in Neuro-Psychopharmacology and Biological Psychiatry* 65: 134–144.

123. Berlett, B.S., and E.R. Stadtman. 1997. Protein oxidation in aging, disease, and oxidative stress. *Journal of Biological Chemistry* 272 (33): 20313–20316.

124. Zhang, Z., et al. 2001. Reactive oxygen species mediate tumor necrosis factor alpha-converting, enzyme-dependent ectodomain shedding induced by phorbol myristate acetate. *The FASEB Journal* 15 (2): 303–305.

125. Song, S., et al. 2003. cDNA microarray analysis of gene expression profiles in human fibroblast cells irradiated with red light. *Journal of Investigative Dermatology* 120 (5): 849–857.

126. Iakymenko, I. and E. Sydoryk. 2001. Regulatory role of low-intensity laser radiation on the status of the antioxidant system. *Ukrains' kyi biokhimichnyi zhurnal* 73 (1): 16–23.

127. Yang, X., et al. 2010. Low energy laser light (632.8 nm) suppresses amyloid-β peptide-induced oxidative and inflammatory responses in astrocytes. *Neuroscience* 171 (3): 859–868.

128. Heo, J.-C., et al. 2019. Photobiomodulation (660 nm) therapy reduces oxidative stress and induces BDNF expression in the hippocampus. *Scientific Reports* 9.

129. Sutalangka, C., et al. 2013. Laser acupuncture improves memory impairment in an animal model of Alzheimer's disease. *Journal of Acupuncture and Meridian Studies* 6 (5): 247–251.

130. Lowenstein, C.J., J.L. Dinerman, and S.H. Snyder. 1994. Nitric oxide: A physiologic messenger. *Annals of Internal Medicine* 120 (3): 227–237.

131. Leung, M.C., et al. 2002. Treatment of experimentally induced transient cerebral ischemia with low energy laser inhibits nitric oxide synthase activity and up-regulates the expression of transforming growth factor-beta 1. *Lasers in Surgery and Medicine: The Official Journal of the American Society for Laser Medicine and Surgery* 31 (4): 283–288.

132. Cantó, C., and J. Auwerx. 2009. Caloric restriction, SIRT1 and longevity. *Trends in Endocrinology & Metabolism* 20 (7): 325–331.

133. St-Pierre, J., et al. 2003. Bioenergetic analysis of peroxisome proliferator-activated receptor γ coactivators 1α and 1β (PGC-1α and PGC-1β) in muscle cells. *Journal of Biological Chemistry* 278 (29): 26597–26603.

134. St-Pierre, J., et al. 2006. Suppression of reactive oxygen species and neurodegeneration by the PGC-1 transcriptional coactivators. *Cell* 127 (2): 397–408.

135. Ransohoff, R.M., et al. 2015. Neuroinflammation: Ways in which the immune system affects the brain. *Neurotherapeutics* 12 (4): 896–909.

136. Hurley, L.L., and Y. Tizabi. 2013. Neuroinflammation, neurodegeneration, and depression. *Neurotoxicity Research* 23 (2): 131–144.

137. Halaris, A. 2018. Neuroinflammation and neurotoxicity contribute to neuroprogression in neurological and psychiatric disorders. *Future Neurology* 13 (2): 59–69.

138. Hamblin, M.R. 2017. Mechanisms and applications of the anti-inflammatory effects of photo-biomodulation. *AIMS Biophysics* 4 (3): 337.
139. Chen, A.C.-H., et al. 2011. Effects of 810-nm laser on murine bone-marrow-derived dendritic cells. *Photomedicine and Laser Surgery* 29 (6): 383–389.
140. Lee, H.I., et al. 2017. Low-level light emitting diode (LED) therapy suppresses inflammasome-mediated brain damage in experimental ischemic stroke. *Journal of Biophotonics* 10 (11): 1502–1513.
141. Fan, S., et al. 2018. LED enhances anti-inflammatory effect of luteolin (3', 4', 5, 7-tetrahydroxyflavone) in vitro. *American Journal of Translational Research* 10 (1): 283.
142. von Leden, R.E., et al. 2013. 808 nm wavelength light induces a dose-D ependent alteration in microglial polarization and resultant microglial induced neurite growth. *Lasers in Surgery and Medicine* 45 (4): 253–263.
143. Song, S., F. Zhou, and W.R. Chen. 2012. Low-level laser therapy regulates microglial function through Src-mediated signaling pathways: Implications for neurodegenerative diseases. *Journal of Neuroinflammation* 9 (1): 219.
144. Esenaliev, R.O., et al. 2018. Nano-pulsed laser therapy is neuroprotective in a rat model of blast-induced neurotrauma. *Journal of Neurotrauma* 35 (13): 1510–1522.
145. Moreira, M.S., et al. 2009. Effect of phototherapy with low intensity laser on local and systemic immunomodulation following focal brain damage in rat. *Journal of Photochemistry and Photobiology B: Biology* 97 (3): 145–151.
146. Zhang, Q., et al. 2014. Low-level laser therapy effectively prevents secondary brain injury induced by immediate early responsive gene X-1 deficiency. *Journal of Cerebral Blood Flow & Metabolism* 34 (8): 1391–1401.
147. Choi, D.-H., et al. 2012. Effect of 710-nm visible light irradiation on neuroprotection and immune function after stroke. *NeuroImmunoModulation* 19 (5): 267–276.
148. Lee, H.I., et al. 2016. Pre-conditioning with transcranial low-level light therapy reduces neuroinflammation and protects blood-brain barrier after focal cerebral ischemia in mice. *Restorative Neurology and Neuroscience* 34 (2): 201–214.
149. Blivet, G., et al. 2018. Neuroprotective effect of a new photobiomodulation technique against Aβ25–35 peptide–induced toxicity in mice: Novel hypothesis for therapeutic approach of Alzheimer's disease suggested. *Alzheimer's & Dementia: Translational Research & Clinical Interventions* 4: 54–63.
150. Cho, G.M., et al. 2018. Photobiomodulation using a low-level light-emitting diode improves cognitive dysfunction in the 5XFAD mouse model of Alzheimer's disease. *The Journals of Gerontology: Series A.*
151. De Taboada, L., et al. 2011. Transcranial laser therapy attenuates amyloid-β peptide neuropathology in amyloid-β protein precursor transgenic mice. *Journal of Alzheimer's Disease* 23 (3): 521–535.
152. Salgado, A.S., et al. 2016. Effects of light emitting diode and low-intensity light on the immunological process in a model of Parkinson's disease. *Medical Research Archives* 4 (8).
153. Pourmemar, E., et al. 2017. Intranasal cerebrolysin attenuates learning and memory impairments in D-galactose-induced senescence in mice. *Experimental Gerontology* 87: 16–22.
154. Obulesu, M., and M.J. Lakshmi. 2014. Apoptosis in Alzheimer's disease: An understanding of the physiology, pathology and therapeutic avenues. *Neurochemical Research* 39 (12): 2301–2312.
155. Da Costa, C.A., and F. Checler. 2011. Apoptosis in Parkinson's disease: Is p53 the missing link between genetic and sporadic Parkinsonism? *Cellular Signalling* 23 (6): 963–968.
156. Desagher, S., and J.-C. Martinou. 2000. Mitochondria as the central control point of apoptosis. *Trends in Cell Biology* 10 (9): 369–377.

157. Gronbeck, K.R., et al. 2016. Application of tauroursodeoxycholic acid for treatment of neuro-logical and non-neurological diseases: Is there a potential for treating traumatic brain injury? *Neurocritical Care* 25 (1): 153–166.

158. Jiang, B., et al. 2004. Catalpol inhibits apoptosis in hydrogen peroxide-induced PC12 cells by preventing cytochrome c release and inactivating of caspase cascade. *Toxicon* 43 (1): 53–59.

159. Duan, R., et al. 2003. Light emitting diode irradiation protect against the amyloid beta 25–35 induced apoptosis of PC12 cell in vitro. *Lasers in Surgery and Medicine: The Official Journal of the American Society for Laser Medicine and Surgery* 33 (3): 199–203.

160. Jiang, W., et al. 2014. Red photon treatment inhibits apoptosis via regulation of bcl-2 proteins and ROS levels, alleviating hypoxic–ischemic brain damage. *Neuroscience* 268: 66–74.

161. Musashi, M., S. Ota, and N. Shiroshita. 2000. The role of protein kinase C isoforms in cell proliferation and apoptosis. *International Journal of Hematology* 72 (1): 12–19.

162. Weinreb, O., et al. 2004. Neuroprotection via pro-survival protein kinase C isoforms associated with Bcl-2 family members. *The FASEB Journal* 18 (12): 1471–1473.

163. Zhang, L., et al. 2008. Low-power laser irradiation inhibiting Aβ25-35-induced PC12 cell apoptosis via PKC activation. *Cellular Physiology and Biochemistry* 22 (1–4): 215–222.

164. Zhang, L., Y. Zhang, and D. Xing. 2010. LPLI inhibits apoptosis upstream of Bax translocation via a GSK-3β-inactivation mechanism. *Journal of Cellular Physiology* 224 (1): 218–228.

165. Zhang, H., S. Wu, and D. Xing. 2012. Inhibition of Aβ25–35-induced cell apoptosis by Low-power-laser-irradiation (LPLI) through promoting Akt-dependent YAP cytoplasmic transloca-tion. *Cellular Signalling* 24 (1): 224–232.

166. Yang, L., et al. 2019. Photobiomodulation preconditioning prevents cognitive impairment in a neonatal rat model of hypoxia-ischemia. *Journal of Biophotonics* 12 (6): e201800359.

167. Yip, K., et al. 2011. The effect of low-energy laser irradiation on apoptotic factors following experimentally induced transient cerebral ischemia. *Neuroscience* 190: 301–306.

168. Quirk, B.J., et al. 2012. Near-infrared photobiomodulation in an animal model of traumatic brain injury: Improvements at the behavioral and biochemical levels. *Photomedicine and Laser Surgery* 30 (9): 523–529.

169. Xuan, W., et al. 2014. Transcranial low-level laser therapy enhances learning, memory, and neuroprogenitor cells after traumatic brain injury in mice. *Journal of Biomedical Optics* 19 (10): 108003.

170. Emery, D.L., et al. 2003. Plasticity following injury to the adult central nervous system: Is recapitulation of a developmental state worth promoting? *Journal of Neurotrauma* 20 (12): 1271–1292.

171. Sanchez, B., et al. 2002. 1, 25-Dihydroxyvitamin D3 increases striatal GDNF mRNA and protein expression in adult rats. *Molecular Brain Research* 108 (1–2): 143–146.

172. Telerman, A., et al. 2011. Induction of hippocampal neurogenesis by a tolerogenic peptide that ameliorates lupus manifestations. *Journal of Neuroimmunology* 232 (1–2): 151–157.

173. Martinowich, K., H. Manji, and B. Lu. 2007. New insights into BDNF function in depression and anxiety. *Nature Neuroscience* 10 (9): 1089.

174. Byrnes, K.R., et al. 2005. Low power laser irradiation alters gene expression of olfactory ensheathing cells in vitro. *Lasers in Surgery and Medicine: The Official Journal of the American Society for Laser Medicine and Surgery* 37 (2): 161–171.

175. Yan, X., et al. 2017. Low-level laser irradiation modulates brain-derived neurotrophic factor mRNA transcription through calcium-dependent activation of the ERK/CREB pathway. *Lasers in Medical Science* 32 (1): 169–180.

176. Ghanbari, A., et al. 2017. Light-emitting diode (LED) therapy improves occipital cortex dam-age by decreasing apoptosis and increasing BDNF-expressing cells in methanol-induced tox-icity in rats. *Biomedicine & Pharmacotherapy* 89: 1320–1330.

177. Ghanbari, A., et al. 2018. Light-emitting diode (LED) therapy attenuates neurotoxicity of methanol-induced memory impairment and apoptosis in the hippocampus. *CNS & Neurological Disorders-Drug Targets (Formerly Current Drug Targets-CNS & Neurological Disorders)* 17 (7): 528–538.
178. Yun, Y.-C., et al. 2017. Laser acupuncture exerts neuroprotective effects via regulation of Creb, Bdnf, Bcl-2, and Bax gene expressions in the hippocampus. *Evidence-Based Complementary and Alternative Medicine* 2017
179. El Massri, N., et al. 2017. Photobiomodulation-induced changes in a monkey model of Parkinson's disease: Changes in tyrosine hydroxylase cells and GDNF expression in the striatum. *Experimental Brain Research* 235 (6): 1861–1874.
180. Mueller, S.G., et al. 2010. Hippocampal atrophy patterns in mild cognitive impairment and Alzheimer's disease. *Human Brain Mapping* 31 (9): 1339–1347.
181. Campbell, S. and G. MacQueen. 2004. The role of the hippocampus in the pathophysiology of major depression. *Journal of Psychiatry & Neuroscience.*
182. Santos, T., et al. 2017. Blue light potentiates neurogenesis induced by retinoic acid-loaded responsive nanoparticles. *Acta Biomaterialia* 59: 293–302.
183. Longo, L. 2017. *Nonsurgical laser treatment (NSLT) of central and peripheral nervous system injuries.* Mary Ann Liebert, Inc. 140 Huguenot Street, 3rd Floor New Rochelle, NY 10801 USA.
184. Xuan, W., et al. 2013. Transcranial low-level laser therapy improves neurological performance in traumatic brain injury in mice: Effect of treatment repetition regimen. *PLoS ONE* 8 (1): e53454.
185. Xuan, W., et al. 2015. Low-level laser therapy for traumatic brain injury in mice increases brain derived neurotrophic factor (BDNF) and synaptogenesis. *Journal of Biophotonics* 8 (6): 502–511.
186. Xuan, W., L. Huang, and M.R. Hamblin. 2016. Repeated transcranial low-level laser therapy for traumatic brain injury in mice: Biphasic dose response and long-term treatment outcome. *Journal of Biophotonics* 9 (11–12): 1263–1272.
187. Oron, A., et al. 2006. Low-level laser therapy applied transcranially to rats after induction of stroke significantly reduces long-term neurological deficits. *Stroke* 37 (10): 2620–2624.
188. Argibay, B., et al. 2019. Light-emitting diode photobiomodulation after cerebral ischemia. *Frontiers in Neurology* 10: 911.
189. Tanaka, Y., et al. 2011. Infrared radiation has potential antidepressant and anxiolytic effects in animal model of depression and anxiety. *Brain Stimulation* 4 (2): 71–76.
190. Shu-Zhi, L., and W. Li-Hua. 1983. Effects of laser guided by optic fiber into rat brain on conditioned avoidance response and brain chemistry. *Lasers in Surgery and Medicine* 2 (3): 231–239.
191. Shu-Zhi, L., and W. Li-Hua. 1982. Effects of a low power laser beam guided by optic fiber on rat brain striatal monoamines and amino acids. *Neuroscience Letters* 32 (2): 203–208.
192. Ahmed, N.A.E.H., et al. 2008. Effect of three different intensities of infrared laser energy on the levels of amino acid neurotransmitters in the cortex and hippocampus of rat brain. *Photomedicine and Laser Surgery* 26 (5): 479–488.
193. Kuo, J.-R., et al. 2015. Deep brain light stimulation effects on glutamate and dopamine concentration. *Biomedical Optics Express* 6 (1): 23–31.
194. Romeo, S., et al. 2017. Fluorescent light induces neurodegeneration in the rodent nigrostriatal system but near infrared LED light does not. *Brain Research* 1662: 87–101.
195. Lombard, A., et al. 1990. Neurotransmitter content and enzyme activity variations in rat brain following in vivo He-Ne laser irradiation. In *Proceedings, round table on basic and applied research in photobiology and photomedicine*, 10–11.

196. Radwan, N.M., et al. 2009. Effect of infrared laser irradiation on amino acid neurotransmitters in an epileptic animal model induced by pilocarpine. *Photomedicine and Laser Surgery* 27 (3): 401–409.

197. Eshaghi, E., et al. 2019. Transcranial photobiomodulation prevents anxiety and depression via changing serotonin and nitric oxide levels in brain of depression model mice: A study of three different doses of 810 nm laser. *Lasers in Surgery and Medicine.*

198. Cassone, M., et al. 1993. Effect of in vivo He-Ne laser irradiation on biogenic amine levels in rat brain. *Journal of Photochemistry and Photobiology B: Biology* 18 (2–3): 291–294.

199. Mann, J.J. 1999. Role of the serotonergic system in the pathogenesis of major depression and suicidal behavior. *Neuropsychopharmacology* 21 (S1): 99S.

200. Jacobsen, J.P., I.O. Medvedev, and M.G. Caron. 2012. The 5-HT deficiency theory of depression: Perspectives from a naturalistic 5-HT deficiency model, the tryptophan hydroxylase 2Arg439His knockin mouse. *Philosophical Transactions of the Royal Society B: Biological Sciences* 367 (1601): 2444–2459.

201. Der-Avakian, A., and A. Markou. 2012. The neurobiology of anhedonia and other reward-related deficits. *Trends in Neurosciences* 35 (1): 68–77.

202. Roy, A., J. De Jong, and M. Linnoila. 1989. Cerebrospinal fluid monoamine metabolites and suicidal behavior in depressed patients: A 5-year follow-up study. *Archives of General Psychiatry* 46 (7): 609–612.

203. Naeser, M.A., et al. 2016. Transcranial, red/near-infrared light-emitting diode therapy to improve cognition in chronic traumatic brain injury. *Photomedicine and Laser Surgery* 34 (12): 610–626.

204. Vincent, J.L., et al. 2007. Intrinsic functional architecture in the anaesthetized monkey brain. *Nature* 447 (7140): 83.

205. Park, H.-J., and K. Friston. 2013. Structural and functional brain networks: From connections to cognition. *Science* 342 (6158): 1238411.

206. Xiao, H., et al. 2015. Structural and functional connectivity in traumatic brain injury. *Neural Regeneration Research* 10 (12): 2062.

207. Kringelbach, M.L., A.L. Green, and T.Z. Aziz. 2011. Balancing the brain: Resting state networks and deep brain stimulation. *Frontiers in Integrative Neuroscience* 5: 8.

208. De La Plata, C.D.M., et al. 2011. Deficits in functional connectivity of hippocampal and frontal lobe circuits after traumatic axonal injury. *Archives of Neurology* 68 (1): 74–84.

209. Johnson, B., et al. 2012. Alteration of brain default network in subacute phase of injury in concussed individuals: Resting-state fMRI study. *NeuroImage* 59 (1): 511–518.

210. Greicius, M.D., et al. 2004. Default-mode network activity distinguishes Alzheimer's disease from healthy aging: Evidence from functional MRI. *Proceedings of the National Academy of Sciences* 101 (13): 4637–4642.

211. Naeser, M.A., et al. 2014. Significant improvements in cognitive performance post-transcranial, red/near-infrared light-emitting diode treatments in chronic, mild traumatic brain injury: Open-protocol study. *Journal of Neurotrauma* 31 (11): 1008–1017.

212. Lim, L. 2018. The Growing Evidence for Photobiomodulation as a Promising Treatment for Alzheimer's Disease. *Journal of Biosciences and Medicines* 6 (12): 100.

213. Sripada, R.K., et al. 2012. Neural dysregulation in posttraumatic stress disorder: Evidence for disrupted equilibrium between salience and default mode brain networks. *Psychosomatic Medicine* 74 (9): 904.

214. Naeser, M.A. and M.R. Hamblin. 2015. *Traumatic brain injury: A major medical problem that could be treated using transcranial, red/near-infrared LED photobiomodulation.* Mary Ann Liebert, Inc. 140 Huguenot Street, 3rd Floor New Rochelle, NY 10801 USA.

215. Naeser, M., et al. 2012. Improved language after scalp application of red/near-infrared light-emitting diodes: Pilot study supporting a new, noninvasive treatment for chronic aphasia. *Procedia-Social and Behavioral Sciences* 61: 138–139.

216. Chao, L.L. 2019. Effects of home photobiomodulation treatments on cognitive and behavioral function, cerebral perfusion, and resting-state functional connectivity in patients with dementia: A pilot trial. *Photobiomodulation, Photomedicine, and Laser Surgery* 37 (3): 133–141.

217. Zomorrodi, R., et al. 2019. pulsed Near Infrared transcranial and Intranasal photobiomodulation Significantly Modulates Neural oscillations: A pilot exploratory study. *Scientific Reports* 9 (1): 6309.

218. Miyauchi, E., et al. 2019. A novel approach for assessing neuromodulation using phase-locked information measured with TMS-EEG. *Scientific Reports* 9.

219. Cohen, M.X. 2017. Where does EEG come from and what does it mean? *Trends in Neurosciences* 40 (4): 208–218.

220. Caruso, G., et al. 1999. Clinical EMG and glossary of terms most commonly used by clinical electromyographers. *Recommendations for the Practice of Clinical Neurophysiology: Guidelines of the International Federation of Clinical Physiology* 189–198.

221. Schomer, D.L. and F.L. Da Silva. 2012. *Niedermeyer's electroencephalography: Basic principles, clinical applications, and related fields*. Lippincott Williams & Wilkins.

222. Stefanics, G., et al. 2009. Attentive anticipation modulates phase-entrainment of human delta EEG oscillations–a single trial analysis. In *Frontiers on* Systems Neuroscience Conference Abstract: 12th Meeting of the Hungarian Neuroscience Society. https://doi.org/10.3389/conf. neuro.

223. Lisman, J.E., and O. Jensen. 2013. The theta-gamma neural code. *Neuron* 77 (6): 1002–1016.

224. Amin, H., and A.S. Malik. 2013. Human memory retention and recall processes. *Neurosciences* 18 (4): 330–344.

225. Spitzer, B. and S. Haegens. 2017. Beyond the status quo: A role for beta oscillations in endogenous content (re) activation. *Eneuro* 4 (4).

226. Omel'Chenko, V., and V. Zaika. 2002. Changes in the EEG-rhythms in endogenous depressive disorders and the effect of pharmacotherapy. *Human Physiology* 28 (3): 275–281.

227. Babiloni, C., et al. 2013. Cortical sources of resting state EEG rhythms are sensitive to the progression of early stage Alzheimer's disease. *Journal of Alzheimer's Disease* 34 (4): 1015–1035.

228. Alfimova, M., and L. Uvarova. 2007. Changes in the EEG spectral power during perception of neutral and emotionally salient words in schizophrenic patients, their relatives and healthy individuals from the general population. *Zhurnal Vysshei Nervnoi Deiatelnosti Imeni IP Pavlova* 57 (4): 426–436.

229. Wacker, J., D.G. Dillon, and D.A. Pizzagalli. 2009. The role of the nucleus accumbens and rostral anterior cingulate cortex in anhedonia: Integration of resting EEG, fMRI, and volumetric techniques. *NeuroImage* 46 (1): 327–337.

230. Ekimova, I., et al. 2016. Changes in sleep characteristics of rat preclinical model of Parkinson's disease based on attenuation of the ubiquitin—Proteasome system activity in the brain. *Journal of Evolutionary Biochemistry and Physiology* 52 (6): 463–474.

231. Clarke, A.R., et al. 2011. Behavioural differences between EEG-defined subgroups of children with attention-deficit/hyperactivity disorder. *Clinical Neurophysiology* 122 (7): 1333–1341.

232. Vlahinić, S., et al. 2020. Analyses of IR stimulation influence on EEG. In *2020 IEEE international instrumentation and measurement technology conference (I2MTC)*. IEEE.

233. Zomorrodi, R., et al. 2019. Modulation of neural oscillation power spectral density with transcranial photobiomodulation. *Brain Stimulation* 12 (2): 457–458.

234. Machado, C., et al. 2018. Effect of low level laser therapy on brain activity assessed by QEEG and QEEGt in normal subjects. *The Internet Journal of Neurology* 20 (1).
235. Mohammed, H.S. 2016. Transcranial low-level infrared laser irradiation ameliorates depression induced by reserpine in rats. *Lasers in Medical Science* 31 (8): 1651–1656.
236. Zomorrodi, R., G. Loheswaran, and L. Lim. 2019. Electroencephalography as the diagnostic adjunct to transcranial photobiomodulation. In *Photobiomodulation in the brain*, 419–426. Elsevier.
237. Wang, X., et al. 2019. Neurophysiological enhancements of the human brain by transcranial photobiomodulation using 1064-nm laser. *Biological Psychiatry* 85 (10): S80.
238. Wang, X., et al. 2019. Transcranial photobiomodulation with 1064-nm laser modulates brain electroencephalogram rhythms. *Neurophotonics* 6 (2): 025013.
239. Wang, X., et al. 2017. Proceedings# 18. Transcranial infrared brain stimulation modulates EEG alpha power. *Brain Stimulation: Basic, Translational, and Clinical Research in Neuromodulation* 10 (4): e67–e69.
240. Vargas, E., et al. 2017. Beneficial neurocognitive effects of transcranial laser in older adults. *Lasers in Medical Science* 32 (5): 1153–1162.
241. Parsaei, F., M.A. Nazari, and S. Heysieattalab. 2019. Deceased delta activity and cross-frequency interaction of resting-state electroencephalographic oscillations in transcranial light emitting diode (LED). *IBRO Reports* 6: S78.
242. Yao, L., et al. 2020. Effects of stimulating frequency of NIR LEDs light irradiation on forehead as quantified by EEG measurements. *Journal of Innovative Optical Health Sciences* 2050025.
243. Jahan, A., et al. 2019. Transcranial near-infrared photobiomodulation could modulate brain electrophysiological features and attentional performance in healthy young adults. *Lasers in Medical Science* 1–8.
244. Grover Jr, F., J. Weston, and M. Weston. 2017. Acute effects of near infrared light therapy on brain state in healthy subjects as quantified by qEEG measures. *Photomedicine and Laser Surgery* 35 (3): 136–141.
245. Berman, M.H., et al. 2017. Photobiomodulation with near infrared light helmet in a pilot, placebo controlled clinical trial in dementia patients testing memory and cognition. *Journal of Neurology and Neuroscience* 8 (1).
246. Berman, M.H., M.R. Hamblin, and P. Chazot. 2017. Photobiomodulation and other light stimulation procedures. In *Rhythmic stimulation procedures in neuromodulation*, 97–129. Elsevier.
247. Zomorrodi, R., et al. 2017. Complementary EEG evidence for a significantly improved Alzheimer's disease case after photobiomodulation treatment. *Alzheimer Dementia* 13 (S7): P621.

Biphasic Dose–response in Photobiomodulation of Neuronal Cells and the Brain

5.1 Introduction

It is generally accepted that there is an optimal dose of light for any particular photo-biomodulation application, and doses of light much lower or much higher than this ideal value may not result in the best therapeutic benefits. In other words, PBM exhibits a biphasic dose–response relationship or hormetic effect, suggesting that the dose must be optimized to produce the optimal PBM effect. This phenomenon has been described as the "Arndt–Schulz law" first reported in 1985 [1]. According to this law, a "biphasic" or "inverted U-shaped" curve can be used to clarify the expected dose–response to light at the cellular level. This law states that light at very low doses has no detectable photo-biomodulation effects (below the threshold), whereas small (but still larger) doses above this threshold produce therapeutic effects; much greater doses of light can lose the benefits and exert inhibitory or even harmful effects [2, 3]. It is noteworthy that the transcranial direct current stimulation (tDCS) technique also appears to exert hormetic effects [4].

In fact, studies suggest that both the irradiance and the fluence are important parameters for the effectiveness of PBM, and they both operate between certain thresholds [5]. It is noteworthy that PBM studies with negative or null results seem to be more often due to over-dosing than to under-dosing [6]. PBM regimens that are based on a protocol involving too frequent repetitions of PBM sessions or a too compressed treatment interval can also result in ineffective results [7]. In other words, selecting the proper PBM regimen is an important factor for optimal therapeutic results [8, 9]. At the cellular level, NO and ROS can both be considered as classical Janus-type mediators that act as beneficial signaling molecules at low concentrations but act as harmful cytotoxic mediators at high concentrations. NO, ROS, ATP, and NF-κB have been shown to respond to PBM in a biphasic dose-dependent manner [10, 11]. Evidence suggests that these molecules may

© The Author(s), under exclusive license to Springer Nature Switzerland AG 2023
F. Salehpour et al., *Photobiomodulation for the Brain*, Synthesis Lectures
on Biomedical Engineering, https://doi.org/10.1007/978-3-031-36231-6_5

be responsible for the overall biphasic dose–response pattern in PBM studies. Herein, we review several reports of biphasic dose–response in PBM of brain/neuronal cells.

5.1.1 In Vitro Studies

5.1.2 Biphasic Response: Power or Irradiance

In 2003, Duan et al. [12] investigated the inhibition of amyloid-beta $(A\beta)_{25-35}$ induced apoptosis in PC12 cells using 640 nm light-emitting diodes (LED) and found a dose-dependent pattern. An irradiance of 0.9 W/m^2 for 60 min significantly decreased apoptotic cells, whereas, 10 W/m^2 for 60 min did not produce the same effects. They suggested that irradiance was the most important factor in their experiment, and when using a series of irradiation times with the same constant irradiance, there was no significant difference [12].

Higuchi et al. [13] in 2007 investigated the effect of 525 nm LED light at different irradiances (0, 0.25, 0.5, 0.75, and 1 mW/cm^2) on the neurite outgrowth of PC12 cells and reported that neurite outgrowth was suppressed at 0.5 and 0.75 mW/cm^2. Furthermore, they showed a biphasic response between neurite outgrowth and the irradiance of light at different wavelengths 470, 525, 880, and 945 nm. The neurite outgrowth at 0.25, 0.5, and 0.75 mW/cm^2 was found to be lower than that at 1 and 1.8 mW/cm^2 for all wavelengths. In addition to this, an irradiance of 2.0 mW/cm^2 showed even greater suppression of neurite outgrowth compared to all of the other irradiances. This suggests that 2 mW/cm^2 of monochromatic LED light can produce damage to PC12 cells [13].

In 2009, Rochkind et al. [14] studied the effects of 780 nm laser PBM at different powers (10, 30, 50, 110, 160, 200, and 250 mW) and durations (1, 4, and 7 min) on the growth and development of embryonic cortical neurons and their fibers. Irradiation with 50 mW for 1 or 4 min significantly accelerated nerve cell sprouting and cell migration within 24 h after seeding. Cultures irradiated with 50 mW for either 4 or 7 min showed thick elongated fibers. Cultures exposed to 1 min of 50 mW irradiation contained much larger neurons than cultures exposed for 4 or 7 min. Also, irradiated cultures with 50 mW for either 1 or 4 min exhibited large-sized neurons with a dense branched interconnected network of neuronal fibers. They suggested that both laser power and irradiation time were important parameters for achieving effective outcomes [14].

Chen et al. [15] in 2014 explored the effect of 808 nm LED pre-treatment with various irradiances (5, 10, 25, and 50 mW/cm^2) on cell viability of cultured neurons treated with $CoCl_2$ (a form of oxidative stress). Among the four irradiances, 25 mW/cm^2 exhibited the best increase of cell viability, and inhibited the damage induced by $CoCl_2$, whereas other irradiances showed non-significant effects on cell viability [15].

5.1.3 Biphasic Response: Time or Fluence

In 2005, Byrnes et al. [16] evaluated the effects of two different doses of 810 nm laser using various irradiation times on the gene expression of olfactory ensheathing cells (OEC). The first group of cells received a fluence of 0.2 J/cm^2 (4 s of light with a power of 127 mW), while the second group received a fluence of 68 J/cm^2 (9 min and 21 s of light with a power of 127 mW). Results showed that PBM of OEC cultures at 0.2 J/cm^2 significantly increased BDNF, collagen, and GDNF gene expression, whereas 68 J/cm^2 did not produce any significant change. Although irradiation with both doses did not result in a significant change in the p75 and GFAP protein expression, a trend toward an increase in GFAP without any increase in p75 expression was found in cultures treated with PBM at 68 J/cm^2. Moreover, both light doses significantly increased OEC proliferation on day 7 post-irradiation, but not on day 3 [16].

Giuliani et al. [17] in 2009 examined the response of PC12 cells to 670 nm pulsed laser at various fluences and irradiation durations. They showed that using constant 20-s of irradiation, PBM at 0.11 J/cm^2 stimulated NGF-induced neurite outgrowth better than 0.22 J/cm^2 on day 4 of culture. Both irradiation times of 20 s (0.11 or 0.22 J/cm^2) and 15 min (5.06 or 10.12 J/cm^2) significantly restored the decrease in MMP caused by H_2O_2-induced oxidative stress. On the other hand, 20 s (0.11 or 0.22 J/cm^2) irradiation slightly but significantly protected cell culture viability, whereas long laser irradiation of 15 min (5.06 or 10.12 J/cm^2) had no protective effect on cell survival [17].

In 2011, Sharma et al. [11] evaluated the effect of 810 nm laser on several cellular processes in mouse cortical neurons. The cells were irradiated in CW mode with different fluences of 0.03, 0.3, 3, 10, and 30 J/cm^2. Results showed the highest efficiency at 3 J/cm^2 for mitochondrial membrane potential (MMP) as well as intracellular ATP and intracellular Ca^{2+} ($[Ca^{2+}]_i$) levels, whereas both low (0.03 and 0.3 J/cm^2) and high (10 J/cm^2) fluences showed lower beneficial effects and a higher dose (30 J/cm^2) induced inhibitory effects. In addition, intracellular NO and mitochondrial ROS levels exhibited two peaks in response to PBM, one at the low dose (0.3 and 3 J/cm^2) and another at high dose (30 J/cm^2), which suggested that neurons have a non-Arndt–Schulz pattern of response at least for ROS and NO production [11].

Saito et al. [18] in 2011 studied the effects of 810 nm laser at two fluences of 5 and 20 J/cm^2 on cell growth factor-induced proliferation and differentiation in PC12 cells. PBM at 20 J/cm^2 significantly decreased the cell numbers after 24 and 48 h. Although there was no significant difference in cell numbers between the 5 J/cm^2 group and the control group, a tendency toward a decrease was observed for the irradiated group. Neurite outgrowth increased with 5 J/cm^2 irradiation after 24 and 48 h, while the neurite outgrowth was significantly higher only after 48 h. Expression of neurofilament and β-tubulin proteins was also increased with both 5 and 20 J/cm^2 after 48 h. However, only PBM at 5 J/cm^2 significantly increased phospho-p38 expression 1 to 3 h post-irradiation [18].

In 2012, Song et al. [19] exposed lipopolysaccharide-activated BV-2 cells or primary microglia to different fluences of 632.8 nm laser (3, 5, 10, 20, 25, and 50 J/cm^2) to determine the optimal dose of PBM. Microglial cytotoxicity was markedly decreased by PBM in a dose-dependent manner. Although laser irradiation with fluences greater than 20 J/cm^2 induced the lowest microglia-mediated neurotoxicity, fluences of 25 and 50 J/cm^2 caused a slight contraction of microglial cells, and thus a dose of 20 J/cm^2 was suggested to be the optimal dose. In addition, PBM at fluences greater than 10 J/cm^2 significantly inhibited the expression of inducible nitric oxide synthase (iNOS) protein, with a maximum effect at 50 J/cm^2 [19].

Fukuzaki et al. [20] in 2013 examined how various fluences of 532 nm laser (10.1, 20.3, and 30.46 × 10^2 J/cm^2) could rescue the γ-secretase inhibitor (GSI)-induced toxic effects in A-172 cells. PBM at 20.3 and 30.46 × 10^2 J/cm^2 significantly increased the cell numbers at 48 h post-irradiation, but not at 24 h. Results also showed the involvement of Akt signaling pathway in the PBM-mediated proliferation, so that irradiation at all fluences significantly increased p-Akt levels, while only fluences of 20.3 and 30.46 × 10^2 J/cm^2 significantly decreased p-PTEN levels. In addition, in the presence of GSI, only a fluence of 30.46 × 10^2 J/cm^2 was able to decrease PTEN expression and to increase Akt expression while keeping the suppression of Aβ [20]. It should be noted that in a similar study in 2015, Fukuzaki et al. [21] reported that among the aforementioned fluences, only a fluence of 30.46 × 10^2 J/cm^2 significantly increased cell proliferation of cultured neural stem/progenitor cells (NSPCs) derived from E10 forebrain in mice [21].

In 2013, von Leden et al. [22] studied the effects of 808 nm laser on microglial polarization and microglial-induced neurite growth using differing fluences (0.2, 4, 10, and 30 J/cm^2). PBM at fluences between 4 and 30 J/cm^2 led to a significant expression of the M1 marker, CD86, in microglia. Also, the CD206 and TIMP1, as markers of the M2 phenotype, were observed at lower fluences of 0.2–10 J/cm^2. PBM had a dose-dependent effect on microglial pro-inflammatory response, so that NO expression was significantly increased in BV2 cells exposed to 4 and 30 J/cm^2 at 24 h post-irradiation. PBM increased ROS in primary microglia in a dose-dependent manner, so that the highest increases were observed in cells exposed to 10 and 30 J/cm^2 at 2 or 24 h post-irradiation. PBM with all doses also increased ROS in BV2 cells, however, a less dose-dependent response was observed in these cells. PBM had no significant effect on TNF-α, IL-1β, and IL-6 expression in primary or BV2 microglia at 2 or 24 h post-irradiation. PBM at low doses significantly up-regulated cytokine and chemokine levels in microglial cultures, so that laser at 0.2 J/cm^2 increased MCP-1 expression in BV2 cells, while laser at 0.2, 4, and 10 J/cm^2 significantly induced TIMP1. Results also demonstrated a dose-dependent effect of PBM on microglial-induced neuronal growth and neurite extension, so that the number of neurites significantly increased at fluences of 0.2 and 30 J/cm^2, while only PBM at 4 J/cm^2 significantly increased the length of the neurites [22].

Meng et al. [23] in 2013 evaluated the neuroprotective effect of 632.8 nm laser at various fluences (0.5, 1, 2, and 4 J/cm^2) on SH-SY5Y cells and primary neurons. Cell viability for both cell types was increased by PBM in a dose-dependent manner, so that only PBM at 2 and 4 J/cm^2 exerted a neuroprotective effect against Aβ toxicity. Furthermore, PBM at fluences of 1, 2, and 4 J/cm^2 significantly increased BDNF protein expression in a biphasic dose-dependent manner, with a maximum response at 2 J/cm^2 [23].

In 2015, Zheng et al. [24] investigated the effects of three different fluences of 658 nm laser (2, 6, and 16 J/cm^2) on NO production in rat dorsal root ganglion neurons. The results showed a biphasic dose–response curve; the neurons exposed to 6 J/cm^2 irradiation produced more NO compared with the control and 2 J/cm^2 irradiation groups, while 16 J/cm^2 irradiation inhibited the NO production [24].

Yan et al. [25] in 2017 investigated the effects of 632.8 nm laser irradiation at various fluences (0.5, 1, 1.9, and 3.8 J/cm^2) on BDNF mRNA transcription in cultured dorsal root ganglion neurons. At both 6 and 24 h post-irradiation, PBM at fluences of 1, 1.9, and 3.8 J/cm^2 significantly increased BDNF mRNA transcripts in a biphasic dose-dependent manner, with a maximal response at 1.9 J/cm^2. The neurite length was also increased at 3 days post-irradiation in a biphasic dose manner with a peak response at 1.9 J/cm^2 [25].

In 2017, Gu et al. [26] also showed that 632.8 nm laser irradiation rescued SH-SY5Y cells against MPP$^+$ toxicity in a dose-dependent manner. Results of cell viability assessment at 24 h post-irradiation showed the highest effects with 2 and 4 J/cm^2 fluences, whereas a fluence of 1 J/cm^2 exhibited minor, but still significant beneficial effects [26].

Zhu et al. [27] in 2017 evaluated how neural stem cells responded to 635 nm laser PBM at various fluences (61, 122, 244, and 366 J/cm^2), and found a dose-dependent response. Cells exposed to a fluence of 61 J/cm^2 demonstrated the highest cell proliferation rate at 24 h post-irradiation. Although there was no significant difference in the cell number between control and 122 J/cm^2 irradiation groups, both higher doses of 244 and 366 J/cm^2 significantly inhibited cell proliferation at 24 h post-irradiation [27].

In 2017, Santos et al. [28] studied the potential effects of 405 nm laser PBM at two fluences of 9 and 18 J/cm^2 on the differentiation of neural stem cells isolated from the hippocampus. PBM at 18 J/cm^2 resulted in a significant transient increase in mitochondrial ROS levels at 30 min post-irradiation, reaching a maximum at 1 h and reverting back to basal levels at 7 h. Cytosolic peroxide levels were also increased 3 h after 18 J/cm^2 irradiation peaking at 7 h, and reverting to basal levels at 36 h. They also showed that blue light induced an increase in mitochondrial ROS with 18 J/cm^2 irradiation in a NADPH oxidase (Nox)-dependent manner, with a significant up-regulation of Nox4 (at 1 h post-irradiation) as well as Nox1 and Rac1 (both at 3 h post-irradiation). In addition, only 18 J/cm^2 irradiation significantly activated β-catenin (at 2 h post-irradiation), induced neuronal differentiation (at 7 days post-irradiation), and up-regulated RARα (at 12 h post-irradiation) [28].

Levchenko et al. [29] in 2018 evaluated lipid metabolic profiles in rat cortical neurons after 808 nm laser irradiation at various fluences (0.3, 3, 10, and 30 J/cm^2). The results showed a significant increase in the lipid contents of the neuronal cell cytoplasm for all applied fluences at 2, 6, and 24 h post-irradiation. Specifically, at 2 h post-irradiation, the lower fluences of 0.3 and 3 J/cm^2 resulted in a slightly lower increase of lipid levels, compared to the fluences of 10 and 30 J/cm^2. However, at the subsequent time points of 6 and 24 h, the lipid contents were found to increase more significantly with 0.3 J/cm^2, compared to all of the other applied fluences. Furthermore, PBM increased the number of lipid droplets (LDs) per cell in a similar manner as reported for the lipid contents, but the results were strikingly different after 0.3 J/cm^2 irradiation. Mitochondrial ROS production was also found to be dependent on irradiation fluence in a nonlinear pattern, with the most significant increase at 0.3 J/cm^2, and a slight increase at 3 and 10 J/cm^2 [29].

In 2019, Rhee et al. [30] investigated the potential effects of 660 nm LED PBM at various fluences (0.78, 1.56, 3.12, 6.24, and 9.36 J/cm^2) on cellular neuronal damage induced by the cardiac glycoside, ouabain. PBM at fluences of 1.56, 3.12, and 6.24 J/cm^2 significantly increased cell survival as well as the ADP/ATP ratio in a biphasic dose-dependent manner, with a maximal response at 3.12 J/cm^2. Only PBM at a fluence of 3.12 J/cm^2 significantly restored the ouabain-induced inhibition of Na/K-ATPase [30].

5.1.4   Biphasic Response: Treatment Sessions

In 2008, Liang et al. [31] evaluated the potential benefits of single and multiple sessions of 670 nm LED irradiation on primary cultured cortical neurons treated with potassium cyanide (KCN). Two sessions of PBM were more effective in decreasing the percentage of apoptotic neurons compared to once a day, whereas three or four times a day did not produce a protective effect. Two and three sessions of PBM exhibited significantly lower ROS production than four sessions of irradiation, while LED irradiation once a day was not effective. Likewise, two and three sessions a day of PBM resulted in a significant reduction of NO levels compared with once or four times a day, both of which were ineffective. Two sessions of PBM resulted in the highest cytochrome c oxidase (CCO) activity and ATP contents, compared with other irradiation regimens. In addition, PBM twice a day for 1, 3, and 5 days significantly increased neuronal ATP contents in a linear manner, with a maximum response after 5 days of irradiation [31].

Choi et al. [32] in 2012 compared the effects of single and multiple sessions of 710 nm LED irradiation on neurite outgrowth in primary cortical neurons following an ischemic insult. PBM once and twice a day significantly increased the mean neurite density, while treatments three and four times a day significantly reduced the neurite density. Also, the mean neurite diameter significantly increased only after PBM once a day. PBM also affected synaptogenesis as shown by changes in synaptic markers, so that the expression levels of GAP43, PSD95, and synaptophysin were significantly increased following PBM once and twice a day. Moreover, PBM twice a day induced MAPK pathway activation with a significant increase in ERK, JNK, and p38 kinase activity at 7 days post-irradiation [32].

5.1.5   In Vivo Studies

5.1.6   Biphasic Response: Power or Irradiance

In 2006, Ilic et al. [33] evaluated the potential adverse effects of various irradiances of 808 nm laser (7.5, 75, and 750 mW/cm^2) when applied transcranially to the brain of mature rats. Their results showed a deficit in neurological function only in the 750 mW/cm^2 group immediately post-irradiation. The deficits were partially resolved after 1 and 7 days. Likewise, PBM with 750 mW/cm^2 caused marked abnormalities in the brain, such as focal areas of subdural necrosis with loss of neuronal tissue [33].

Ahmed et al. [34] in 2006 studied the effects of near-infrared (NIR) laser PBM using various output powers and wavelengths (90 mW for 830 nm; 190 and 500 mW for 808 nm) on the axonal conduction of hippocampal and cortical tissues in the rat brain. The 90 mW group exhibited a decrease in cortical aspartate, glutamate, and glutamine, and an increase in glycine, whereas in the hippocampus they observed an increase in aspartate, glutamate, and GABA. The 190 mW group showed an increase in aspartate accompanied by a decrease in glutamine only in the cortices. Animals irradiated with 500 mW irradiance showed a decrease in aspartate, glutamate, and taurine in the cortex, and a decrease in hippocampal GABA. They suggested that daily PBM with 90 mW produced the most noticeable inhibitory effect in the cortex after 7 days.

In 2010, Uozumi et al. [35] examined the effect of 808 nm laser irradiation at three irradiances (0.8, 1.6, and 3.2 W/cm^2) on cerebral blood flow (CBF) and NO levels in brain tissue during NIR laser irradiation. The results showed that among the tested irradiances, laser irradiation at both 1.6 and 3.2 W/cm^2 could improve CBF most effectively. Both irradiances of 0.8 and 1.6 W/cm^2 resulted in an immediate increase in NO levels after the start of the laser irradiation, nevertheless, animals in the 0.8 W/cm^2 irradiation group showed a small increase in NO levels, only 15% of that seen in the 1.6 W/cm^2 irradiation group [35].

De Taboada et al. [36] in 2011 investigated the effects of transcranial 808 nm laser irradiation at various irradiances (0.56, 2.83, and 5.66 W/cm^2) on the neurobehavioral and neurochemical profiles in an Aβ-induced Alzheimer's disease (AD) mouse model and found a biphasic dose–response for all measured parameters. Along with an improvement in spatial memory, PBM at all irradiances significantly reduced brain, plasma, and cerebrospinal fluid Aβ levels, decreased the expression of the inflammatory markers (IL-1β, TNF-α, and TGF-β), increased soluble AβPP and CTFβ levels, and improved brain mitochondrial function and ATP levels. Irradiation at 2.83 W/cm^2 showed the maximum effectiveness for all the parameters [36].

In 2012, Khuman et al. [37] reported the effects of 800 nm laser irradiation on a mouse model of traumatic brain injury (TBI). None of the fluences between 60 and 210 J/cm^2 changed motor recovery and performance after a controlled cortical impact (CCI). Although doses of 30, 105, 120, and 210 J/cm^2 did not show strong effects on the hidden platform trials in the Morris water maze, a fluence of 120 J/cm^2 had positive effects on the probe trial score [37].

Kuo et al. [38] in 2014 evaluated concentration changes in glutamate and dopamine in the striatum of intact rats following intracranial 840 nm laser irradiation at various powers (2, 5, and 10 mW). Results showed that intracranial PBM at all three power levels caused a clear decrease in glutamate concentration in a dose-dependent manner, with a maximal response at 5 mW. On the other hand, only intracranial PBM at 10 mW power significantly increased the secretion of dopamine, suggesting a different response of neurotransmitters to NIR light stimulation [38].

In 2015, Oueslati et al. [39] studied the neuroprotective effects of transcranial 808 nm PBM at various irradiances (5, 10, 20, and 30 mW/cm^2) in a rat model of Parkinson's disease (PD). They showed that only low irradiances (5 and 10 mW/cm^2) decreased α-syn-induced akinesia, while higher light irradiances resulted in a significant reduction in contralateral motor function Moreover, although none of the irradiances significantly changed α-syn-induced dopaminergic neuronal loss in the substantia nigra (SNc) or the dopaminergic fiber denervation in the striatum, the irradiance of 5 mW/cm^2 showed a tendency to alleviate α-syn-induced toxicity. Four weeks of PBM at the low irradiances of 2.5 and 5 mW/cm^2 significantly suppressed α-syn-induced motor impairment. However, chronic treatment with 5 mW/cm^2, but not 2.5 mW/cm^2, suppressed α-syn-induced nigral neuronal degeneration as well as striatal fiber denervation [39].

Meyer et al. [40] in 2016 utilized a rabbit small clot embolic stroke model to investigate the best dosage regimen of transcranial 808.5 nm laser irradiation. PBM with 111 mW/cm^2 resulted in a significant improvement in behavioral measures at 24 h post-embolization, compared to either irradiances of 55.6 or 333 mW/cm^2 [40].

In 2016, Reinhart et al. [41] explored whether intracranial 670 nm LED at low (333 nW) or high (0.16 mW) powers had any effect on apomorphine-induced turning behavior in a rat PD model induced by 6-hydroxydopamine (6-OHDA). The results showed that PBM at both powers significantly reduced rotational behavior in animals at day 21 post-surgery, while only the 0.16 mW power exhibited a beneficial effect at day 14 post-surgery. However, none of the power values mitigated 6-OHDA-induced Tyrosine Hydroxylase-positive (TH+) cell loss in the SNc [41].

Mohammed [42] in 2016 compared the effects of three powers of 804 nm laser irradiation (80, 200, and 400 mW) on the depression-like behavior of rats. The lowest power tested (80 mW) significantly increased animal activity and reduced the immobility period in the forced swim test (FST). On the other hand, the power of 200 mW did not significantly affect the animal behavior in the FST, while the 400 mW exhibited inhibitory effects [42].

5.1.7 Biphasic Response: Time or Fluence

In 2002, Leung et al. [43] evaluated the effects of 660 nm laser at various fluences (2.64, 13.2, and 26.4 J/cm^2) on NOS activity and TGF-β1 expression at 4 days post-stroke. PBM at all fluences significantly decreased NOS-specific activity, while this reduction was more pronounced for 13.2 and 26.4 J/cm^2. The decrease in the expression of each of the three NOS isoforms (eNOS, nNOS, and iNOS) following PBM showed a similar trend to the total NOS activity. For all applied fluences, the decreases were most pronounced with eNOS, followed by nNOS, and least with iNOS. In addition, all fluences effectively up-regulated the expression of TGF-β1 [43].

Moreira et al. [44] in 2009 studied the effects of laser PBM with two fluences and two wavelengths (3 or 5 J/cm^2 of 660 nm; 3 or 5 J/cm^2 of 780 nm) on local and systemic immunomodulatory effects following cryogenic brain injury in rats. PBM with 5 J/cm^2 of 660 nm or 3 J/cm^2 of 780 nm significantly decreased brain levels of IL-1β at 24 h post-injury compared to 6 h. In contrast, PBM at 3 and 5 J/cm^2 of 660 nm or 5 J/cm^2 of 780 nm significantly increased blood TNF-α and IL-6 levels at 24 h post-injury compared to 6 h [44].

In 2010, Lapchak and De Taboada [45] assessed the effect of pulsed 808 nm laser at two cortical fluences (4.5 and 31.5 J/cm^2) on cortical ATP content using the rabbit small clot embolic stroke model. Both fluences significantly increased cortical ATP content at 3 h post-embolization, of note, the increase was larger for 31.5 J/cm^2 (by 221%) than 4.5 J/cm^2 (by 157%) [45].

Yip et al. [46] in 2011 investigated the effects of 660 nm laser irradiation at various fluences (2.64, 13.2, and 26.4 J/cm^2) on the expression and activity of a number of anti- and pro-apoptotic proteins after cerebral ischemia in rats. PBM at all fluences significantly increased Akt, pAkt, Bcl-2, and pBAD levels at 4 days post-injury. All fluences of PBM significantly reduced caspase-3 activity, however, only irradiation at 2.64 and 13.2 J/cm^2 resulted in a significant decrease in caspase-9 activity after ischemia–reperfusion injury [46].

In 2012, Rojas et al. [47] studied the effects of 660 nm LED irradiation at various fluences on the rat cortical metabolic capacity and memory retention. Their preliminary results showed that among the low fluences (1 and 5 J/cm^2), only 5 J/cm^2 significantly increased the rate of oxygen consumption in the prefrontal cortex in vivo. In addition, among all the fluences (5.4, 10.8, 16.2, and 21.6 J/cm^2), only 10.8 J/cm^2 showed statistical significant results for memory extinction, as well as an improvement in CCO activity in the prefrontal cortex [47].

El Massri et al. [48] in 2016 used a MPTP mouse PD model to explore whether the neuroprotective effects of 670 nm LED irradiation in the SNc were dose-dependent. Only PBM at the higher fluence of 4 J/cm^2, but not at a lower fluence of 2 J/cm^2, significantly mitigated the loss of TH+ cells. Likewise, a fluence of 4 J/cm^2, unlike 2 J/cm^2, abrogated the MPTP-induced increase in caudate-putamen complex (CPu) astrocytes after 14 days of irradiation. They suggested that the neuroprotective effect of PBM was dose-dependent. With a higher MPTP dose leading to increased toxicity, higher doses of light were required to protect the neurons and reduce astrogliosis [48].

In 2017, Salehpour et al. [49] studied the effects of laser PBM using various cortical fluences and wavelengths (4 or 8 J/cm^2 for 660 nm; 4 or 8 J/cm^2 for 810 nm) on the D-galactose-induced cognitive impairment in mice, and the resulting mitochondrial dysfunction and apoptosis. For both wavelengths of 660 and 810 nm, only a fluence of 8 J/cm^2 significantly increased spatial and episodic-like memories, improved brain mitochondrial function, and ameliorated apoptosis. However, PBM at a lower fluence of 4 J/cm^2 was also able to decrease ROS production in the brain tissue [49].

Blivet et al. [50] in 2018 evaluated the neuroprotective effects of combined red/NIR PBM at various fluences (2.1, 4.2, 8.4, and 16.8 J/cm^2) against Aβ–induced toxicity in mice. PBM at all fluences significantly improved short-term memory in the Y-maze test with a maximum effect at 16.8 J/cm^2. Enhanced long-term memory after PBM showed a biphasic dose–response curve, with a peak at 8.4 J/cm^2. Also, PBM at a fluence of 8.4 J/cm^2 most effectively reduced lipid peroxidation as well as inflammatory markers of GFAP and TNF-α, as compared to other fluences tested [50].

In 2019, Eshaghi et al. [51] investigated the therapeutic effects of three different fluences of 810 nm laser irradiation (4, 8, and 16 J/cm^2) on neurobehavioral function in a mouse stress model. Both depression and anxiety-like behavior showed a biphasic dose–response curve after PBM, with a maximum response at 8 J/cm^2. Moreover, PBM at both fluences of 8 and 16 J/cm^2 increased the levels of NO and serotonin in both hippocampus and prefrontal cortex regions, with a peak response at 8 J/cm^2. Nevertheless, PBM at all fluences significantly decreased serum cortisol levels [51].

Zinchenko et al. [52] in 2019 investigated the optimum fluence of 1267 nm laser to produce neuroprotective effects in a mouse Aβ-induced AD model. Their results showed that both fluences of 32 and 39 J/cm^2 significantly reduced the accumulation of Aβ in the mouse brain, as compared to lower fluences of 18 and 25 J/cm^2, however, the positive effect of 39 J/cm^2 was associated with some thermal injury to the dura matter and arachnoid membranes [52].

5.1.8   Biphasic Response: Treatment Sessions

In 2013, Huisa et al. [53] aimed to determine whether multiple 808.5 nm laser PBM treatments would have further improvement in neurobehavioral outcomes in a rabbit stroke model. Two and three PBM treatments were compared against a single PBM treatment alone. The double treatment group received PBM (7.5 mW/cm^2) at 3 and 5 h and the triple treatment group (20 mW/cm^2) at 2, 3, and 4 h post-embolization. Behavioral analysis was performed 24 h post-embolization using a dichotomized behavioral score. The results showed that triple treatment led to a 91% or 245% improvement in stroke outcomes when compared with a single treatment or sham treatment, respectively. They suggested that additional PBM treatments would provide further behavioral improvement when applied during the acute ischemic stroke phase [53].

A series of studies by Xuan et al. investigated the effects of different treatment repetition regimens of transcranial 810 nm laser PBM on neurobehavioral function and neurochemical markers in a mouse CCI model. At 4 h post-injury, mice first received PBM with either a single treatment, 3 daily treatments, or 14 daily treatments. In 2013, Xuan et al. [9] published the first study showing that three daily treatments resulted in a better improvement in neurological severity score (NSS) compared to a single treatment, over a period of 4 weeks post-injury. Results also showed that animals receiving one and three PBM treatments had smaller lesion size (at 28 days post-injury) and less neuronal degeneration (at 14 days post-injury) than in sham and 14-PBM treatment groups. In addition, three times PBM stimulated more neurogenesis in the brain (as shown via BrdU-positive cells) than the other treatment regimens [9].

In 2014, Xuan et al. [54] compared the effects of a single treatment and three daily treatments of transcranial 810 nm laser PBM in a mouse CCI-TBI model. Results showed that one and (to a greater extent) three daily PBM sessions enhanced neurological performance as assessed by wire grip and motor test especially at 3 and 4 weeks post-injury. Likewise, one and three times PBM improved spatial learning and memory as measured by the Morris water maze (MWM) test. In addition, both treatment protocols significantly decreased caspase-3 expression in the lesion region (at day 4 post-injury), increased neurogenesis in the dentate gyrus (DG) and subventricular zone (SVZ) (at days 7 and 28 post-injury), and up-regulated migrating neuroprogenitor cells and neuronal differentiation in the DG and SVZ (at days 7 and 28 post-injury); however, three daily treatments had the best effect on all measures [54].

In 2014, Xuan et al. [55] used their CCI mouse model of severe TBI treated with one or three daily transcranial 810 nm laser sessions and measured the expression of genes for the neurotrophin (BDNF) and synaptogenesis (synapsin) in different brain areas. Both one and three times PBM improved NSS (at 14, 21, and 28 days post-injury) and increased expression of BDNF in the DG and SVZ areas. Nevertheless, only three daily treatments significantly increased the expression of synapsin-1 in SVZ and cortex lesion area at 28 days post-injury [55].

In 2016, Xuan et al. [8] compared the effects of the 3 daily and 14 daily sessions of transcranial 810 nm laser PBM in the mouse CCI-TBI model. Results from the 14 daily treatment group showed worse neurological function than the no-treatment control group at 2 weeks, but the NSS started to improve gradually during the next 6 weeks, and by 8 weeks was meaningfully better than the no-treatment group, but still worse than the three daily treatment group. Although 14 daily PBM significantly increased GFAP (a marker of activated glial cells) compared to both no-treatment and 3 daily PBM groups at 4 weeks, the GFAP had fallen to low levels in both 3 and 14 daily PBM groups by 8 weeks. They suggested that because of the temporary induction of reactive gliosis (high

GFAP), an excessive number of PBM sessions could temporarily suppress the process of brain repair, but then the inhibitory effect disappeared, and brain repair could restart [8].

In 2016, Meyer et al. [40] compared the neuroprotective effects of single and triple PBM treatments in a rabbit stroke model. In the single PBM group, laser treatment was initiated 2 h post-embolization, whereas, the triple treatment group received PBM at 2, 3, and 4 h post-embolization. Overall, three times PBM resulted in significantly improved behavioral results at 24 h post-embolization compared to single PBM protocol [40].

Basha et al. [56] in 2016 aimed to investigate whether single or multiple pre-treatment sessions of 670 nm LED PBM would have different effects on the activity of Na^+–K^+ ATPase, Ca^{2+} ATPase, and CCO in the brains of rats that were exposed to bright fluorescent light. They compared the neuroprotective effects of 1, 15, and 30 daily PBM pre-treatment sessions against fluorescent light-induced neural damage. The results from PBM pre-treatment groups showed that there was a time-dependent change in the enzyme activity of Na^+–K^+ ATPase and Ca^{2+} ATPase, with a maximum response in the 30 daily irradiations group. Also, 15 and 30 daily PBM pre-treatments significantly increased CCO activity as compared to both the non-treatment and once-daily treatment groups [56].

Xu et al. in 2017 [57] investigated the effects of 808 nm laser PBM on depression-like behavior in a mouse stress model. PBM was immediately performed after spatial restraint stress once daily for a duration of 28 days. Subsequently, depression-like behavior was assessed by the FST and tail suspension tests (TST) at different time points. The results from both FST and TST showed no significant difference in immobility time between PBM and non-treatment groups until 14 days post-stress. After this time point, the immobility time in mice subjected to PBM was significantly decreased compared with the stress group, and these differences even were maintained at 21 and 28 days post-stress [57].

In 2019, Ganeshan et al. [58] investigated an effective PBM pre-treatment regimen for inducing neuroprotection in a mouse PD model. Mice received irradiation to the body but not the head with 670 nm LEDs for 2, 5, or 10 days, followed by injection of the neurotoxin MPTP over 2 consecutive days. Results showed that only 10 days pre-treatment significantly attenuated MPTP-induced loss of midbrain TH+ dopaminergic cells. Nonetheless, all treatment regimens significantly mitigated the MPTP-induced increase in FOS + cells in the CPu [58].

In 2019, Wang et al. [59] aimed to explore an optimal PBM regimen in a rat global cerebral ischemia model. Transcranial 808 nm laser PBM was initiated at 6 h post-injury and continued every day for 1, 2, 3, or 4 days. According to their results, the density of hippocampal CA1 pyramidal neurons was markedly increased in the 2 and 3 daily PBM groups. In addition, all treatment regimens significantly increased the number of surviving neurons, with a maximum effect with 3 and 4 PBM regimens. Likewise, although one and two daily PBM sessions significantly decreased the number of apoptotic cells based on TUNEL labeling, the protective effects of three and four daily treatments were more pronounced [59].

References

1. Mester, E., A.F. Mester, and A. Mester. 1985. The biomedical effects of laser application. *Lasers in surgery and medicine* 5 (1): 31–39.
2. Huang, Y.-Y., et al., *Biphasic dose response in low level light therapy.* Dose-response, 2009. 7(4): p. dose-response. 09–027. Hamblin.
3. Huang, Y.-Y., et al., *Biphasic dose response in low level light therapy–an update.* Dose-Response, 2011. 9(4): p. dose-response. 11–009. Hamblin.
4. Fresnoza, S., et al. 2014. Nonlinear dose-dependent impact of D1 receptor activation on motor cortex plasticity in humans. *Journal of Neuroscience* 34 (7): 2744–2753.
5. Sommer, A.P., et al. 2001. Biostimulatory windows in low-intensity laser activation: Lasers, scanners, and NASA's light-emitting diode array system. *Journal of clinical laser medicine & surgery* 19 (1): 29–33.
6. Zein, R., W. Selting, and M.R. Hamblin. 2018. Review of light parameters and photobiomodulation efficacy: Dive into complexity. *Journal of biomedical optics* 23 (12): 120901.
7. Hamblin, M.R. 2019. Photobiomodulation for traumatic brain injury in mouse models. In *Photobiomodulation in the Brain*, 155–168. Elsevier.
8. Xuan, W., L. Huang, and M.R. Hamblin. 2016. Repeated transcranial low-level laser therapy for traumatic brain injury in mice: Biphasic dose response and long-term treatment outcome. *Journal of biophotonics* 9 (11–12): 1263–1272.
9. Xuan, W., et al. 2013. Transcranial low-level laser therapy improves neurological performance in traumatic brain injury in mice: Effect of treatment repetition regimen. *PLoS ONE* 8 (1): e53454.
10. Hamblin, M.R., et al., *Low-level light therapy: Photobiomodulation.* 2018: SPIE Press Bellingham.
11. Sharma, S.K., et al. 2011. Dose response effects of 810 nm laser light on mouse primary cortical neurons. *Lasers in surgery and medicine* 43 (8): 851–859.
12. Duan, R., et al. 2003. Light emitting diode irradiation protect against the amyloid beta 25–35 induced apoptosis of PC12 cell in vitro. *Lasers in Surgery and Medicine: The Official Journal of the American Society for Laser Medicine and Surgery* 33 (3): 199–203.
13. Higuchi, A., et al. 2007. Visible light regulates neurite outgrowth of nerve cells. *Cytotechnology* 54 (3): 181–188.
14. Rochkind, S., et al. 2009. Increase of neuronal sprouting and migration using 780 nm laser phototherapy as procedure for cell therapy. *Lasers in Surgery and Medicine: The Official Journal of the American Society for Laser Medicine and Surgery* 41 (4): 277–281.
15. Chen, X., et al., *Effect and mechanism of 808 nm light pretreatment of hypoxic primary neurons.* International Journal of Photoenergy, 2014. **2014**.
16. Byrnes, K.R., et al. 2005. Low power laser irradiation alters gene expression of olfactory ensheathing cells in vitro. *Lasers in Surgery and Medicine: The Official Journal of the American Society for Laser Medicine and Surgery* 37 (2): 161–171.
17. Giuliani, A., et al. 2009. Low infra red laser light irradiation on cultured neural cells: Effects on mitochondria and cell viability after oxidative stress. *BMC complementary and alternative medicine* 9 (1): 8.
18. Saito, K., et al. 2011. Effect of diode laser on proliferation and differentiation of pc12 cells. *The Bulletin of Tokyo Dental College* 52 (2): 95–102.
19. Song, S., F. Zhou, and W.R. Chen. 2012. Low-level laser therapy regulates microglial function through Src-mediated signaling pathways: Implications for neurodegenerative diseases. *Journal of neuroinflammation* 9 (1): 219.

20. Fukuzaki, Y., et al. 2013. 532 nm low-power laser irradiation recovers γ-secretase inhibitor-mediated cell growth suppression and promotes cell proliferation via Akt signaling. *PLoS ONE* 8 (8): e70737.
21. Fukuzaki, Y., et al. 2015. 532 nm low-power laser irradiation facilitates the migration of GABAergic neural stem/progenitor cells in mouse neocortex. *PLoS ONE* 10 (4): e0123833.
22. von Leden, R.E., et al. 2013. 808 nm Wavelength Light Induces a Dose-D ependent Alteration in Microglial Polarization and Resultant Microglial Induced Neurite Growth. *Lasers in surgery and medicine* 45 (4): 253–263.
23. Meng, C., Z. He, and D. Xing. 2013. Low-level laser therapy rescues dendrite atrophy via upregulating BDNF expression: Implications for Alzheimer's disease. *Journal of Neuroscience* 33 (33): 13505–13517.
24. Zheng, L.-Q., et al. 2015. Modulating nitric oxide levels in dorsal root ganglion neurons of rat with low-level laser therapy. *Optoelectronics Letters* 11 (3): 233–236.
25. Yan, X., et al. 2017. Low-level laser irradiation modulates brain-derived neurotrophic factor mRNA transcription through calcium-dependent activation of the ERK/CREB pathway. *Lasers in medical science* 32 (1): 169–180.
26. Gu, X., et al. 2017. Photoactivation of ERK/CREB/VMAT2 pathway attenuates MPP+-induced neuronal injury in a cellular model of Parkinson's disease. *Cellular signalling* 37: 103–114.
27. Zhu, W., et al. 2017. 3D printing scaffold coupled with low level light therapy for neural tissue regeneration. *Biofabrication* 9 (2): 025002.
28. Santos, T., et al. 2017. Blue light potentiates neurogenesis induced by retinoic acid-loaded responsive nanoparticles. *Acta biomaterialia* 59: 293–302.
29. Levchenko, S.M., et al. 2018. Near-infrared irradiation affects lipid metabolism in neuronal cells, inducing lipid droplets formation. *ACS chemical neuroscience* 10 (3): 1517–1523.
30. Rhee, Y.-H., et al. 2019. Effect of photobiomodulation therapy on neuronal injuries by ouabain: The regulation of Na, K-ATPase; Src; and mitogen-activated protein kinase signaling pathway. *BMC neuroscience* 20 (1): 19.
31. Liang, H.L., et al. 2008. Near-infrared light via light-emitting diode treatment is therapeutic against rotenone-and 1-methyl-4-phenylpyridinium ion-induced neurotoxicity. *Neuroscience* 153 (4): 963–974.
32. Choi, D.-H., et al. 2012. Effect of 710 nm visible light irradiation on neurite outgrowth in primary rat cortical neurons following ischemic insult. *Biochemical and biophysical research communications* 422 (2): 274–279.
33. Ilic, S., et al. 2006. Effects of power densities, continuous and pulse frequencies, and number of sessions of low-level laser therapy on intact rat brain. *Photomedicine and Laser Therapy* 24 (4): 458–466.
34. Ahmed, N.A.E.H., et al. 2008. Effect of three different intensities of infrared laser energy on the levels of amino acid neurotransmitters in the cortex and hippocampus of rat brain. *Photomedicine and laser surgery* 26 (5): 479–488.
35. Uozumi, Y., et al. 2010. Targeted increase in cerebral blood flow by transcranial near-infrared laser irradiation. *Lasers in surgery and medicine* 42 (6): 566–576.
36. De Taboada, L., et al. 2011. Transcranial laser therapy attenuates amyloid-β peptide neuropathology in amyloid-β protein precursor transgenic mice. *Journal of Alzheimer's disease* 23 (3): 521–535.
37. Khuman, J., et al. 2012. Low-level laser light therapy improves cognitive deficits and inhibits microglial activation after controlled cortical impact in mice. *Journal of neurotrauma* 29 (2): 408–417.
38. Kuo, J.-R., et al. 2015. Deep brain light stimulation effects on glutamate and dopamine concentration. *Biomedical optics express* 6 (1): 23–31.

39. Oueslati, A., et al. 2015. Photobiomodulation suppresses alpha-synuclein-induced toxicity in an AAV-based rat genetic model of Parkinson's disease. *PLoS ONE* 10 (10): e0140880.

40. Meyer, D.M., Y. Chen, and J.A. Zivin. 2016. Dose-finding study of phototherapy on stroke outcome in a rabbit model of ischemic stroke. *Neuroscience letters* 630: 254–258.

41. Reinhart, F., et al. 2016. Intracranial application of near-infrared light in a hemi-parkinsonian rat model: The impact on behavior and cell survival. *Journal of neurosurgery* 124 (6): 1829–1841.

42. Mohammed, H.S. 2016. Transcranial low-level infrared laser irradiation ameliorates depression induced by reserpine in rats. *Lasers in medical science* 31 (8): 1651–1656.

43. Leung, M.C., et al. 2002. Treatment of experimentally induced transient cerebral ischemia with low energy laser inhibits nitric oxide synthase activity and up-regulates the expression of transforming growth factor-beta 1. *Lasers in Surgery and Medicine: The Official Journal of the American Society for Laser Medicine and Surgery* 31 (4): 283–288.

44. Moreira, M.S., et al. 2009. Effect of phototherapy with low intensity laser on local and systemic immunomodulation following focal brain damage in rat. *Journal of photochemistry and photobiology B: Biology* 97 (3): 145–151.

45. Lapchak, P.A., and L. De Taboada. 2010. Transcranial near infrared laser treatment (NILT) increases cortical adenosine-5′-triphosphate (ATP) content following embolic strokes in rabbits. *Brain research* 1306: 100–105.

46. Yip, K., et al. 2011. The effect of low-energy laser irradiation on apoptotic factors following experimentally induced transient cerebral ischemia. *Neuroscience* 190: 301–306.

47. Rojas, J.C., A.K. Bruchey, and F. Gonzalez-Lima. 2012. Low-level light therapy improves cortical metabolic capacity and memory retention. *Journal of Alzheimer's Disease* 32 (3): 741–752.

48. El Massri, N., et al. 2016. The effect of different doses of near infrared light on dopaminergic cell survival and gliosis in MPTP-treated mice. *International Journal of Neuroscience* 126 (1): 76–87.

49. Salehpour, F., et al. 2017. Transcranial low-level laser therapy improves brain mitochondrial function and cognitive impairment in D-galactose–induced aging mice. *Neurobiology of aging* 58: 140–150.

50. Blivet, G., et al. 2018. Neuroprotective effect of a new photobiomodulation technique against Aβ25–35 peptide–induced toxicity in mice: Novel hypothesis for therapeutic approach of Alzheimer's disease suggested. *Alzheimer's & Dementia: Translational Research & Clinical Interventions* 4: 54–63.

51. Eshaghi, E., et al., *Transcranial photobiomodulation prevents anxiety and depression via changing serotonin and nitric oxide levels in brain of depression model mice: A study of three different doses of 810 nm laser.* Lasers in surgery and medicine, 2019.

52. Zinchenko, E., et al. 2019. Pilot study of transcranial photobiomodulation of lymphatic clearance of beta-amyloid from the mouse brain: Breakthrough strategies for non-pharmacologic therapy of Alzheimer's disease. *Biomedical Optics Express* 10 (8): 4003–4017.

53. Huisa, B.N., et al. 2013. Incremental treatments with laser therapy augments good behavioral outcome in the rabbit small clot embolic stroke model. *Lasers in medical science* 28 (4): 1085–1089.

54. Xuan, W., et al. 2014. Transcranial low-level laser therapy enhances learning, memory, and neuroprogenitor cells after traumatic brain injury in mice. *Journal of biomedical optics* 19 (10): 108003.

55. Xuan, W., et al. 2015. Low-level laser therapy for traumatic brain injury in mice increases brain derived neurotrophic factor (BDNF) and synaptogenesis. *Journal of biophotonics* 8 (6): 502–511.

56. Mathangi, D., and R. Shyamala. 2016. Effect of LED photobiomodulation on fluorescent light induced changes in cellular ATPases and Cytochrome c oxidase activity in Wistar rat. *Lasers in medical science* 31 (9): 1803–1809.
57. Xu, Z., et al. 2017. Low-level laser irradiation improves depression-like behaviors in mice. *Molecular neurobiology* 54 (6): 4551–4559.
58. Ganeshan, V., et al. 2019. Pre-conditioning with remote photobiomodulation modulates the brain transcriptome and protects against MPTP insult in mice. *Neuroscience* 400: 85–97.
59. Wang, R., et al. 2019. Photobiomodulation for global cerebral ischemia: Targeting mitochondrial dynamics and functions. *Molecular neurobiology* 56 (3): 1852–1869.

Light Delivery Approaches for Brain Photobiomodulation

6

6.1 Transcranial Photobiomodulation

Non-invasive delivery of photons from an external light source to the head and thence into the brain tissue is generally referred to as transcranial photobiomodulation (PBM). In this approach, light must pass through several types of tissue, such as the scalp, skull, periosteal, meningeal, subdural space, arachnoid mater, subarachnoid space, and pia mater, successively, until reaching the cortical surface. Hair can also act as a significant attenuator of light in the visible and near-infrared (NIR) wavelengths, and its barrier role should be taken into account when other parts of head (not the forehead) are irradiated. Owing to exponential attenuation of light during when passing through the aforementioned tissue layers, the maximum fluence (still only a relatively small fraction of the incident light on the scalp) will be delivered to those neurons located in the outermost layers of the cerebral cortex. Thus, there will be an additional gradient of light flux within the cerebral cortex, so that usually only a very few neurons will absorb a sufficient light dose in transcranial PBM [1]. It is worth noting that up to now, the transcranial method has been the most widely explored approach for light delivery in brain PBM therapy, but its effectiveness in reaching the deeper structures of the brain with biostimulatory/therapeutic light doses (e.g., limbic system and brainstem) remains an open question [2]. However, it would not be possible for biostimulatory dose levels to reach the deeper regions without delivering an overdose to the surface tissue, due to the exponential reduction in the transmission of light through the brain tissue [3]. Nevertheless, it is noteworthy that modern-day PBM can use high-power lasers or light-emitting diodes (LEDs) (> 5 W, class IV) instead of more old-fashioned low-power devices (<500 mW, class III), while still delivering a higher fluence to the deeper layers of the cortex without any thermal damage [4]. These LED devices do not have any safety concerns, which is an important issue that must be taken into account when delivering light to the head. Generally speaking, transcranial PBM is either applied using an in-contact mode (hand-held, strapped-on, or helmet) or in a non-contact mode (projection of light from a distance). Some researchers have suggested that diffuse light reflection from the tissue surface can be minimized if the probe is held in firm contact with the skin/scalp [5].

In the majority of the clinical trials of transcranial brain PBM, hand-held probes have initially been applied to a single site on the scalp. After completion of the irradiation, the light device is turned off and the probe is moved and positioned on the next spot and the procedure is repeated [6]. In the single-spot irradiation method, although the irradiation procedure is more controllable in terms of heat generation than a whole-head irradiation method, the total treatment period may be longer because of one by one irradiation of each target spot. Nowadays, there are several companies that have designed light-emitting helmets or "buckets" to irradiate the whole area of the scalp [7–13]. The BrainTHOR helmet has been also introduced for investigational use to treat traumatic brain injury (TBI), concussion, and chronic traumatic encephalopathy by Thor Photomedicine Ltd. (Chesham, Bucks, UK). These wearable PBM devices consist of hundreds of separate diodes distributed equally inside the helmet to deliver a large total amount of light (many thousands of Joules) to the head. The development of these devices has been in response to the growing realization that modern health care intervention service delivery is shifting from face-to-face office-based treatments to telemedicine and cloud-based at home applications [14]. Irradiation of specific areas of the brain using light headsets that target the nodes of the default mode network: (a) mesial prefrontal cortex, (b) precuneus, (c) posterior cingulate cortex, and (d) inferior parietal lobe, has also been described for treating neurodegenerative diseases such as Alzheimer's disease (AD) as well as boosting brain function in healthy individuals [15, 16]. Lately, a robotic-based laser device has also been applied for brain PBM in clinical settings [17, 18].

In some of the clinical trials using brain PBM, the 10–20 system for electroencephalogram (EEG) electrode placement, and more rarely the Brodmann areas, has been used to define the target irradiation spots. For example, the Fp1 and Fp2 spots refer to the left and right ventromedial prefrontal cortex (vmPFC), respectively, and Fpz refers to the rostral medial PFC. Moreover, the F3 and F4 as well as the F7 and F8 sites refer to the left and right dorsolateral PFC (dlPFC), and left and right lateral frontal area, respectively. A recent simulation study has shown that when LED probes were positioned on the F3 and F4 sites over the scalp, the highest light energy was received by the caudal dlPFC, rostral dorsolateral inferior PFC, and rostral dorsolateral superior PFC areas [19]. It must be noted that the frontal sinus in the skull is an optically void area. Mathematical simulation of light propagation within a human head model has revealed that penetration and scattering of light in the brain are affected by the frontal sinus [20], therefore, this should be taken into account when areas close to the frontal sinus such as Fp1, Fp2, and Fpz are irradiated. A review of the literature has shown that the frontal region is the most often irradiated area in clinical trials of brain PBM. Anatomically speaking, the frontal lobes are the largest and most developed brain regions in humans, and extend from the central sulcus to the frontal pole. The frontal lobes are involved in control of several important body functions such as motor control, emotional expression, task planning, problem-solving, memory, language and speech, reasoning and judgment, and sexual behavior. Given that the frontal lobe is located in the anterior part of the head, it is often the most susceptible region to brain injury. Damage to the frontal lobe can often affect the ability to make good choices, personality, facial expression, and interpretation. The frontal lobes are subdivided into two main areas, including the PFC and the motor cortex [21].

The motor cortex is composed of the premotor cortex and primary motor cortex. The primary motor cortex (Brodmann area 4) is located on the anterior wall of the central sulcus and is responsible for planning and executing movements, in association with other motor areas and subcortical brain regions. It contains large nerve cells, known as Betz cells, which send nerve signals down to the body to control voluntary movement of the skeletal muscles. The premotor cortex (Brodmann area 6) is located anterior to the primary motor cortex, and on the lateral surface of the cerebral hemisphere, and functions to modify movements [22–25].

The PFC is located in the anterior part of the frontal lobe on the lateral, medial, and basal surfaces of the brain. The PFC amounts to more than a quarter of the total cerebral cortex in the human brain. It encompasses the Brodmann areas 8–13, 24, 25, 32, and 44–47 [26]. The PFC is closely connected with subcortical and other cortical structures involved in memory, sensory perception, and emotions, including the thalamus, basal ganglia, hypothalamus, amygdala, hippocampus, and the temporal and parietal lobes of the cortex. It is crucial for the higher cognitive skills, and the regulation of behavior, personality, attention, and planning of complex cognitive tasks [23]. The PFC can be subdivided into several major anatomical and functional regions, including the dorsolateral and the ventromedial regions, dorsomedial regions, anterior cingulate cortex (ACC), and the orbitofrontal region. The dorsolateral region provides cognitive support for language, speech, and logical reasoning. Moreover, emotional behavior has been attributed to the orbitofrontal and ventromedial regions [23, 24].

The dlPFC, Brodmann areas 45–49, constitutes the largest part of the frontal cortex. The dlPFC is a functional structure rather than a localized anatomical structure. The dlPFC is located in the middle of the frontal gyrus, superior to the orbitofrontal cortex (OFC) and rostral to the frontal eye fields. The dlPFC is connected to motor control centers, such as the basal ganglia, premotor cortex, supplementary motor area, cingulate cortex, hippocampus, thalamus, OFC, and parietal cortex. It is also connected to the vmOFC. The dlPFC is involved in cognitive functions, such as planning, executive function, working memory, risky or moral decision-making, speech, and logical reasoning [24, 27].

The vmPFC is located below the cerebral hemispheres, inferior to the dorsomedial PFC, and encompasses the whole area of the PFC, both in the ventral and medial positions approximately below the genu of the corpus callosum. Brodmann areas 24, 25, 32, 12, 11, and 10 together comprise the vmPFC zone [28, 29]. The vmPFC receives input from the five senses and is extensively interconnected with the amygdala, ACC, temporal lobe, olfactory system, dorsomedial thalamus, and the anterior, medial, and ventral regions of the striatum. The main outputs from vmPFC are connected to many different brain regions, including the lateral hypothalamus, temporal lobe, amygdala, periaqueductal gray matter, hippocampus, nucleus accumbens, cingulate cortex, and superior temporal cortex. By means of this large network of connections, the vmPFC can receive and process a wide variety of sensory inputs [30, 31]. Within the vmPFC, there are anatomically and/or functionally distinct sub-regions aligned with their functional domains. The anterior/pregenual vmPFC and the ventral striatum are involved with decision-making, the posterior/subgenual vmPFC and the amygdala are associated with emotion, and the anterior/pregenual vmPFC and the dorsomedial PFC, precuneus and temporoparietal cortex are responsible for social cognition [30, 32]. More precisely, the anterior/perigenous subregion of the vmPFC is believed to be involved in positive valence, such as rewards, whereas the posterior/subgenuous subregion is linked to negative valence such as threats and fear [24]. This region is implicated in the inhibition of emotional responses, processing of fear response, social cognition, episodic and semantic memory, decision-making, prospection, and self-control [31, 33]. Moreover, functional magnetic resonance imaging (fMRI) studies have shown that a dysfunction of this region has been associated with a number of brain disorders, including addiction, depression, bipolar disorder, posttraumatic stress disorder (PTSD), schizophrenia, anxiety disorder, and attention-deficit/hyperactivity disorder [32, 34, 35].

The OFC is part of the PFC within the frontal lobe, which is located on the ventral surface of the frontal lobe directly above the orbits (in which the eyes are positioned) and extends posteriorly from the frontal pole to the insula, and ventrally from the rostral sulcus on the medial wall to the ventrolateral convexity to form the frontal base of the brain. Its blood supply is provided by orbitofrontal arteries or frontobasal arteries, both of which are branches of the middle cerebral artery. This region is anatomically composed of Brodmann areas 11–13, and 47. Moreover, it is anatomically connected to the limbic system and the anterior temporal lobe cortex. The OFC receives projections from the amygdala, the temporal association cortex, the visual system, somatosensory cortex, medial dorsal nucleus of the thalamus, and hypothalamus, and is involved in taste and smell [36]. The OFC is involved in reward and punishment learning, emotional control, emotionally-guided behavior, decision-making, and social and emotional processing [24, 37]. Moreover, the right OFC mediates the conscious perception of smell [38]. An OFC lesion in humans leads to increased impulsivity, difficulties in executive function, impairment of motivation, arousal, and prediction of rewards [39].

6.2   Intracranial or Deep Brain Photobiomodulation

Neurons and circuits within the globus pallidus internus (GPi) and the subthalamic nucleus (STN) (two deep structures of the brain) are involved in Parkinson's disease (PD) pathophysiology. The restricted ability of light to penetrate through the head to the deeper brain structures may limit the delivery of a sufficient light dose to the midbrain neurons. Over the past decade, there have been attempts to develop new effective techniques for delivering light energy to deeper brain tissues, such as the substantia nigra pars compacta (SNpc), using an intracranial approach, similar to that used in deep brain stimulation (DBS) procedures [40, 41]. Intracranial delivery of light via an implanted optical fiber to target specific neuroanatomical structures is a novel technology that could be used both for PBM [42] as well as for optogenetic studies [43]. A Monte Carlo study showed that when a fiber-optic source was implanted in the third ventricular region, it had a 20-fold higher efficiency in delivering light to the SNc region, compared to a non-invasive transcranial method. It was found that only 1% of 670 nm LED light reached the mouse SNc at a depth of 5 mm from the light source in the standard transcranial approach [44].

Anatomically speaking, the SN is a large cluster of pigmented neurons located dorsal to the cerebral peduncles, ventral to the midbrain tegmentum, and inferior to the red nucleus. It is the largest nucleus in the midbrain and is a major output structure to the basal ganglia. The SN plays an important role in reward-seeking, motor planning, addiction, and movement. The term SN means black substance in Latin because of the high level of the dark-colored neuromelanin pigment in dopaminergic neurons. Anatomical studies have divided the SN into two distinct parts, with different connections and functions: (1) the superior and anterior pars reticulata (SNpr); (2) the inferior and posterior pars compacta (SNpc). The SNpc is the main output to the basal ganglia, while the SNpr is the main input, primarily from the striatum, and is also an important processing center. Its main projections are to the ventral lateral and ventral anterior nuclei of the thalamus and superior colliculus. The SNpr and the internal globus pallidus are separated by the white matter of the internal capsule, and its neurons are mostly GABAergic [1, 19, 20]. Cells of the SNpc synthesize the neurotransmitter dopamine, some of which is converted into melanin. Dopaminergic neurons of this region project to the dorsal striatum through the nigrostriatal pathway, as well as to the lateral and medial pallidum of the basal ganglia system, the SNpr, and the subthalamic nucleus. These receive inhibitory inputs from the collateral axons of the SNpr [21–23]. This region is mainly involved in control of movement. Pathological changes to the dopaminergic neurons of the SNpc have been associated with PD, schizophrenia, sleep disturbance, and clinical depression [24, 25]. Given this, the SNpc is a promising treatment target for patients suffering from PD, and the results of early animal studies suggest that deep brain PBM may be a novel treatment approach.

The John Mitrofanis laboratory in Australia has studied PD in animal models, and after reporting the implantation of an optical fiber into the mouse brain [40], they repeated this minimally invasive procedure in Macaque monkeys with PD [42]. Their preliminary investigations demonstrated the neuroprotective effects of intracranial PBM, without showing any toxic adverse effects around the implant site within the midbrain [45]. Moreover, it has been shown by others that the use of an optical fiber with 0.4 mm diameter, to deliver 638 nm light irradiation at an irradiance of 100–600 mW/mm^2 and pulsed wave (PW) of 20–60 Hz, resulted in a maximum temperature rise of only about 0.1–2.5 °C. Moreover, histological examination did not show any light-induced cellular or tissue damage when a power of 200 mW/mm^2 was continuously applied [46]. In agreement with these findings, another study reported that after 1 h of irradiation with an optical fiber connected to 720 nm LEDs with a power of 180 μW/cm^2, the increase in temperature at the tip of the fiber was negligible, being less than 0.1 °C [47].

6.3   Intranasal Photobiomodulation

Recently, intranasal PBM has been proposed to provide active irradiation of the prefrontal area and some of the limbic structures in the brain, as an alternative to the limitations of transcranial PBM [19, 44]. Intranasal PBM is a therapeutic approach using the nostrils, in which one or two small portable probes each equipped with a laser or LED are inserted. Systemic effects on the blood (cells and/or proteins) could possibly explain the neuro-therapeutic effects reported after nose-based intranasal irradiation [48, 49]. Blood capillaries are plentiful within the lining of the nasal cavity, and the blood flow is comparatively slow. A decrease in blood viscosity [50], an increase in hemorheology [50], and an increase in blood coagulability [51] have been reported following intranasal PBM, and these improvements in blood flow parameters have been found to be correlated with enhanced cognitive function [52] and mood states [53]. Overall, the systemic effects of blood irradiation in intranasal PBM could have potential neuroprotective effects in the brain [48, 49, 54, 55].

6.3.1   Intranasal Photobiomodulation from the Nostrils

The nasal cavity is a large air-filled space above and behind the nose, which fills the space between the base of the skull and the roof of the mouth. Anatomically, the roof of the nasal cavity is formed by frontal and nasal bones in the anterior, the cribriform plate of the ethmoid in the middle, and the body of the sphenoid in the posterior. The floor is the upper surface of the hard palate. The anterior wall is formed by the nasal surface of the ethmoid labyrinth, posterior wall by the perpendicular plate of palatine bone, and the nasal surface of the maxilla in the lower anterior part. There is a close relationship between the nasal cavity and the cranial cavity, both in anatomy and physiology, thus allowing a unique and short path to deliver some drugs to the brain. Given that the blood–brain barrier (BBB) can prevent drugs from easily penetrating to the central nervous system (CNS), when a rapid onset of action is required to treat brain disorders, an intranasal delivery route could be considered due to its ability to bypass the BBB [56]. The bilateral nasal cavities in human have a total surface area of around 150–160 cm^2, and a total volume of 15–20 mL. The nasal cavity is divided into two symmetrical cavities by the nasal septum, whose inner sides are lined with mucosa. Each of the two nasal cavities opens at the face through the nostrils and extends posterior to the nasopharynx. The nasal cavity is subdivided into three areas, based on the anatomical and histological characteristics, including the nasal vestibule, respiratory region (also called the conchae), and the olfactory region. The most anterior part of the nasal cavity is the nasal vestibule with a surface area of approximately 0.6 cm^2. This region is covered by stratified squamous and keratinized epithelial cells equipped with vibrissae or nasal hairs. The respiratory region, the largest part of the nasal cavity, is full of blood capillaries and receives abundant blood flow with a surface area of about 130 cm^2. This region is divided into superior, middle, and inferior turbinates by projections from the lateral wall. These turbinates are responsible for moisturizing and warming up inhaled air. The olfactory region is located on the roof of the nasal cavity, which is composed of the superior nasal concha and the upper third of the septum, and contains olfactory cells, which act as receptors for the sense of smell. The neuroepithelium of the olfactory region is the only portion of the CNS that is directly exposed to the external environment. This area also plays the most important part in the transportation of drugs to the CNS [56].

Nasal administration has been acknowledged to be a useful route to deliver drugs for the treatment of respiratory-related complications, such as asthma, colds, coughs, and sinusitis [57]. The nose also provides direct access to the brain for delivering exogenous drug molecules, making the nasal system an attractive non-invasive approach [58]. The nasal cavity has a relatively rich supply of blood vessels within its mucosal layer. Recently, attention has been given to the nasal delivery route for systemic drugs, which can then be distributed via the blood circulation to the brain [59]. Intranasal PBM involves a simple process of placing a small laser diode or LED unilaterally or bilaterally into the nostril(s). Research on the applications of intranasal PBM is becoming more and more common, and a range of portable intranasal light sources are now available on the market, designed for both medical and home use. To all intents and purposes, intranasal PBM is an inexpensive, non-invasive, patient-friendly, ambulatory home-based approach, which needs no trained medical personnel, and gives better patient comfort and compliance. In addition, portable intranasal PBM devices available on the market are powered by a lightweight, rechargeable battery, making them an ideal option for a self-operated device. The available devices often contain red diodes with wavelengths of 600–680 nm and/ or NIR diodes emitting 800–850 nm. Such devices are typically customizable to select the treatment duration, and continuous or pulsed irradiation mode [49]. The PW irradiation mode is considered to provide extra benefits for intranasal PBM. The first advantage of PW is that the dark interval between the pulses will allow the tissue to cool down to some degree, without which the light would produce too much heat if in continuous contact with the nasal cavity tissue. For another advantage, recent studies have shown some beneficial neurobiological effects of specific pulse frequencies of the light, including 10–Hz, 40–Hz, and 100 Hz [60]. In patients with cognitive or mood disorders, PBM at 10-Hz PW was shown to have neurotherapeutic effects [9, 16, 61]. As a third advantage, the 40-Hz pulse rate could increase brain gamma waves, which have been shown to be associated with a decrease in amyloid-β (Aβ) levels and tau phosphorylation within the brain. This was shown in a study using an animal AD model with illumination in the visible spectrum [62]. In this regard, portable intranasal PBM applicators with a PW of 40-Hz have been tested to improve concentration and cognitive function in AD patients [63] and also to modulate neural oscillations in healthy adults [15]. For the last advantage, PD patients treated with a 60-Hz frequency DBS demonstrated significant improvements in motor function and swallowing symptoms [64]. Nevertheless, intranasal PBM at 60-Hz PW mode should be further studied in clinical trials with a focus on neurodegenerative diseases.

Commercially available portable devices involve a light source to be placed into the nostril. In this approach, the light is supposed to be absorbed mainly superficially by the nasal mucosa and surrounding tissues owing to the position of the light source at the inside tip of the nose [49]. In an intranasal simulation experiment with 810 nm light, it was demonstrated that the vmPFC and vmOFC obtained the greatest portion of light among the CNS structures, but still only a tiny fraction of the primary photons (~0.001%) reach the brain. That is to say, only 0.014 and 0.025 J/cm^2 reached vmPFC and vmOFC, respectively, in terms of total light fluence for a commercially available device. More experiments showed that the photon flux at the amygdala and hippocampus was weaker than that of vmPFC and vmOFC by two orders of magnitude. Overall, such results suggest that the nostril-based intranasal PBM is inadequate to deliver light to brain structures, especially limbic structures, and therefore the beneficial effects must be systemic in nature [19].

The systemic effects of nostril-based irradiation with intranasal PBM on hematologic cells and/or blood elements are possible mechanisms of action that require further investigation [48, 54, 55]. Therefore, intranasal PBM may achieve the same therapeutic benefits as the intravenous blood irradiation approach, but the former is much superior in terms of being non-invasive and relatively low-cost [55]. For example, intranasal laser irradiation has been shown to be as effective as intravascular blood irradiation in increasing local CBF and brain function in patients with cerebral infarction [55]. Also, 10 days of remote pre-conditioning PBM (670 nm, 4.5 J/cm^2; administered to the dorsal and hind limbs) were recently shown to be protective against MPTP-induced neuropathology in mice by modulating various molecular pathways within the brain, including oxidative stress response pathways, cell signaling, increased migration (including CXCR4 + stem cell and adipocytokine signaling), and modulation of the blood–brain barrier [65]. These data provide further evidence for the possibility of blood absorption of light to be considered as the main mechanism responsible for nostril-based intranasal PBM protective effects.

The nasal cavity has abundant arterial blood supply from both the internal and external carotid arteries for the purpose of direct blood irradiation. Anatomically, the anterior and posterior ethmoidal arteries stem from the internal carotid artery. The ethmoidal arteries are branches of the ophthalmic artery that pass through the cribriform plate into the nasal cavity and supply the upper nasal septum and lateral walls. The external carotid artery, however, branches into the sphenopalatine artery, greater palatine artery, superior labial artery, and lateral nasal arteries. The sphenopalatine artery is the terminal branch of the maxillary artery and supplies a substantial portion of the septum and lateral wall. The sphenopalatine artery penetrates the posterior nasal cavity through the sphenopalatine foramen. A branch of the facial artery, known as the superior labial artery, supplies the frontal part of the nose and the nasal septum. Furthermore, the lateral nasal arteries mainly supply the nasal vestibule [66, 67]. In the anterior portion of the nose, an area known as Little's area ends the branches of these arteries making anastomoses with one other, and creating the plexus of Kiesselbach. The Kiesselbach plexus is a gathering of arterial anastomoses, very rich in vascular tissue, located in the anteroinferior quadrant of the nasal septum above the septal cartilage. The nasal veins accompany the arteries, which drain into the pterygoid plexus, the cavernous sinus, and the ophthalmic artery [67, 68].

It is believed that nostril-based intranasal PBM increases blood oxygenation and leads to increased levels of ATP in different tissues, including the brain. PBM using visible or NIR light results in partial photochemical dissociation of hemoglobin-ligand complexes in the blood (e.g., O_2, carbon dioxide, and nitric oxide [NO]) [69–72]. This leads to substantial oxygenation of local tissue that is followed by a decrease in oxygen saturation of capillaries (SpO_2) [73–75]. Hemoglobin also absorbs 660 nm photons, which may intensify the effect of laser irradiation on blood lymphocytes [76]. In addition, the released NO (because NO is a potent vasodilator of microcirculation) enhances the perfusion and oxygenation of tissues. This suggests that nostril-based intranasal PBM is a promising and potentially effective tool for treating hypoxic-ischemic brain injury, as well as neurodegenerative and neuropsychiatric disorders. As a matter of fact, in depressed patients, less NO is released from the vascular endothelium [77]. Similarly, hypertension increases oxidative stress in the endothelial cells reducing the bioavailability of released NO, which is associated with impaired cognitive function [78]. On the contrary, NO release from endothelium or platelets is stimulated by nostril-based intranasal PBM, resulting in increased cerebrovascular circulation. Another reported mechanism for the systemic effects of PBM on the blood circulation has been attributed to conformational changes occurring in the membrane properties of red blood cells (RBCs) caused by light-induced hydrogen bond disruption. This leads to an increased dissociation of water molecules from the RBC membrane layers causing structural changes in membrane proteins and in lipid bilayer fluidity [69, 79–82]. Consequently, the alteration in membrane ion pumps results in increased

RBC deformability, ATP content, normalization of osmotic properties, and membrane electrokinetic potential [83–85]. The latter effect would also reduce RBC aggregation improving hemorheology and circulation. In general, RBC aggregations cause rouleaux, resembling coin stacks that physically interfere with the capillary flow [82, 86, 87]. It has also been demonstrated that RBCs absorbing 810 nm laser light have an increased activity of ATPase and show modifications in their membrane proteins [88]. It is, therefore, proposed that the beneficial long-term effect of PBM on blood viscosity and circulation could be caused by RBC membrane alterations [69, 72, 88].

Along with the changes in tissue oxygenation and changes in RBC structure, the systemic effect of nostril-based intranasal PBM might be directly linked to the regulation of hemostasis mechanisms following blood irradiation. PBM causes a dose-dependent reversible suppression of platelet activity in response to certain agonists and reduces enzymatic activity in the arachidonic acid cascade [89–91], which can prevent platelet apoptosis and extend their lifespan in some pathological conditions [92–94]. Some evidence suggests that patients with depression have a relatively lower mitochondrial respiratory rate leading to reduced ATP levels in platelets [95]. Because the central mechanism of PBM involves stimulation of cellular energy metabolism, the long-term systemic effects seen in depression might be caused by the irradiation of circulating platelet mitochondria via the intranasal route [96].

Various studies on immune cells have shown that direct tissue exposure to red/NIR light may discourage polymorphonuclear (PMN) diapedesis to inflammatory sites and may also diminish the oxidative burst [97–99]. Paradoxically, the remote neuroprotective effects of blood-based intranasal PBM may be explained by the regulation of reactive oxygen species (ROS) [100]. ROS play a central role in the regulation of signal transduction pathways and gene expression. The function of ROS is vital in reprogramming the macrophage polarization from an M1 phenotype to an M2 phenotype, which releases tissue-related anti-inflammatory mediators [101, 102]. This mechanism is consistent with a previous report of a spinal cord injury rat model, in which 810 nm PBM altered the polarization state of the macrophage/microglia to an M2 phenotype, increased the expression of anti-inflammatory cytokines such as interleukin-4 (IL-4) and interleukin-13 (IL-13) [103], yet suppressed pro-inflammatory interleukin-6 (IL-6) [104] resulting in alternative macrophage activation.

The nose is also abundantly supplied by specialized nerves, which can be functionally divided into special and general innervation. The role of the nerve supply in the nasal cavity is to transfer information about temperature, chemicals, odors, pain, and touch to the CNS. The branches of the trigeminal nerve including the ophthalmic nerve and the maxillary nerve innervate the posterior part of the nose, including the olfactory and respiratory regions, respectively. These branches transmit chemosensory information from the nasal cavity to the brain, indicating their involvement in olfaction. The ophthalmic branch of the trigeminal nerve provides sensory innervation to the anterior part of the nose, including the vestibule region. General sensory innervation in the nasal cavity is provided by branches of the trigeminal nerve. The external skin of the nose is innervated by the trigeminal nerve. The nasociliary nerve, a branch of the ophthalmic nerve, supplies sensation to the top half of the nasal cavity and septum. The nasopalatine nerve, a branch of maxillary nerve, supplies sensation to the lateral nasal wall and septum. Moreover, the autonomic nervous system, parasympathetic and sympathetic pathways, control the vessels, nasal resistance, nasal secretions, and mucus gland production of the nose and nasal cavity [56, 105]. In addition to the systemic effects of intranasal PBM on blood components, several other reports of its potential mechanisms have suggested that it can influence the olfactory nerve and bulb, olfactory endothelium, autonomic nervous system, and lymphatic system [106, 107]. Systemic activation of mesenchymal stem cells/marrow stromal cells (MSCs) in the nasal bone marrow [50] and olfactory ensheathing cells (another type of stem cell) in the nasal mucosa [48] have also been suggested as possible receptors for PBM effects.

6.3.2   Intranasal Photobiomodulation from the Nasal Cavity and from the Nasal Submucosal Space

As described previously, the use of nostril-based portable applicators is likely to provide the deeper structures of the brain with only a tiny amount of photon energy [19]. In addition, restricted light penetration through the skull continues to be a fundamental obstacle to the stimulation of subcortical neurons by transcranial PBM. Only 2% of 1064 nm laser light has been reported to be able to pass across the human supraorbital frontal bone [108]. In a cadaver model, nearly 0.5% of 633 nm, and 2.1% of 830 nm LED light were shown to penetrate 1 cm of frontal skull and its overlying tissue [109]. It has been also reported that with a high-power laser device 2.9% of 810 nm light penetrated through 3 cm of scalp, skull and brain tissue [110]. Because of these measurements, the substantial attenuation of light flux would result in an unsatisfactory intracranial light dose and subsequent poor photostimulation of subcortical gray and white matter.

Recently, it has been proposed that implantable intranasal PBM devices could be employed to address the limitations of both portable devices and the transcranial irradiation techniques. Especially for patients who need a long-term therapeutic regimen, the advent of implantable intranasal light devices could offer a chance to avoid frequent treatment visits. Recently, the implantation of miniaturized LEDs in accessible areas of the nose within the submucosal pockets has become feasible. The procedure for implanting (or removing) a submucosal LED would only require an ambulatory surgical procedure with local anesthesia. Nevertheless, the relative proximity to the cribriform plate may partially restrict the submucosal positioning of miniaturized LEDs in practice [49].

The cribriform plate is a part of the ethmoid bone between the anterior cranial fossa and the nasal cavity, separating the brain from the underlying hollow space of the nose.

Anatomically, it is positioned anteroposteriorly from the crista galli until the planum sphenoidale, lying below the PFC forming the roof of nasal cavity. It has a thickness of about 1 mm with small perforations on almost half of its surface to allow olfactory nerve fibers to enter the nasal cavity through a bony protection [111]. The olfactory bulb is a neural extension of the brain lying superior to the cribriform plate and inferior to the basal frontal lobe [112]. Given that olfactory bulb conveys smell information to the brain, it is essential for the sense of smell. The olfactory bulb has a multi-layered cellular architecture (from surface to the center) including glomerular layer, external plexiform layer, mitral cell layer, internal plexiform layer, and granule cell layer [113]. The second-order neurons are mitral cells at the glomerular layer of the bulb, which are connected to the olfactory nerve fibers. The axons of mitral cells project to the olfactory cortex through the olfactory tract, a bundle of afferent nerve fibers, by which the olfactory information is transferred to different brain regions, including the amygdala, piriform cortex, OFC, and hippocampus. The olfactory tract lies in the olfactory sulcus on the inferior surface of the frontal lobe and its destruction results in anosmia [114]. Olfactory dysfunction is associated with the likelihood of certain neurodegenerative diseases, such as mild cognitive impairment (MCI), dementia, PD, and neuropsychiatric complications of PD [115]. In adult mammals, it is believed that neurogenesis occurs only in the olfactory bulb and the hippocampus dentate gyrus (DG); while in humans, only in the hippocampus [116]. In a TBI model of mice, it has been shown that 810 nm transcranial PBM significantly induced neurogenesis and upregulated migratory neuroprogenitor cells in the DG and the subventricular zone [117]. Although no study has yet confirmed induction of neurogenesis in the olfactory bulb after PBM, the possibility for neural stem cells to be triggered by direct irradiation of this region should be considered.

Using fiber-optic technology, it is proposed that wavelengths of 650 nm and 850 nm may potentially penetrate deeply through the nasal cavity [118]. The blue and green spectrum of visible light (400–540 nm) has penetration values of less than 0.1% through the human skull bone, which is even less significant considering all the tissues involved [119]. Thus, despite the generally unsuitable nature of these wavelengths for transcranial PBM of humans, their use in intranasal PBM could be considered due to their beneficial biological effects. A 532 nm laser has a soft tissue penetration depth of 0.8 mm [120]. Therefore, when implanted in close proximity to the cribriform plate, it would be possible for 532 nm green light to irradiate nerve fiber tissue within 1 mm of the bony plate. Nevertheless, it is unlikely that this wavelength would completely irradiate the whole olfactory bulb, which has a thickness of 3 mm. Studies have shown the in vitro increase of ATP levels and cell proliferation with 532 nm green laser, which would likely be due to absorption by the mitochondrial complex III (cytochromes b, c1, and c) [121]. It has also been shown that irradiation with a 532 nm laser increased the migration of GABAergic neural stem/progenitor cells into deeper layers of the mouse neocortex [121]. In addition, it has been proposed that 420 nm blue laser could effectively increase the synthesis of ATP, possibly by regulating mitochondrial complex I (NADH-dehydrogenase) [122].

As mentioned previously, the existence of very small perforations in the cribriform plate could allow for a portion of the photons emitted from immediately below the plate to directly irradiate the olfactory nerve fibers, the olfactory bulb, and the PFC. The PFC has abundant connections with the thalamus, basal ganglia, hypothalamus, amygdala, and hippocampus, which are involved in memory, sensory perception, and emotion. The vmPFC is situated at the bottom of the cerebral hemisphere inferior to the dorsomedial PFC and refers to the whole ventral and medial PFC region below the genu of the corpus callosum [123]. Damage to the vmPFC region leads to severe impairments in personality and behavior, social decision-making, incapability to learn from previous mistakes, difficulty in planning, mood and anxiety disorders, despite the fact that the intellectual abilities generally remain normal [31, 33]. As mentioned earlier in this chapter, the OFC is located in the ventral surface of the frontal lobe, adjacent and superior to the orbits, extending posteriorly from the frontal pole to the insula, and ventrally from the rostral sulcus, on the medial wall, to the ventrolateral convexity to form the frontal base of the brain [36]. The OFC is responsible for classical learning (rewards and punishment), decision-making, social processing, and actions that are driven by emotions [37, 124].

The olfactory epithelium is a specialized epithelial tissue with a thickness of about 60 μm that covers the nasal cavity above the superior concha and is identified by the cribriform plate, which is roughly 7 cm distant from the nostrils. It consists of four distinct cell types, including basal cells, supporting cells, brush cells, and olfactory sensory neurons. Basal cells are a population of stem cells localized on or near the basal lamina of the olfactory epithelium with the ability to divide and differentiate into other olfactory cells, resulting in turnover of the olfactory epithelium every 6–8 weeks. Supporting cells, analogous to neural glial cells, are scattered among the receptor cells and possess numerous microvilli and secretory granules, which empty their contents onto the mucosal surface [125]. Brush cells are microvilli-like columnar cells with their basal surface in contact with the afferent nerve endings of the trigeminal nerve. Olfactory sensory neurons are bipolar neurons located in the mucosa of the superior conchae and superior part of the nasal septum [125, 126]. Because of its position within the nose in close contact with the cribriform plate, the olfactory epithelium could be suggested as a possible location for implantation of an intranasal PBM probe. It is important to keep in mind that the sense of smell is facilitated by the sensitive cells in the epithelium [126], thus the use of this tissue as a place for light probe implantation should be approached with caution. Nevertheless, if implanted at the level of the cribriform surface, the olfactory bulb and vmPFC would effectively receive irradiation. Research on a cadaver head suggests that intra-nasal delivery of red/NIR light can theoretically illuminate anteromedial and posteromedial parts of the OFC [127]. When the light source is positioned in the close proximity to the cribriform plate, Monte Carlo simulation of vmPFC energy deposition demonstrated showed it was 46-fold, and 658-fold greater than when the light source was inserted in the mid-nose, and nostril, respectively [19]. In addition, the placement of the light source just below the cribriform plate resulted in at least two orders of magnitude more light fluence to the vmPFC and vmOFC compared with the dlPFC. The study also indicated that while limbic system structures, such as the amygdala and hippocampus, may receive very little light energy from a source located in the nostril, both mid-nose and cribriform plate light source positions would deliver a slightly higher dose (~0.01% of primary photons) to those areas, but this would still not be enough [19].

6.3.3   Intranasal Photobiomodulation from the Sphenoid Sinus

A sphenoid sinus implant might be possible by introducing an optical fiber into the sinus through the nasal cavity [127–129]. As an example, under direct endoscopic visualization, the tip of the optical fiber connected to a handheld laser/LED source could be inserted and placed within the sphenoid bone as an indwelling implant [130]. Nevertheless, this procedure would entail a complicated otorhinolaryngology (ENT) surgery under general anesthesia to insert the fiber into a delicate bony structure that is not easily accessible [49]. There are two sphenoidal sinuses within the body of the sphenoid bone which both open to the roof of the nasal cavity through the sphenoethmoidal recess on the anterior wall of the sinuses [131]. Sphenoid sinuses are located inferior to the ethmoid sinus roof, cavernous sinus, optic nerve, the olfactory nerve, sella turcica, and supraposterior to the nasal cavities and posterior ethmoid air cells, and anterior to the contents of the middle cranial fossa, and medial to the cavernous sinus and the cranial cavity [132]. The sphenoid sinus is adjacent to major limbic system structures, such as the pituitary gland, amygdala, hypothalamus, and hippocampus.

The pituitary gland is suspended from the hypothalamus via the pituitary stalk or infundibulum and is located in the hypophysial fossa of the sphenoid bone surrounded by the sella turcica [133]. Its anatomical connections are as follows: anteriorly and inferiorly to the sphenoid sinus; posteriorly to the posterior intercavernous sinus, dorsum sellae, basilar artery and the pons, superiorly to the diaphragma sellae and optic chiasm; and laterally to the cavernous sinus [133]. Since the pituitary gland regulates the majority of endocrine glands throughout the body, the pituitary is referred to as the master gland of the entire body. Damage to the pituitary gland leads to hormonal imbalance resulting in a wide variety of disorders in many different areas of the body. The pituitary gland has three main parts or lobes: anterior, intermediate, and posterior. The anterior pituitary or adenohypophysis is glandular tissue that arises from hypophyseal (Rathke's) pouch and secretes hormones under the control of release or inhibition hormones derived from the hypothalamus. It has three regions: the pars distalis, the pars intermedia, and the pars tuberalis. However, the posterior pituitary or neurohypophysis is neural tissue. Indeed it is an extension of the neurons of the paraventricular and supraoptic nuclei of the hypothalamus, whose axons descend within the infundibulum and end at the posterior pituitary [134], and it stores hormones secreted by the hypothalamic nuclei. The anterior lobe of the pituitary gland obtains its arterial blood supply from the superior hypophyseal artery, a branch of the internal carotid artery, while the posterior pituitary gland and infundibulum receive a rich blood supply from the superior hypophyseal artery, infundibular artery, and inferior hypophyseal artery [133, 134].

The amygdala is located anteriorly to the hippocampus, medially within the temporal lobes, and laterally and slightly posteriorly to the sphenoid sinus. The amygdala sends and receives projections to and from various structures of the brain, including dorsomedial thalamus, thalamic reticular nucleus, hypothalamus, nuclei of the trigeminal and facial nerves, locus coeruleus, ventral tegmental area, and the laterodorsal tegmental nucleus [135]. That is why the amygdala is considered to be vital in a variety of behavioral functions and emotional responses, such as fear, anxiety, aggression, memory, facial expression perception, and decision-making [136–138]. There are around 13 nuclei in the amygdala, categorized into three subdivisions, including centromedial complex, basolateral complex, and corticomedial complex. The basolateral complex or deep nuclei comprise the lateral nucleus, the basal nucleus, and the accessory basal nucleus. It receives the majority of sensory information from the HIP and primary auditory cortex and projects to the medial dorsal nucleus of the thalamus, the basal nucleus (of Meynart), the ventral striatum, and the nucleus accumbens. The primary function of the basolateral complex is to stimulate the fear response and consolidate fear memory, while the lateral nucleus has a major role in plasticity. It also mediates response to stress, feeding, and drinking behavior [137, 139]. Damage to this complex prevents emotion-enhanced memory [140]. The corticomedial nuclear group is superficial structure located at the surface of the brain. This group includes the cortical nucleus (anterior and posterior), the nucleus of the lateral olfactory tract, and the bed nucleus. These neurons receive afferents of olfaction and project to the ventromedial nucleus of the hypothalamus and participate in behavior associated with hunger and eating [137]. The medial nucleus receives input from the olfactory bulb and olfactory cortex, and mediates the sense of smell and pheromone-processing [139]. The centromedial nuclear group is located in the dorsomedial portion of the amygdaloid complex, and comprises the central (CeA), medial, and the amygdaloid part of the bed nucleus of stria terminalis. The CeA is the major output of the amygdala and is involved in receiving and processing pain signals [138]. Moreover, the CeA mediates autonomic components of emotion, such as alteration of heart rate, blood pressure, and respiratory rate mainly via output pathways to the brain stem and lateral hypothalamus [137, 139].

The hypothalamus is the ventral part of the diencephalon and is located at the base of the brain, adjacent to the pituitary gland on either side of the third ventricle. The hypothalamic sulcus separates the hypothalamus superiorly from the dorsal thalamus. It makes up less than 1% of brain volume and weighs only about 5 g. The hypothalamus is interconnected with the brainstem and reticular formation, limbic structures including the amygdala and septum, and parts of the autonomous nervous system [141, 142]. The hypothalamus receives many inputs from the brainstem, the most notable being from the nucleus of the solitary tract, the locus coeruleus, and the ventrolateral medulla. It is bounded superiorly by the hypothalamic sulcus, rostrally by the lamina terminalis, later-ally by the substantia innominata, and caudally by the medial edge of the posterior limb of the internal capsule [141]. Arterial blood to the hypothalamus is provided by the anterior and posterior branches of the circle of Willis, as well as the hypothalamic branches of the superior hypophyseal artery. The hypothalamus is engaged in the regulation of the endocrine system, body temperature, food and water intake, reproduction, sexual behav-ior, circadian rhythm, fatigue, sleep, emotional response, and memory function [143]. The hypothalamus has three main regions: the supraoptic or anterior region located internal to the optic chiasm, the middle or tuberal region located at the levels of tubercinereum, and the mammillary or posterior region located at the level of the mammillary bodies [142]. Each region contains different nuclei. The supraoptic/anterior region comprises four nuclei including the supraoptic, paraventricular, suprachiasmatic, and anterior nuclei. The nuclei in the supraoptic/anterior region are largely involved in the secretion of var-ious hormones, such as corticotropin-releasing hormone (CRH), thyrotropin-releasing hormone (TRH), gonadotropin-releasing hormone (GnRH), oxytocin, vasopressin, and somatostatin. This region also regulates the body temperature, via sweating, and circadian rhythms. The major nuclei of the middle/tuberal region are the ventromedial and arcuate nuclei. The ventromedial nucleus controls appetite and food intake, while the arcuate nucleus regulates the growth and development of the body via the release of the growth hormone-releasing hormone (GHRH) [141]. The posterior/mammillary region comprises mammillary nuclei and the posterior hypothalamic nuclei. The posterior hypothalamic nucleus regulates body temperature through shivering and hindering sweat production, while the mammillary nuclei are involved in memory function, particularly short-term memory [141].

The hippocampus is a convex structure formed by the gray matter within the parahip-pocampal gyrus situated at the temporal lobe, beneath the cerebral cortex, lateral to the inferior horns of the lateral ventricles. The sphenoid sinus meets the anterior portion of the hippocampus superoposteriorly [123]. The hippocampus blood supply is provided by the posterior cerebral artery, which is also reinforced by the anterior choroidal artery. The veins of the hippocampus drain into the basal vein [144]. The hippocampus is involved in several higher cognitive functions, such as learning, memory (particularly long-term memory), spatial navigation, regulation of hypothalamic functions, and emotions [144, 145]. The hippocampus receives direct inputs from the olfactory bulb hence it is impor-tant in olfaction. Therefore, damage to this part of the brain leads to a serious long-term effect on different types of memory. Several studies have linked hippocampal volume loss or atrophy to (AD), depression, schizophrenia, and MCI [144, 146]. The hippocampus consists of three distinct zones, including the dentate gyrus (DG), cornu ammonis (CA) fields, and the subiculum [146]. The DG is a "V" shaped cortical structure, located in the most proximal part of the hippocampal formation and considered to be the hippocampal input region. It receives major inputs from the EC via the perforant path and minor pro-jections from the presubiculum and parasubiculum, brain stem, and hypothalamus, and itself projects fibers to the CA3. It has three layers, including molecular layer, granule cell layer, and polymorphic layer. The most superficial molecular layer lies nearby the hippocampal fissure and contains the dendrites of the dentate granule cells. The principal granular layer mainly contains granule cells. The polymorphic layer lies within the granule cell layer and contains the unmyelinated axons called mossy fibers, and often refers to the cellular component of the hilus [146]. The DG is responsible for the formation of episodic memories and plays a role in stress and depression [147, 148]. The cornu ammonis (CA) fields (or hippocampus proper) contain pyramidal cells, which are often subdivided into four subfields (CA1–CA4). The CA1 subfield is located near to the subiculum and projects to the entorhinal cortex (EC) and subiculum and receives inputs from the CA3 subfield. CA2 is a small region found between CA1 and CA3. CA3 is the largest in the hippocampus, which is closest to the DG and is inserted into the hilus. CA3 is considered as the pacemaker of the hippocampus. CA3 receives fibers from the DG, EC, the medial septum, and the diagonal band of Broca. Whereas CA3 pyramidal cells project back via the Schaffer collaterals mostly to the CA2 and CA1 subfields as well as the DG hilus. The CA4 subfield (also called the hilus or hilar region) is within the hilus of the DG. It primarily receives fibers from the DG and also small projections from the pyramidal cells in CA3, and in turn projects back into the DG [149]. The dorsal hippocampus is primarily involved in cognitive functions, such as learning and memory, while the ventral hippocampus is mainly concerned with stress and emotion [150]. The subiculum is the most inferior part of the hippocampal formation, positioned between the EC and the CA1 subfield in the dorsal part of the parahippocampal gyrus. The subiculum is divided into three subdivisions: the subiculum proper, the presubiculum, and the parasubiculum. The subiculum receives inputs from CA1 and EC pyramidal neurons and other cortices, as well as several subcortical structures. Given that the subiculum projects to many cortical and subcortical regions particularly various hypothalamic nuclei, it is considered to be the major output structure of the hippocampus [151]. It plays a crucial role in the facilitation of hippocampal-cortical interactions and mediates a wide range of neurocognitive functions, such as memory, motivation, reward, spatial information processing, and the temporal control of behavior [152].

It could be feasible to use sphenoid sinus as a potential location for a light source implant, which could provide the limbic structures with therapeutic doses of light [128]. Pitzschke et al. [44] investigated transsphenoidal light delivery and photon distribution in a human cadaver head and also proposed the possibility of pars compacta (SNpc) irradiation for PD. They first coupled an optical fiber-based light diffuser to a laser diode with a wavelength of 671 nm or 808 nm and then inserted it into the nasal cavity under endoscopic guidance, where it was implanted in the sphenoidal sinus to irradiate the SNpc. Their results revealed that 0.03% of 671 nm and 0.36% of 808 nm light emitted by the optical fiber could reach the SNpc [44]. With an appropriate total power of the light source, such percentages of transmitted light could probably provide the SNpc with a satisfactory photostimulatory fluence. The SNpc is a central outlet to the basal ganglia circuit. Dopaminergic neurons of the SNpc project to the dorsal striatum through the nigrostriatal pathway, to the lateral and medial pallidum of the basal ganglia system, to the pars reticulata (SNpr), and the subthalamic nucleus [153–155]. The SN plays a central role in reward-seeking, motor planning, and movement [156, 157]. With transsphenoidal irradiation, it is expected that a significant fraction of the primary light energy could reach the pituitary gland, and probably, even the amygdala, hypothalamus, and the anterior tail of hippocampus before reaching the SNpc [49].

6.4   Intraoral, Intraaural, and Transocular Photobiomodulation Therapies

One serendipitous finding during intraoral PBM for a dental problem was a reduction in clinical signs in a single PD patient after 660 nm laser PBM. The laser probe was directed to the posterior regions of the cranium/upper neck, and PBM was applied for ~2–3 min. It was suggested, in this case, that unexpected primary or scattered photons had a chance of reaching PD-related brain structures via the deep oral cavity [158]. Nevertheless, recent Monte Carlo simulation modeling has indicated that when a light source at 671 nm was positioned in the oral cavity (pointing toward the SNpc), the percentage of received photons at the SNpc was only 0.0005% [44]. It is obvious that this level of light is much lower than that required for beneficial PBM effects. In order to achieve a sufficient fluence at the human SNpc, a combination of PBM via the oral cavity and sphenoid sinus was also suggested by this simulation study [44]. It is clear that further clinical trials are needed to evaluate whether intraoral PBM has any effect in neurodegenerative diseases.

Bright LED light (at 448 nm) applied through the ear canal was shown to penetrate the skull and reach the temporal lobes of the human brain [159]. Intraaural PBM to treat brain conditions has been proposed, in order to treat seasonal affective disorder [160, 161] as well as modulation of brain function [159]. Moreover, recently, in a Boston VA Hospital double-blind study, the possible procognitive benefits of a combination of transcranial and intranasal PBM along with intraaural irradiation of the brain using LED cluster heads were investigated for Gulf War illness in veterans [12].

In a speculative study, researchers have sought to determine whether the eyes could be potential beneficial routes for visible and NIR light to reach the SN in humans. As a first step, they irradiated the rat head with polychromatic light and measured the wavelength range and intensity of light reaching the SN using a microoptical probe implanted into mesencephalic dopaminergic region. Results showed that the eyes could be a gateway for light transmission to the SN, since covering the eyes with aluminum foil reduced light intensity in the brain by half. Moreover, they suggested that only wavelengths above 600 nm reached the SN, with a transmission peak at 709 nm. In the next step, using computed tomography (CT) and magnetic resonance imaging (MRI) of a human head, they found that the distance between the eye surface and the SN inside the mesencephalon was about 8 cm long, of which almost 5.6 cm was composed of transparent fluids. They also proposed that the eye and the superior orbital fissure could be possible gateways for external light to reach the mesencephalon in humans [162]. The potential neurotherapeutic benefits of the CerebroLite transcranial extraocular 1068 nm PBM device have recently been suggested by Berman and Nichols in dementia patients with PD. The device consists of a transcranial helmet with two additional LED arrays covering the eyes. It is assumed that irradiation of the eyes with these LED arrays may have some beneficial effects on the deep neurons behind the eyes [14, 163].

6.4.1 Remote Photobiomodulation for Brain

The possibility that PBM shone onto a remote part of the body might elicit systemic effects that stimulate protection or repair of distant injured tissues has been of great interest to investigators in recent years [164]. Indeed, the observation that localized treatment of a specific area of the body can produce positive effects in other remote tissues has been reported many times using the technique known as remote ischemic preconditioning [165, 166]. In the context of brain PBM, although light irradiation is mostly shone on the head for localized cerebral pathology, and its effects are often considered to be limited to the irradiated area, there are several reports suggesting that the brain might benefit remotely from light irradiation of different parts of the body, in a systemic manner (also called abscopal effect) [167–169]. Apparently, the first observation of the beneficial effects of indirect PBM therapy on the recovery of the injured peripheral and CNS was reported by Rochkind and his colleagues in 1989 [170]. In a rat model, daily laser PBM (632.8 nm, 7.6 J/cm^2) over the right crushed sciatic nerve in a model of bilateral crush injury resulted in inhibition of bilateral degeneration of motor neurons in the spinal cord segments and also led to an increased compound action potential in the left non-PBM irradiated crushed sciatic nerve. Their findings suggested that indirect PBM may help to preserve conduction in the injured peripheral nerve, and these systemic effects on the nerve recovery could be long-lasting [170]. So far, there have been a number of other convincing examples of the systemic effects of PBM in animal models of AD [171, 172] and PD [65, 166, 173].

In a pioneering study, using the 5xFAD transgenic mouse model of AD, Farfara et al. [171] performed abscopal PBM by placing a fiber-optic directly on the middle portion of the medial part of the tibia and treated the animals daily for 2 months. The results showed a significant reduction in the hippocampal burden of Aβ deposits, along with improvement in cognitive function at 6 months of age. Additionally, they conducted a conceptual in vitro study to determine the possible mechanisms and found that remote PBM may induce the mobilization of bone marrow-derived mesenchymal stem cells (MSCs) and immune cells which most likely migrate to the brain, lessening the progression of the AD [171, 174]. In another remarkable study, Blivet et al. [172] applied PBM (625/850 nm) once a day for 7 days either on the top of the head or to the center of the abdomen (or both) in an Aβ mouse model of AD. Overall, their results showed memory restoration and improvements in key markers of AD (Aβ and p-Tau), oxidative stress, apoptosis (Bax/Bcl-2), and neuroinflammation following both transcranial and remote tissue PBM. They suggested that abscopal PBM to the abdomen may change the gut microbiota with the release of yet unidentified circulating mediators producing a neuroprotective action on the brain [172]. It is worth mentioning that an abdominal light device has been recently introduced by Chinese researchers for remote PBM to the belly in AD patients [175].

Daniel Johnstone and his group in Australia [173] was the first to report the neuroprotective effects of remote PBM in a MPTP mouse model of PD. Their initial results showed that PBM of the animal body preserved functional dopaminergic neurons in the SNc against Parkinsonian insults with 50 mg/kg MPTP, comparable to the transcranial irradiation method [173]. To explain the observed systemic effects, they suggested that remote PBM may modulate some immune cells (e.g., macrophages) [104] and affect inflammatory mediators (e.g., IFN-γ, TNF-α, IL-4, and IL-10) [176], resulting in robust neuroprotective effects. Their subsequent experimental studies revealed some of the cellular signaling pathways that were actually involved in remote PBM-induced neuroprotection, including upregulation of stem cell-related CXCR4 signaling, adipocytokine signaling, oxidative stress response pathways involving nuclear factor-erythroid 2-related factor 2 (NRF2), and also promotion of cell proliferation and migration [65]. In their most recent study [166], they also showed that remote tissue conditioning, whether ischemia of a limb or PBM of the torso using 670 nm LEDs, could induce protection of the SNpc in PD mice, although combining these two treatment modalities provided no additive or synergistic effects [166].

There is also another possibility that irradiation of superficial structures of the head (scalp, skull, and meningeal layers) with transcranial PBM may produce some benefits to the brain tissue despite the light not actually reaching the brain. In the transcranial PBM method, photons must not only penetrate the scalp, and skull but also the periosteal, meningeal, subdural space, arachnoid mater, subarachnoid space, and pia mater, all surrounded by a large amount of fluid, including blood and water. Therefore, along with scattering, all of these layers absorb photons and therefore could induce systemic effects, which should be taken into account when using the transcranial approach. For example, since the calvarial bone marrow of the skull has large numbers of stem cells, and there is also significant blood flow in the scalp and the skull, photon absorption by these tissues before the light actually reaches the brain cortex, might contribute to the systemic protective effects [177]. Specifically, it has been suggested that bone marrow-derived stem cells, in particular MSCs, are the main candidate for mediating remote PBM-induced neuroprotection [178, 179] as they can easily transmigrate across the BBB [173, 180]. PBM has been shown to enhance the proliferation of MSCs [181]. Also, PBM of the bone marrow stimulates the proliferation and mobilization of MSCs accompanied by increased localization of MSCs at sites of injury [182]. NO is a well-known vasodilator, which regulates blood flow. It has been suggested that PBM could act on cerebral superficial blood vessels and alter the elasticity and resistance of the cerebrovascular endothelium, at least partly through the NO pathway [177]. The improved cerebral blood flow induced by transcranial PBM could affect global blood circulation, with a subsequent increase in cerebral blood flow. This evidence supports the assumption that blood absorption of light could be the mechanism responsible for the protective effects of remote PBM. Because the cytoskeleton is an initiator of signal transduction pathways, it has also been suggested that systemic PBM could modulate the neuronal cytoskeleton and produce neuroprotection via photon activation of photoreceptors in the membrane, including opsins, nicotinamide adenine dinucleotide phosphate (NADPH), or the transient receptor potential vanilloid 1 (TRPV1) ion channel, leading to effects on tau protein and the microtubules via ROS [183, 184]. Nevertheless, it should be noted that the systemic mechanism underlying the phenomenon of remote PBM has not yet been widely investigated, so further studies are warranted.

6.4.2   Transcatheter Brain Photobiomodulation

In recent years, transcatheter intracerebral laser PBM therapy has been proposed as a promising method for revascularization of cerebral blood vessels and regeneration of cerebral tissue in some brain disorders [185–187]. The procedure consists of threading a thin, flexible fiber-optic all the way from the femoral artery to the middle cerebral artery under local anesthesia to deliver laser light energy into the brain. It has been reported that the penetration depth of 632.8 nm laser light (with a power of 25 mW) with the transcatheter intracerebral approach is 20–40 mm in cerebral tissue [188]. Thus, it could be supposed that by using this technique, laser light from the tip of the fiber could irradiate not only the vascular wall but also volumes of surrounding cerebral tissue. Clinical studies have shown that transcatheter intracerebral PBM therapy could restore the cerebral distal arterial and capillary blood supply, reduce hypoxia, and stimulate metabolic functions in the cerebral tissue in the early, less severe stages of ischemic brain disease [187] as well as in advanced stages of neurodegenerative diseases, such as AD and PD [185]. In addition, it has been shown in experimental animal studies that transcatheter PBM therapy using 632.8 nm laser could stimulate physiological angiogenesis leading to rapid opening of arterial and capillary collateral vessels [188].

6.4.3   Intravascular Photobiomodulation for Brain

Intravascular or intravenous laser PBM is a type of systemic laser therapy that relies on direct irradiation of the blood. This technique consists of inserting a sterile optical fiber through a special puncture needle into a blood vessel, commonly the cubital vein in the wrist [189]. Because of the dynamic nature of blood flow, intravascular PBM could irradiate the whole blood and all the blood components including platelets, erythrocytes, leukocytes, and lipids, through circulating throughout the body. The early procedures used a He–Ne laser (632.8 nm) with an output power of 1–3 mW and an irradiation time of 20–60 min. In fact, Russian researchers began to develop intravascular PBM for various medical conditions during the 1970s [189]. Afterwards, several Western studies brought a better understanding of the action mechanisms of intravascular PBM to clarify some of the results previously reported by the Russians [190]. It should be noted that the methodology and findings of much of the Russian research have been comprehensively reviewed elsewhere [191]. However, other researchers from Western countries and China have also reported the efficacy of intravascular laser PBM for various cerebrovascular disorders such as cerebral ischemia [192], stroke [193, 194], and AD [195]. Indeed, the underlying action mechanisms for intravascular PBM are similar to those involved in intranasal PBM, where blood plays a key role in light absorption. Intravascular PBM could act on the membrane of blood cells, degranulate mast cells, and modulate lipids or other blood components. Upon blood irradiation, photodissociation of NO from hemoglobin could increase the oxygenation of the blood. Intravascular PBM could also activate cellular metabolism in blood components and affect endothelial cells resulting in vasodilation [190].

6.4.4   Laser Acupuncture for Brain

Laser acupuncture is a non-invasive stimulation method, which in contrast to the Chinese needle acupuncture method, patients do not feel any painful sensations during the procedure. Because of the non-invasive nature of this modality, there is no need to puncture the skin and also there is no risk of adverse effects including infection or bleeding complications. Nevertheless, some mild and transient adverse effects such as dizziness, fatigue, and headaches have been reported by individuals undergoing laser acupuncture [196]. Evidence suggests that laser irradiation to the acupoints HT9, LR1, HT3, and KI10 could produce neuroprotective effects in a rat model of cerebral ischemia [197]. Also, laser acupuncture to the HT7 acupoint has been shown to induce neuroprotection in animal models of AD [198] and autism [199, 200]. Furthermore, PBM using laser acupuncture has been tested clinically in stroke [201] and TBI [202] patients, as well as healthy volunteers [203–205]. In addition to the above-mentioned peripheral acupoints, acupuncture meridian points of Governing Vessel (GV) on the skull, namely GV 20, GV 21, GV 22, GV 23, and GV 24 located in midline, inferior to occipital protuberance up to near center-front hairline have been employed to treat patients with TBI [202], stroke [206], and coma [207]. Of note, recently, Nicholas Alexander Wise has provided a review of all the animal and clinical studies evaluating the neuroprotective effects of laser acupuncture for the brain [208].

6.4.5 Whole-Body Photobiomodulation for Brain

Whole-body laser or LED PBM therapy has also shown to have neuroprotective benefits in various animal models of brain disorders such as AD [209, 210], PD [211], stroke [212], as well as ophthalmological neurotoxicity models induced by methanol [213, 214] or fluorescent light [47, 215]. It has been shown that 710 nm LED light when applied directly on top of the animal cage could activate cellular immunity, reduce microglial activation, and decrease brain infarct size, as well as improve neurological scores in a rat stroke model [212]. In another report, long-term irradiation of white fluorescent light onto the animal body induced a reduction in dopaminergic neurons in the mouse SNc, while 710 nm LED did not produce any detrimental effects [47]. In addition, whole-body PBM therapy using 1072 nm LEDs has been shown to improve working memory in middle-aged mice [216] and to mitigate Aβ plaque deposition in AD mice [209]. Recently, the NovoThor "light-pod" device has been introduced by Thor Photomedicine (Chesham, Bucks, UK) for whole-body irradiation in humans. The NovoThor consists of hundreds of combined 660/850 nm LEDs with a total optical output power of 500W. The use of this apparatus has been suggested to relieve muscle fatigue and pain and to improve muscle performance, as well as improve weight loss in combination with exercise [217, 218]. It is possible that this light bed could also be effective for preconditioning or post-conditioning to benefit a wide variety of brain conditions.

References

1. Bhattacharya, M., and A. Dutta. 2019. Computational modeling of the photon transport, tissue heating, and cytochrome C oxidase absorption during transcranial near-infrared stimulation. *Brain sciences* 9 (8): 179.
2. Johnstone, D.M., et al. 2016. Turning on lights to stop neurodegeneration: The potential of near infrared light therapy in Alzheimer's and Parkinson's disease. *Frontiers in neuroscience* 9: 500.
3. Caldieraro, M.A., and P. Cassano. 2019. Transcranial and systemic photobiomodulation for major depressive disorder: A systematic review of efficacy, tolerability and biological mechanisms. *Journal of affective disorders* 243: 262–273.
4. Henderson, T.A. 2016. Multi-watt near-infrared light therapy as a neuroregenerative treatment for traumatic brain injury. *Neural regeneration research* 11 (4): 563.
5. Ohshiro, T. 2012. The proximal priority theory: An updated technique in low level laser therapy with an 830 nm GaAlAs laser. *Laser therapy* 21 (4): 275–285.
6. Lapchak, P.A. 2012. Transcranial near-infrared laser therapy applied to promote clinical recovery in acute and chronic neurodegenerative diseases. *Expert review of medical devices* 9 (1): 71–83.
7. Hamilton, C.L., et al. 2019. "Buckets": Early Observations on the Use of Red and Infrared Light Helmets in Parkinson's Disease Patients. *Photobiomodulation, photomedicine, and laser surgery* 37 (10): 615–622.

8. Salehpour, F., M.R. Hamblin, and J.O. DiDuro. 2019. Rapid Reversal of Cognitive Decline, Olfactory Dysfunction, and Quality of Life Using Multi-Modality Photobiomodulation Therapy: Case Report. *Photobiomodulation, photomedicine, and laser surgery* 37 (3): 159–167.

9. Berman, M.H., et al., *Photobiomodulation with near infrared light helmet in a pilot, placebo controlled clinical trial in dementia patients testing memory and cognition.* Journal of neurology and neuroscience, 2017. **8**(1).

10. Hipskind, S.G., et al., *Pulsed transcranial red/near-infrared light therapy using light-emitting diodes improves cerebral blood flow and cognitive function in veterans with chronic traumatic brain injury: a case series.* Photomedicine and laser surgery, 2018.

11. Poiani, G.d.C.R., et al., *Photobiomodulation using low-level laser therapy (LLLT) for patients with chronic traumatic brain injury: a randomized controlled trial study protocol.* Trials, 2018. **19**(1): p. 17.

12. Gefvert, B., *Medical Lasers/Neuroscience: Photobiomodulation and the brain: Traumatic brain injury and beyond.*

13. Litscher, G., *Brain photobiomodulation—preliminary results from regional cerebral oximetry and thermal imaging*, 2019, Multidisciplinary Digital Publishing Institute.

14. Berman, M.H., and T.W. Nichols. 2019. Treatment of Neurodegeneration: Integrating Photobiomodulation and Neurofeedback in Alzheimer's Dementia and Parkinson's: A Review. *Photobiomodulation, photomedicine, and laser surgery* 37 (10): 623–634.

15. Zomorrodi, R., et al. 2019. pulsed Near Infrared transcranial and Intranasal photobiomodulation Significantly Modulates Neural oscillations: A pilot exploratory study. *Scientific reports* 9 (1): 6309.

16. Saltmarche, A.E., et al. 2017. Significant improvement in cognition in mild to moderately severe dementia cases treated with transcranial plus intranasal photobiomodulation: Case series report. *Photomedicine and laser surgery* 35 (8): 432–441.

17. https://www.helpmychronicpain.com/blog/brain-laser-transcranial-photobiomodulation-improves-brain-blood-flow-and-energy-production-video.

18. https://www.youtube.com/watch?v=gJ7RDXbwOj8.

19. Cassano, P., et al. 2019. Selective photobiomodulation for emotion regulation: Model-based dosimetry study. *Neurophotonics* 6 (1): 015004.

20. Okada, E., et al. 2010. Theoretical and experimental investigation of the influence of frontal sinus on the sensitivity of the NIRS signal in the adult head. In *Oxygen Transport to Tissue XXXI*, 231–236. Springer.

21. Netter, F.H. and S. Colacino, *Atlas of human anatomy.* 1989: Ciba-Geigy Corporation.

22. Carpenter, M.B. and J. Sutin, *Human neuroanatomy.* 1983: Williams & Wilkins.

23. Fuster, J.n.M., *The Prefrontal Cortex—An Update: Time Is of the Essence.* Neuron, 2001. **30**(2): p. 319–333.

24. Hiser, J., and M. Koenigs. 2018. The multifaceted role of the ventromedial prefrontal cortex in emotion, decision making, social cognition, and psychopathology. *Biological psychiatry* 83 (8): 638–647.

25. Miller, B.L. and J.L. Cummings, *The human frontal lobes: Functions and disorders.* 2017: Guilford Publications.

26. Murray, E.A., S.P. Wise, and K.S. Graham, *The evolution of memory systems: ancestors, anatomy, and adaptations.* 2016: Oxford University Press.

27. Cieslik, E.C., et al., *Is there "one" DLPFC in cognitive action control? Evidence for heterogeneity from co-activation-based parcellation.* Cerebral cortex (New York, N.Y. : 1991), 2013. **23**(11): p. 2677–2689.

28. Bechara, A. 2004. Disturbances of Emotion Regulation After Focal Brain Lesions. In *International Review of Neurobiology*, 159–193. Academic Press.

29. Finger, E.C., et al. 2008. Abnormal ventromedial prefrontal cortex function in children with psychopathic traits during reversal learning. *Archives of general psychiatry* 65 (5): 586–594.
30. Myers-Schulz, B., and M. Koenigs. 2012. Functional anatomy of ventromedial prefrontal cortex: Implications for mood and anxiety disorders. *Molecular psychiatry* 17 (2): 132–141.
31. Motzkin, J.C., et al. 2015. Ventromedial prefrontal cortex is critical for the regulation of amygdala activity in humans. *Biological psychiatry* 77 (3): 276–284.
32. Roy, M., D. Shohamy, and T.D. Wager. 2012. Ventromedial prefrontal-subcortical systems and the generation of affective meaning. *Trends in cognitive sciences* 16 (3): 147–156.
33. Bechara, A. 2004. Disturbances of emotion regulation after focal brain lesions. *International Review of Neurobiology* 62 (159): 93.
34. Etkin, A., and T.D. Wager. 2007. Functional neuroimaging of anxiety: A meta-analysis of emotional processing in PTSD, social anxiety disorder, and specific phobia. *American Journal of Psychiatry* 164 (10): 1476–1488.
35. Price, J.L., and W.C. Drevets. 2010. Neurocircuitry of mood disorders. *Neuropsychopharmacology* 35 (1): 192.
36. Rudebeck, P.H., et al. 2013. Effects of amygdala lesions on reward-value coding in orbital and medial prefrontal cortex. *Neuron* 80 (6): 1519–1531.
37. Rolls, E.T., and F. Grabenhorst. 2008. The orbitofrontal cortex and beyond: From affect to decision-making. *Progress in neurobiology* 86 (3): 216–244.
38. Li, W., et al. 2010. Right orbitofrontal cortex mediates conscious olfactory perception. *Psychological science* 21 (10): 1454–1463.
39. Kuusinen, V., et al., *Orbitofrontal Lesion Alters Brain Dynamics of Emotion-Attention and Emotion-Cognitive Control Interaction in Humans.* Frontiers in Human Neuroscience, 2018. **12**(437).
40. Moro, C., et al. 2014. Photobiomodulation inside the brain: A novel method of applying near-infrared light intracranially and its impact on dopaminergic cell survival in MPTP-treated mice. *Journal of neurosurgery* 120 (3): 670–683.
41. Johnstone, D.M., et al. 2014. The potential of light therapy in Parkinson's disease. *Chrono-Physiology and Therapy* 4: 1.
42. Darlot, F., et al. 2016. Near-infrared light is neuroprotective in a monkey model of P arkinson disease. *Annals of neurology* 79 (1): 59–75.
43. Tye, K.M., and K. Deisseroth. 2012. Optogenetic investigation of neural circuits underlying brain disease in animal models. *Nature Reviews Neuroscience* 13 (4): 251.
44. Pitzschke, A., et al. 2015. Red and NIR light dosimetry in the human deep brain. *Physics in Medicine & Biology* 60 (7): 2921.
45. Moro, C., et al. 2017. No evidence for toxicity after long-term photobiomodulation in normal non-human primates. *Experimental brain research* 235 (10): 3081–3092.
46. Stujenske, J.M., T. Spellman, and J.A. Gordon. 2015. Modeling the spatiotemporal dynamics of light and heat propagation for in vivo optogenetics. *Cell reports* 12 (3): 525–534.
47. Romeo, S., et al. 2017. Fluorescent light induces neurodegeneration in the rodent nigrostriatal system but near infrared LED light does not. *Brain research* 1662: 87–101.
48. Hennessy, M., and M.R. Hamblin. 2016. Photobiomodulation and the brain: A new paradigm. *Journal of optics* 19 (1): 013003.
49. Salehpour, F., et al., *Therapeutic potential of intranasal photobiomodulation therapy for neurological and neuropsychiatric disorders: a narrative review.* Reviews in the Neurosciences, 2019.
50. Liu, T.C.-Y., et al., *Randomized, double-blind, and placebo-controlled clinic report of intranasal low-intensity laser therapy on vascular diseases.* International Journal of Photoenergy, 2012. **2012**.

51. Gao, X., P. Zhi, and X. Wu, *Low-energy semiconductor laser intranasal irradiation of the blood improves blood coagulation status in normal pregnancy at term.* Nan fang yi ke da xue xue bao= Journal of Southern Medical University, 2008. **28**(8): p. 1400–1401.

52. Elwood, P.C., J. Pickering, and J.E. Gallacher. 2001. Cognitive function and blood rheology: Results from the Caerphilly cohort of older men. *Age and Ageing* 30 (2): 135–139.

53. Gao, Z., L. Zhang, and C. Qin, *The Relationship between Hemorheological Changes and the Anxiety and Depression Symptoms in Schizophrenia.* Chinese Journal of Hemorheology, 2004. **1**.

54. Caldieraro, M.A., et al. 2018. Long-term near-infrared photobiomodulation for anxious depression complicated by Takotsubo cardiomyopathy. *Journal of clinical psychopharmacology* 38 (3): 268–270.

55. Xiao, X., et al. 2005. Effects of low power laser irradiation in nasal cavity on cerebral blood flow perfusion of patients with brain infarction. *Chinese Journal of Physical Medicine* 27 (7): 418–420.

56. Mygind, N., and R. Dahl. 1998. Anatomy, physiology and function of the nasal cavities in health and disease. *Advanced drug delivery reviews* 29 (1–2): 3–12.

57. Marianecci, C., et al. 2017. Drug delivery in overcoming the blood–brain barrier: Role of nasal mucosal grafting. *Drug design, development and therapy* 11: 325.

58. Frey, W. 1991. Neurologic agents for nasal administration to the brain. *World Intellectual Property Organization* 5: 89.

59. Jiang, Y., Y. Li, and X. Liu. 2015. Intranasal delivery: Circumventing the iron curtain to treat neurological disorders. *Expert opinion on drug delivery* 12 (11): 1717–1725.

60. Salehpour, F., et al. 2018. Brain photobiomodulation therapy: A narrative review. *Molecular neurobiology* 55 (8): 6601–6636.

61. Morries, L.D., P. Cassano, and T.A. Henderson. 2015. Treatments for traumatic brain injury with emphasis on transcranial near-infrared laser phototherapy. *Neuropsychiatric disease and treatment* 11: 2159.

62. Iaccarino, H.F., et al. 2016. Gamma frequency entrainment attenuates amyloid load and modifies microglia. *Nature* 540 (7632): 230.

63. Chao, L.L. 2019. Effects of home photobiomodulation treatments on cognitive and behavioral function, cerebral perfusion, and resting-state functional connectivity in patients with dementia: A pilot trial. *Photobiomodulation, photomedicine, and laser surgery* 37 (3): 133–141.

64. Xie, T., et al. 2015. Low-frequency stimulation of STN-DBS reduces aspiration and freezing of gait in patients with PD. *Neurology* 84 (4): 415–420.

65. Ganeshan, V., et al. 2019. Pre-conditioning with remote photobiomodulation modulates the brain transcriptome and protects against MPTP insult in mice. *Neuroscience* 400: 85–97.

66. MacArthur, F.J., and G.W. McGarry. 2017. The arterial supply of the nasal cavity. *European Archives of Oto-Rhino-Laryngology* 274 (2): 809–815.

67. Moore, K.L. and A. Dalley, *Clinically oriented Anatomy. Williams and Wilkins*, 1999, Lippincott.

68. Rajagopal, M., and J. Paul. 2005. Applied anatomy and physiology of the airway and breathing. *Indian Journal of Anaesthesia* 49 (4): 251–256.

69. Komorowska, M., et al. 2002. Erythrocyte response to near-infrared radiation. *Journal of Photochemistry and Photobiology B: Biology* 68 (2–3): 93–100.

70. Lohr, N.L., et al. 2009. Enhancement of nitric oxide release from nitrosyl hemoglobin and nitrosyl myoglobin by red/near infrared radiation: Potential role in cardioprotection. *Journal of molecular and cellular cardiology* 47 (2): 256–263.

71. Vladimirov, Y.A., A. Osipov, and G. Klebanov. 2004. Photobiological principles of therapeutic applications of laser radiation. *Biochemistry (Moscow)* 69 (1): 81–90.

72. Walski, T., et al. 2015. Near infrared light induces post-translational modifications of human red blood cell proteins. *Photochemical & Photobiological Sciences* 14 (11): 2035–2045.
73. Asimov, M., A. Korolevich, and E. Konstantinova. 2007. Kinetics of oxygenation of skin tissue exposed to low-intensity laser radiation. *Journal of Applied Spectroscopy* 74 (1): 133–139.
74. Gisbrecht, A., et al. *Estimation of the quantum efficiency of the photodissociation of HbO2 and HbCO.* in *19th International Conference and School on Quantum Electronics: Laser Physics and Applications.* 2017. International Society for Optics and Photonics.
75. Yesman, S., et al. 2016. Local changes in arterial oxygen saturation induced by visible and near-infrared light radiation. *Lasers in medical science* 31 (1): 145–149.
76. Stadler, I., et al. 2000. In vitro effects of low-level laser irradiation at 660 nm on peripheral blood lymphocytes. *Lasers in Surgery and Medicine: The Official Journal of the American Society for Laser Medicine and Surgery* 27 (3): 255–261.
77. Chrapko, W., et al. 2006. Alteration of decreased plasma NO metabolites and platelet NO synthase activity by paroxetine in depressed patients. *Neuropsychopharmacology* 31 (6): 1286.
78. Bomboi, G., et al. 2010. Alzheimer's disease and endothelial dysfunction. *Neurological sciences* 31 (1): 1–8.
79. Natzle, W.C., et al. 1981. Dissociative ionization of water induced by single-photon vibrational excitation. *The Journal of Physical Chemistry* 85 (20): 2882–2884.
80. Natzle, W.C., and C.B. Moore. 1985. Recombination of hydrogen ion (H+) and hydroxide in pure liquid water. *The Journal of Physical Chemistry* 89 (12): 2605–2612.
81. Szymborska-Małek, K., M. Komorowska, and M. Gąsior-Głogowska, *Effects of Near Infrared Radiation on DNA. DLS and ATR-FTIR Study.* Spectrochimica Acta Part A: Molecular and Biomolecular Spectroscopy, 2018. **188**: p. 258–267.
82. Chludzińska, L., et al. 2005. Near-infrared radiation protects the red cell membrane against oxidation. *Blood Cells, Molecules, and Diseases* 35 (1): 74–79.
83. Itoh, T., et al. 1996. The protective effect of low power He-Ne laser against erythrocytic damage caused by artificial heart-lung machines. *Hiroshima journal of medical sciences* 45: 15–22.
84. Walski, T., et al., *Individual osmotic fragility distribution: a new parameter for determination of the osmotic properties of human red blood cells.* BioMed research international, 2014. **2014**.
85. Wang, H., et al. 2016. The hematologic effects of low intensity 650 nm laser irradiation on hypercholesterolemia rabbits. *American journal of translational research* 8 (5): 2293.
86. Komorowska, M., et al. 2001. Erythrocyte response to near infrared radiation. *Cellular and Molecular Biology Letters* 6 (2): 212–212.
87. Mi, X., et al., *A comparative study of 632.8 and 532 nm laser irradiation on some rheological factors in human blood in vitro.* Journal of Photochemistry and Photobiology B: Biology, 2004. **74**(1): p. 7–12.
88. Kujawa, J., et al., *Effect of low-intensity (3.75–25 J/cm2) near-infrared (810 nm) laser radiation on red blood cell ATPase activities and membrane structure.* Journal of clinical laser medicine & surgery, 2004. **22**(2): p. 111–117.
89. Drohomirecka, A., et al. 2018. Low-level light therapy reduces platelet destruction during extracorporeal circulation. *Scientific reports* 8 (1): 16963.
90. G. Brill, B.S., GE Brill, I. Tamarin, R. Dardik, VF Kirichuk, N. Savion, D. Varon, A, *Blood irradiation by He-Ne laser induces a decrease in platelet responses to physiological agonists and an increase in platelet cyclic GMP.* Platelets, 2000. **11**(2): p. 87–93.
91. Rola, P., et al., *Low-level laser irradiation exerts antiaggregative effect on human platelets independently on the nitric oxide metabolism and release of platelet activation markers.* Oxidative medicine and cellular longevity, 2017. **2017**.

92. Yang, J., et al. 2016. Low-level light treatment ameliorates immune thrombocytopenia. *Scientific reports* 6: 38238.

93. Zhang, Q., et al., *Noninvasive low-level laser therapy for thrombocytopenia*. Science translational medicine, 2016. **8**(349): p. 349ra101–349ra101.

94. Zhang, Q., M. Lu, and M.X. Wu. *Prolonging shelf-life of platelets by low-level laser*. in *Mechanisms of Photobiomodulation Therapy XIII*. 2018. International Society for Optics and Photonics.

95. Hroudová, J., et al. 2013. Mitochondrial respiration in blood platelets of depressive patients. *Mitochondrion* 13 (6): 795–800.

96. Sommer, A.P. and M.A. Trelles, *Light pumping energy into blood mitochondria: a new trend against depression?*, 2014, Mary Ann Liebert, Inc. 140 Huguenot Street, 3rd Floor New Rochelle, NY 10801 USA.

97. de Lima, F.M., et al. 2011. Dual Effect of low-level laser therapy (LLLT) on the acute lung inflammation induced by intestinal ischemia and reperfusion: Action on anti-and pro-inflammatory cytokines. *Lasers in Surgery and Medicine* 43 (5): 410–420.

98. Oliveira, M.C., Jr., et al. 2014. Low level laser therapy reduces acute lung inflammation in a model of pulmonary and extrapulmonary LPS-induced ARDS. *Journal of Photochemistry and Photobiology B: Biology* 134: 57–63.

99. Walski, T., et al. 2018. Low-Level Light Therapy Protects Red Blood Cells Against Oxidative Stress and Hemolysis During Extracorporeal Circulation. *Frontiers in Physiology* 9: 647.

100. Karu, T.I., L.V. Pyatibrat, and N.I. Afanasyeva. 2005. Cellular effects of low power laser therapy can be mediated by nitric oxide. *Lasers in Surgery and Medicine: The Official Journal of the American Society for Laser Medicine and Surgery* 36 (4): 307–314.

101. Cheon, S.Y., et al. 2017. Regulation of microglia and macrophage polarization via apoptosis signal-regulating kinase 1 silencing after ischemic/hypoxic injury. *Frontiers in molecular neuroscience* 10: 261.

102. Zhang, Y., et al. 2013. ROS play a critical role in the differentiation of alternatively activated macrophages and the occurrence of tumor-associated macrophages. *Cell research* 23 (7): 898.

103. Song, J.W., et al. 2017. Low-level laser facilitates alternatively activated macrophage/microglia polarization and promotes functional recovery after crush spinal cord injury in rats. *Scientific reports* 7 (1): 620.

104. Byrnes, K.R., et al. 2005. Light promotes regeneration and functional recovery and alters the immune response after spinal cord injury. *Lasers in Surgery and Medicine: The Official Journal of the American Society for Laser Medicine and Surgery* 36 (3): 171–185.

105. Prendergast, P.M. 2013. Neurologic anatomy of the nose. In *Advanced aesthetic rhinoplasty*, 17–23. Springer.

106. Liu, T.C.-Y., et al. *Mechanism of Itranasal Low Intensity Laser Irradiation Therapy*. in *World Symposium on TCM Acupuncture and Moxibustion, Tarragona, Spain*. 2007.

107. Liu, T.C.-Y., et al. 2010. Applications of intranasal low intensity laser therapy in sports medicine. *Journal of Innovative Optical Health Sciences* 3 (01): 1–16.

108. Barrett, D.W., and F. Gonzalez-Lima. 2013. Transcranial infrared laser stimulation produces beneficial cognitive and emotional effects in humans. *Neuroscience* 230: 13–23.

109. Jagdeo, J.R., et al. 2012. Transcranial red and near infrared light transmission in a cadaveric model. *PLoS ONE* 7 (10): e47460.

110. Henderson, T.A., and L.D. Morries. 2015. Near-infrared photonic energy penetration: Can infrared phototherapy effectively reach the human brain? *Neuropsychiatric disease and treatment* 11: 2191.

111. Erdem, G., et al. 2004. A radiological anatomic study of the cribriform plate compared with constant structures. *Rhinology* 42 (4): 225–229.

112. Masurkar, A. and W. Chen, *Olfactory Bulb Physiology.* 2009.

113. Zhou, G., et al., *Characterizing functional pathways of the human olfactory system.* Elife, 2019. **8**.

114. Sarnat, H.B., *Development of olfaction and taste in the human fetus and neonate*, in *Fetal and Neonatal Physiology.* 2017, Elsevier. p. 1411–1420. e3.

115. Wilson, R.S., L. Yu, and D.A. Bennett. 2010. Odor identification and mortality in old age. *Chemical senses* 36 (1): 63–67.

116. Bergmann, O., K.L. Spalding, and J. Frisén. 2015. Adult neurogenesis in humans. *Cold Spring Harbor perspectives in biology* 7 (7): a018994.

117. Xuan, W., et al. 2014. Transcranial low-level laser therapy enhances learning, memory, and neuroprogenitor cells after traumatic brain injury in mice. *Journal of biomedical optics* 19 (10): 108003.

118. Zubia, J., and J. Arrue. 2001. Plastic optical fibers: An introduction to their technological processes and applications. *Optical Fiber Technology* 7 (2): 101–140.

119. Litscher, D. and G. Litscher, *Laser therapy and stroke: quantification of methodological requirements in consideration of yellow laser.* International Journal of Photoenergy, 2013. **2013**.

120. Te Alexis, E. 2006. The next generation in laser treatments and the role of the GreenLight high-performance system laser. *Reviews in urology* 8 (Suppl 3): S24.

121. Fukuzaki, Y., et al. 2013. 532 nm low-power laser irradiation recovers γ-secretase inhibitor-mediated cell growth suppression and promotes cell proliferation via Akt signaling. *PLoS ONE* 8 (8): e70737.

122. Karu, T. 1988. Molecular mechanism of the therapeutic effect of low-intensity laser radiation. *Lasers Life Sci* 2 (1): 53–74.

123. Netter, F.H., *Atlas of Human Anatomy E-Book: including NetterReference. com Access with Full Downloadable Image Bank.* 2017: Elsevier Health Sciences.

124. Clark, D.L., N.N. Boutros, and M.F. Mendez, *The brain and behavior: an introduction to behavioral neuroanatomy.* 2010: Cambridge university press.

125. Purves, D., et al., *The olfactory epithelium and olfactory receptor neurons.* Neuroscience. 2nd Edition. Purves D, Augustine GJ, Fitzpatrick D (Eds). Sinauer Associates, Sunderland, MA, 2001.

126. Choi, R., and B.J. Goldstein. 2018. Olfactory epithelium: Cells, clinical disorders, and insights from an adult stem cell niche. *Laryngoscope investigative otolaryngology* 3 (1): 35–42.

127. DiMauro, T.M., et al., *Intranasal Insert for OFC Neuroprotection*, 2018, Google Patents.

128. DiMauro, T.M., et al., *Intranasal red light probe for treating alzheimer's disease*, 2014, Google Patents.

129. Pfleiderer, M., Y.S. Tardy, and B. Lovisa, *Transnasal Delivery of Low Level Light Via the Sphenoidal Sinus to Irradiate the Substantia Nigra*, 2017, Google Patents.

130. Dimauro, T.M., et al., *Intranasal red light probe for treating alzheimer's disease*, 2008, Google Patents.

131. Wiebracht, N.D., and L.A. Zimmer. 2014. Complex anatomy of the sphenoid sinus: A radiographic study and literature review. *Journal of Neurological Surgery Part B: Skull Base* 75 (06): 378–382.

132. Budu, V., et al. 2013. The anatomical relations of the sphenoid sinus and their implications in sphenoid endoscopic surgery. *Romanian Journal of Morphology and Embryology* 54 (1): 13–16.

133. Amar, A.P., and M.H. Weiss. 2003. Pituitary anatomy and physiology. *Neurosurgery Clinics* 14 (1): 11–23.

134. Barkhoudarian, G. and D.F. Kelly, *Chapter 1 - The Pituitary Gland: Anatomy, Physiology, and its Function as the Master Gland*, in *Cushing's Disease*, E.R. Laws, Editor. 2017, Academic Press. p. 1–41.
135. Gloor, P. 1978. Inputs and outputs of the amygdala: What the amygdala is trying to tell the rest of the brain. In *Limbic mechanisms*, 189–209. Springer.
136. Ehrlich, I., et al. 2009. Amygdala inhibitory circuits and the control of fear memory. *Neuron* 62 (6): 757–771.
137. LeDoux, J. 2007. The amygdala. *Current biology* 17 (20): R868–R874.
138. Sah, P., et al. 2003. The amygdaloid complex: Anatomy and physiology. *Physiological reviews* 83 (3): 803–834.
139. Garrett, A., and K. Chang. 2008. The role of the amygdala in bipolar disorder development. *Development and Psychopathology* 20 (4): 1285–1296.
140. McGaugh, J.L. 2000. Memory–a century of consolidation. *Science* 287 (5451): 248–251.
141. Parent, A.D. and E. Perkins, *Chapter 30 - The Hypothalamus*, in *Fundamental Neuroscience for Basic and Clinical Applications (Fifth Edition)*, D.E. Haines and G.A. Mihailoff, Editors. 2018, Elsevier. p. 442–456.e1.
142. Freeman, J.L. 2003. The anatomy and embryology of the hypothalamus in relation to hypothalamic hamartomas. *Epileptic disorders* 5 (4): 177–186.
143. Parent, A. and E. Perkins, *Hypothalamus*. Fundamental neuroscience for basic and clinical applications. 4th ed. Philadelphia: Elsevier, 2013: p. 417–30.
144. Anand, K.S., and V. Dhikav. 2012. Hippocampus in health and disease: An overview. *Annals of Indian Academy of Neurology* 15 (4): 239–246.
145. Insausti, R. and D.G. Amaral, *Hippocampal formation*, in *The Human Nervous System: Second Edition*. 2003, Elsevier Inc. p. 871–914.
146. Insausti, R., and D.G. Amaral. 2012. Chapter 24 - Hippocampal Formation. In *The Human Nervous System (Third Edition)*, ed. J.K. Mai and G. Paxinos, 896–942. San Diego: Academic Press.
147. Sahay, A., M.R. Drew, and R. Hen. 2007. Dentate gyrus neurogenesis and depression. *Progress in brain research* 163: 697–822.
148. Anacker, C., et al., *190. Neurogenesis Inhibits Stress-Responsive Cells in the Ventral Dentate Gyrus*. Biological Psychiatry, 2018. **83**(9): p. S76.
149. Adam Samuels, B., E.D. Leonardo, and R. Hen, *Hippocampal subfields and major depressive disorder*. Biological psychiatry, 2015. **77**(3): p. 210–211.
150. Fanselow, M.S., and H.-W. Dong. 2010. Are the dorsal and ventral hippocampus functionally distinct structures? *Neuron* 65 (1): 7–19.
151. O'Mara, S. 2005. The subiculum: What it does, what it might do, and what neuroanatomy has yet to tell us. *Journal of anatomy* 207 (3): 271–282.
152. O'mara, S.M., et al., *Roles for the subiculum in spatial information processing, memory, motivation and the temporal control of behaviour.* Progress in Neuro-Psychopharmacology and Biological Psychiatry, 2009. **33**(5): p. 782–790.
153. Cragg, S.J., et al. 2004. Synaptic release of dopamine in the subthalamic nucleus. *European Journal of Neuroscience* 20 (7): 1788–1802.
154. Hajos, M., and S. Greenfield. 1994. Synaptic connections between pars compacta and pars reticulata neurones: Electrophysiological evidence for functional modules within the substantia nigra. *Brain research* 660 (2): 216–224.
155. Lavoie, B., Y. Smith, and A. Parent. 1989. Dopaminergic innervation of the basal ganglia in the squirrel monkey as revealed by tyrosine hydroxylase immunohistochemistry. *Journal of Comparative Neurology* 289 (1): 36–52.

156. Delong, M.R., M.D. Crutcher, and A.P. Georgopoulos. 1983. Relations between movement and single cell discharge in the substantia nigra of the behaving monkey. *Journal of Neuroscience* 3 (8): 1599–1606.

157. Ilango, A., et al. 2014. Similar roles of substantia nigra and ventral tegmental dopamine neurons in reward and aversion. *Journal of Neuroscience* 34 (3): 817–822.

158. Burchman, M. 2011. Using photobiomodulation on a severe Parkinson's patient to enable extractions, root canal treatment, and partial denture fabrication. *J Laser Dent* 19: 297–300.

159. Sun, L., et al. 2016. Human brain reacts to transcranial extraocular light. *PLoS ONE* 11 (2): e0149525.

160. Timonen, M., et al. 2012. Can transcranial brain-targeted bright light treatment via ear canals be effective in relieving symptoms in seasonal affective disorder?–A pilot study. *Medical hypotheses* 78 (4): 511–515.

161. Jurvelin, H., et al. 2014. Transcranial bright light treatment via the ear canals in seasonal affective disorder: A randomized, double-blind dose-response study. *BMC Psychiatry* 14 (1): 288.

162. Romeo, S., et al., *Eyes as gateways for environmental light to the substantia nigra: relevance in Parkinson's disease.* The Scientific World Journal, 2014. **2014**.

163. Maggio, R., et al., *Parkinson's Disease and Light: The Bright and the Dark Sides.* Brain research bulletin, 2019.

164. Kim, B., et al. 2017. Remote tissue conditioning—An emerging approach for inducing body-wide protection against diseases of ageing. *Ageing research reviews* 37: 69–78.

165. Kharbanda, R.K., T.T. Nielsen, and A.N. Redington. 2009. Translation of remote ischaemic preconditioning into clinical practice. *The Lancet* 374 (9700): 1557–1565.

166. Kim, B., et al. 2018. Remote tissue conditioning is neuroprotective against MPTP insult in mice. *IBRO reports* 4: 14–17.

167. Gordon, L.C., and D.M. Johnstone. 2019. Remote photobiomodulation: An emerging strategy for neuroprotection. *Neural Regeneration Research* 14 (12): 2086.

168. Gordon, L., et al. 2019. Remote photobiomodulation as a neuroprotective intervention—harnessing the indirect effects of photobiomodulation. In *Photobiomodulation in the Brain*, 139–154. Elsevier.

169. Johnstone, D.M., J. Mitrofanis, and J. Stone. 2015. Targeting the body to protect the brain: Inducing neuroprotection with remotely-applied near infrared light. *Neural regeneration research* 10 (3): 349.

170. Rochkind, S., et al. 1989. Systemic effects of low-power laser irradiation on the peripheral and central nervous system, cutaneous wounds, and burns. *Lasers in surgery and medicine* 9 (2): 174–182.

171. Farfara, D., et al. 2015. Low-level laser therapy ameliorates disease progression in a mouse model of Alzheimer's disease. *Journal of Molecular Neuroscience* 55 (2): 430–436.

172. Blivet, G., et al. 2018. Neuroprotective effect of a new photobiomodulation technique against Aβ25–35 peptide–induced toxicity in mice: Novel hypothesis for therapeutic approach of Alzheimer's disease suggested. *Alzheimer's & Dementia: Translational Research & Clinical Interventions* 4: 54–63.

173. Johnstone, D., et al. 2014. Indirect application of near infrared light induces neuroprotection in a mouse model of parkinsonism–an abscopal neuroprotective effect. *Neuroscience* 274: 93–101.

174. Oron, A., and U. Oron. 2016. Low-level laser therapy to the bone marrow ameliorates neurodegenerative disease progression in a mouse model of Alzheimer's disease: A minireview. *Photomedicine and laser surgery* 34 (12): 627–630.

175. Zhang, J., et al. 2019. Illumination with 630 nm Red Light Reduces Oxidative Stress and Restores Memory by Photo-Activating Catalase and Formaldehyde Dehydrogenase in SAMP8 Mice. *Antioxidants & redox signaling* 30 (11): 1432–1449.

176. Muili, K.A., et al. 2012. Amelioration of experimental autoimmune encephalomyelitis in C57BL/6 mice by photobiomodulation induced by 670 nm light. *PLoS ONE* 7 (1): e30655.

177. Hamilton, C., et al. 2018. Exploring the use of transcranial photobiomodulation in Parkinson's disease patients. *Neural regeneration research* 13 (10): 1738.

178. Tuby, H., L. Maltz, and U. Oron. 2009. Implantation of low-level laser irradiated mesenchymal stem cells into the infarcted rat heart is associated with reduction in infarct size and enhanced angiogenesis. *Photomedicine and laser surgery* 27 (2): 227–233.

179. Tuby, H., L. Maltz, and U. Oron. 2011. Induction of autologous mesenchymal stem cells in the bone marrow by low-level laser therapy has profound beneficial effects on the infarcted rat heart. *Lasers in surgery and medicine* 43 (5): 401–409.

180. Matsushita, T., et al. 2011. Mesenchymal stem cells transmigrate across brain microvascular endothelial cell monolayers through transiently formed inter-endothelial gaps. *Neuroscience letters* 502 (1): 41–45.

181. Fekrazad, R., et al. 2016. Effect of photobiomodulation on mesenchymal stem cells. *Photomedicine and laser surgery* 34 (11): 533–542.

182. Oron, U., et al. 2014. Autologous bone-marrow stem cells stimulation reverses post-ischemic-reperfusion kidney injury in rats. *American journal of nephrology* 40 (5): 425–433.

183. Liebert, A.D., et al., *Neuroprotective effects against POCD by photobiomodulation: evidence from assembly/disassembly of the cytoskeleton.* Journal of experimental neuroscience, 2016. **10**: p. JEN. S33444.

184. Liebert, A.D., B.T. Bicknell, and R.D. Adams. 2014. Protein conformational modulation by photons: A mechanism for laser treatment effects. *Medical hypotheses* 82 (3): 275–281.

185. Maksimovich, I.V. 2019. Intracerebral Transcatheter Laser Photobiomodulation Therapy in the Treatment of Binswanger's Disease and Vascular Parkinsonism: Research and Clinical Experience. *Photobiomodulation, photomedicine, and laser surgery* 37 (10): 606–614.

186. Maksimovich, I.V. 2015. Dementia and cognitive impairment reduction after laser transcatheter treatment of Alzheimer's disease. *World Journal of Neuroscience* 5 (03): 189.

187. Maksimovich, I.V. 2019. Transcatheter intracerebral photobiomodulation in ischemic brain disorders: Clinical studies (Part 2). In *Photobiomodulation in the Brain*, 529–543. Elsevier.

188. Maksimovich, I.V. 2019. Transcatheter intracerebral photobiomodulation in degenerative brain disorders: Clinical studies (Part 1). In *Photobiomodulation in the Brain*, 515–528. Elsevier.

189. Meneguzzo, D.T., et al., *Intravascular Laser Irradiation of Blood*, in *Handbook of Low-Level Laser Therapy*. 2016, Pan Stanford. p. 969–988.

190. Mikhaylov, V. 2015. The use of Intravenous Laser Blood Irradiation (ILBI) at 630–640 nm to prevent vascular diseases and to increase life expectancy. *Laser therapy* 24 (1): 15–26.

191. Moskvin, S.V., and A.V. Kochetkov. 2019. Russian low level laser therapy techniques for brain disorders. In *Photobiomodulation in the Brain*, 545–572. Elsevier.

192. Xiaoa, X., et al. 2000. A single photon emission computed tomography study of the therapy of intravascular low intensity laser irradiation on blood for brain infarction. *Laser Therapy* 13 (1): 110–113.

193. Chang, J.Y., et al. 2019. Effects of Intravascular Laser Irradiation of Blood on Cognitive Function in a Stroke Survivor with Hyperhomocysteinemia: Dual Recuperations in Thalamus and Serum Homocysteine. *Biomed J Sci & Tech Res* 16: 11864–11868.

194. Yang, W.-H., S.-P. Lin, and S.-T. Chang, *Case report: Rapid improvement of crossed cerebellar diaschisis after intravascular laser irradiation of blood in a case of stroke.* Medicine, 2017. **96**(2).

195. Arakelyan, H.S. 2005. TREATMENT OF ALZHEIMER'S DISEASE WITH A COMBINATION OF LASER, MAGNETIC FIELD AND CHROMO LIGHT (COLOUR) THERAPIES: A DOUBLE-BLIND CONTROLLED TRIAL BASED ON A REVIEW AND OVERVIEW OF THE ETIOLOGICAL PATHOPHYSIOLOGY OF ALZHEIMER'S DISEASE. *LASER THERAPY* 14 (1): 19–28.

196. Chon, T.Y., et al. 2019. Laser acupuncture: A concise review. *Medical acupuncture* 31 (3): 164–168.

197. Na, C.-S., et al. 2015. Low-level Green and Red Laser Treatment of Shaochong (HT9)· Dadun (LR1) and Shaohai (HT3)· Yingu (KI10) Acupoints in a Rat Model of Focal Cerebral Ischemia. *Transactions on Electrical and Electronic Materials* 16 (2): 65–69.

198. Sutalangka, C., et al. 2013. Laser acupuncture improves memory impairment in an animal model of Alzheimer's disease. *Journal of acupuncture and meridian studies* 6 (5): 247–251.

199. Khongrum, J., and J. Wattanathorn. 2015. Laser acupuncture improves behavioral disorders and brain oxidative stress status in the valproic acid rat model of autism. *Journal of acupuncture and meridian studies* 8 (4): 183–191.

200. Khongrum, J., and J. Wattanathorn. 2017. Laser acupuncture at HT7 improves the cerebellar disorders in valproic acid-rat model of autism. *Journal of acupuncture and meridian studies* 10 (4): 231–239.

201. Naeser, M.A., et al. 1995. Laser acupuncture in the treatment of paralysis in stroke patients: A CT scan lesion site study. *American journal of acupuncture* 23: 13–13.

202. Naeser, M.A., et al. 2011. Improved cognitive function after transcranial, light-emitting diode treatments in chronic, traumatic brain injury: Two case reports. *Photomedicine and laser surgery* 29 (5): 351–358.

203. Raith, W., et al., *Near-infrared spectroscopy for objectifying cerebral effects of laser acupuncture in term and preterm neonates.* Evidence-Based Complementary and Alternative Medicine, 2013. **2013**.

204. Quah-Smith, I., et al. 2013. Differential brain effects of laser and needle acupuncture at LR8 using functional MRI. *Acupuncture in Medicine* 31 (3): 282–289.

205. Lv, J., et al. 2016. The brain effects of laser acupuncture at thirteen ghost acupoints in healthy individuals: A resting-state functional MRI investigation. *Computerized Medical Imaging and Graphics* 54: 48–54.

206. Deadman, P., M. Al-Khafaji, and K. Baker, *A manual of acupuncture.* 1998: Journal of Chinese Medicine Publications East Sussex, UK.

207. Frost, E. 1976. Acupuncture for the comatose patient. *American Journal of Acupuncture* 4: 45–48.

208. Wise, N.A. 2019. Action at a distance: Laser acupuncture and the brain. In *Photobiomodulation in the Brain*, 489–501. Elsevier.

209. Grillo, S., et al. 2013. Non-invasive infra-red therapy (1072 nm) reduces β-amyloid protein levels in the brain of an Alzheimer's disease mouse model, TASTPM. *Journal of Photochemistry and Photobiology B: Biology* 123: 13–22.

210. Han, M., et al. 2018. Near infra-red light treatment of Alzheimer's disease. *Journal of Innovative Optical Health Sciences* 11 (01): 1750012.

211. Vos, M., et al. 2013. Near-infrared 808 nm light boosts complex IV-dependent respiration and rescues a Parkinson-related pink1 model. *PLoS ONE* 8 (11): e78562.

212. Choi, D.-H., et al. 2012. Effect of 710 nm visible light irradiation on neurite outgrowth in primary rat cortical neurons following ischemic insult. *Biochemical and biophysical research communications* 422 (2): 274–279.

213. Ghanbari, A., et al. 2017. Light-emitting diode (LED) therapy improves occipital cortex damage by decreasing apoptosis and increasing BDNF-expressing cells in methanol-induced toxicity in rats. *Biomedicine & Pharmacotherapy* 89: 1320–1330.
214. Ghanbari, A., et al., *Light-Emitting Diode (LED) Therapy Attenuates Neurotoxicity of Methanol-Induced Memory Impairment and Apoptosis in The Hippocampus.* CNS & Neurological Disorders-Drug Targets (Formerly Current Drug Targets-CNS & Neurological Disorders), 2018. **17**(7): p. 528–538.
215. Mathangi, D., and R. Shyamala. 2016. Effect of LED photobiomodulation on fluorescent light induced changes in cellular ATPases and Cytochrome c oxidase activity in Wistar rat. *Lasers in medical science* 31 (9): 1803–1809.
216. Michalikova, S., et al. 2008. Emotional responses and memory performance of middle-aged CD1 mice in a 3D maze: Effects of low infrared light. *Neurobiology of learning and memory* 89 (4): 480–488.
217. Avci, P., et al. 2013. Low-level laser therapy for fat layer reduction: A comprehensive review. *Lasers in surgery and medicine* 45 (6): 349–357.
218. Paolillo, F.R., et al. 2017. Effects of phototherapy plus physical training on metabolic profile and quality of life in postmenopausal women. *Journal of Cosmetic and Laser Therapy* 19 (6): 364–372.

Preconditioning with Photobiomodulation in Neuronal Cells and the Brain

7

7.1 Introduction

Preconditioning is based on the concept that low levels of various types of external stress could produce a protective response against subsequent high levels of stress to cells and tissue. In other words, preconditioning is a potent endogenous protective response that can activate various endogenous signaling pathways resulting in tolerance against more severe insults [1]. This paradigm has received increasing attention in medicine, where ischemic preconditioning has been the most well-known and investigated intervention [2]. Mitochondria, reactive oxygen species (ROS), hypoxia-inducible factor-1α (HIF-1α), and heat-shock factors (HSFs) have all been implicated as possible mechanisms involved in the preconditioning paradigm [3].

Photobiomodulation (PBM) preconditioning refers to the irradiation of the body with low levels of light in advance of some insult, which could subsequently damage the cells and tissue. Over the last decade, several pre-clinical studies have shown that PBM can exert preconditioning effects in medical conditions, such as pain, heart attack, wound healing, and brain damage [3]. In the context of neurons and the brain, evidence suggests that PBM has the potential to be used as a strategy for prevention, and not just treatment of central nervous system damage, such as neurotoxicity [4–8], ischemia [9, 10], and Parkinson's disease (PD) [11, 12]. PBM-induced nitric oxide (NO) production plays an important role in the regulation of blood pressure and increasing the cerebral circulation and blood flow [13]. Besides improving the mitochondrial electron transport chain and ATP production, PBM can also alter the cellular redox state and intracellular signaling molecules, such as Ca^{2+}, NO, and ROS, resulting in regulation of various transcription factors such as nuclear factor kappa B (NF-κB), redox factor-1 (Ref-1), activator protein-1 (AP-1), HIF-1, HIF-like factor as well as some cytoprotective gene products including antioxidant enzymes, antiapoptotic proteins, and heat-shock proteins (HSPs) [14, 15], all

F. Salehpour et al., *Photobiomodulation for the Brain*, Synthesis Lectures on Biomedical Engineering, https://doi.org/10.1007/978-3-031-36231-6_7

of which have been proposed to be involved in PBM preconditioning mechanisms. Herein, we review studies of PBM preconditioning in neuronal cell culture in vitro as well as in animal models of brain insults.

7.2   Preconditioning in Vitro Studies

In 2005, Wong-Riley et al. [6] studied whether PBM preconditioning with light emitting diodes (LEDs) before exposure to potassium cyanide (KCN) could improve the positive effect of PBM on KCN toxicity in primary cultured visual neurons. Preconditioning with 670 nm LED (4 J/cm^2) once a day for both 3 or 5 days provided additional improvement in the level of cytochrome c oxidase (CCO) activity in darkly reactive metabolic neurons, but not in the lightly reactive group, compared to those that received PBM (twice a day) during their 5-day exposure to 10 or 100 μM of KCN [6].

In 2006, Liang et al. [7] explored the effect of PBM preconditioning on rat visual cortical neurons to test if PBM could suppress apoptotic cell death induced by higher concentrations of KCN. The primary neurons were cultured and pre-treated with a 670 nm LED (30 J/cm^2) and were then exposed to KCN (100 or 300 μM). Collectively, the results showed that preconditioning significantly protected neurons against apoptosis by a mechanism most likely involving the decrease of reactive oxygen species (ROS) production, down-regulation of pro-apoptotic proteins (e.g., Bax and caspase-3), and activation of antiapoptotic proteins (e.g., Bcl-2) [7].

In 2008, Ying et al. [4] tested whether PBM preconditioning could be an effective preventative strategy to rescue cultured neurons from neurotoxins linked to PD-like disease, namely the pesticide rotenone and 1-methyl-4-phenylpyridinium [MPP$^+$]. They showed that PBM preconditioning with 670 nm LED (4 J/cm^2) for 2 days before exposure to rotenone (200 nM) or MPP$^+$ (250 μM) significantly suppressed apoptosis in both striatal and cortical neurons. PBM preconditioning also significantly reversed the MPP$^+$-induced decrease in ATP in striatal neurons [4].

In 2009, Lim et al. [5] showed that preconditioning with 635 nm laser (18 J/cm^2) before 200 μM of sodium nitroprusside (SNP) treatment can increase SH-SY5Y cell viability to about 60% that of the control, along with a decrease in nuclear fragmentation and membrane breakdown. PBM preconditioning also inhibited mitochondrial-dependent apoptosis by decreasing the Bax/Bcl-2 ratio and the activity of caspase-9 and caspase-3. The results also suggested that PBM pre-treatment could protect against neuronal death induced by peroxynitrite (ONOO$^-$) formation from NO and superoxide, by promoting the scavenging of ROS and ONOO$^-$ [5].

In 2010, Yang et al. [8] examined whether PBM preconditioning with 632.8 nm laser light could abrogate the oxidative and inflammatory responses induced by 5 μM of amyloid-beta (Aβ) peptide in primary astrocytes. They found that PBM preconditioning (16.2 J/cm^2) could suppress Aβ-induced superoxide production, reduce colocalization between NADPH oxidase gp91^{phox} and p47^{phox} subunits, phosphorylation of cPLA$_2$, as well as the expression of interleukin (IL)-1β and inducible NO synthase (iNOS) [8].

In 2014, Chen et al. [16] investigated the effect of 808 nm LED preconditioning on cobalt chloride (CoCl$_2$)-induced hypoxic damage in primary mouse cortical neurons. PBM preconditioning (2 J/cm^2) for 3 days before exposure to 100 μM of CoCl$_2$ significantly increased cell viability and improved the morphology of the neurons, as well as increasing COX activity and ATP content [16].

In 2014, Jiang et al. [17] investigated the optimal PBM preconditioning time window for reducing apoptosis in oxygen–glucose deprivation (OGD) treated PC12 cells. Their results indicated that 4–6 h of continuous 660 nm LED pre-treatment was the best irradiation time before OGD for decreasing apoptosis, raising mitochondrial membrane potential (MMP), and decreasing ROS levels. The data also showed continuing cell protection in the PBM preconditioning group lasting until 8 h post-irradiation, along with down-regulation of Bax and up-regulation of Bcl-2 levels, indicating the long-term protective effects of PBM preconditioning on apoptosis [17].

7.3 Preconditioning in Vivo Studies

In 2009, Mirzaii-Dizgah et al. [18] examined the effects of 830 nm laser preconditioning on naloxone-induced withdrawal signs in morphine-dependent rats. Animals were irradiated immediately, 15, or 30 min prior to naloxone injection. Results indicated that transcranial PBM preconditioning (12.5 J/cm^2 at the scalp) applied immediately or 15 min before naloxone injection, significantly attenuated the naloxone-precipitated withdrawal symptoms as shown by a decreased total withdrawal score [18].

In 2010, Uozumi et al. [13] reported the first evidence suggesting that PBM preconditioning could mitigate cerebral ischemia and subsequent re-perfusion injury in a mouse model. An 808 nm laser irradiation (1.6 W/cm^2) was applied to the left hemisphere transcranially for 30 min before transient bilateral common carotid artery occlusion (BCCAO). Pre-treatment significantly improved residual cerebral blood flow during BCCAO in both the irradiated and non-irradiated hemispheres of the mouse brain and also reduced the number of apoptotic cells in the hippocampus [13].

In 2016, Mathangi et al. [19] reported the protective effects of whole-body PBM preconditioning on fluorescent light-induced changes to cellular enzymes in the rat brain. Each rat was irradiated with 670 nm LED light (9 J/cm^2 at the skin surface) prior to fluorescent light exposure (with a power of 18 W) for 1 (acute), 15 (sub-chronic), or 30 days (chronic). PBM preconditioning resulted in a significant increase in the enzyme activity of Na$^+$–K$^+$ ATPase, Ca^{2+} ATPase, and CCO at all the time points examined [19].

In 2016, Reinhart et al. [12] tested whether the timing of PBM treatment could have a different effect on the behavioral and structural measures of damage in a Parkinson's disease mouse model. A 670 nm LED PBM (2 J/cm^2) was applied transcranially either

2 days before, at the same time, or 2 days after the MPTP injection. According to the results, PBM was equally effective in decreasing locomotor impairment and protecting cells against a toxic insult whether applied as preconditioning, simultaneous, or post-treatment after MPTP injection. These findings suggested both fast-acting and long-lasting effects of PBM in this model [12].

In 2016 [9] and 2017 [10], Lee et al. examined the effects of 610 nm LED pre-conditioning on two different models of cerebral ischemia in mice, the photothrombotic [9] and middle cerebral artery occlusion/reperfusion [10] models. In both studies, each mouse received 2 J/cm^2 of PBM, twice a day for 2 days prior to the infliction of the ischemic event. The results from both behavioral and molecular measurements suggested that PBM preconditioning could induce tolerance to a cerebral ischemic insult in the photothrombotic model via decreasing infarct volume and edema as well as mitigating inflammatory responses [9]. While in the occlusion/reperfusion model, PBM precondi-tioning could attenuate brain damage by upregulating endothelial nitric oxide synthase (eNOS) phosphorylation through the PI3K/Akt pathway [10].

In 2018, Kim et al. [11] sought to determine whether either PBM or mild ischemia preconditioning could provide protective effects in the brain against the Parkinsonian neu-rotoxin MPTP. They showed that remote tissue preconditioning, whether by ischemia of a limb or by PBM of the torso using 670 nm LEDs, induced protection of the substan-tia nigra pars compacta (SNpc) in the brain, although combining these two treatment modalities provided no synergistic or additive effects [11].

In 2019, Ganeshan et al. [20] aimed to determine the most effective precondition-ing regimen for remote tissue PBM to induce neuroprotection in a mouse PD model. A 670 nm LED irradiation (4.5 J/cm^2) was applied to the dorsum and hind limbs once a day for 2, 5, or 10 days prior to MPTP injection. PBM for 2 and 5 days only increased the Fos$^+$ cell number in the CPu, however, 10 days of irradiation increased TH$^+$ cell num-bers in the SNc and Fos$^+$ cell numbers in the CPu, as well as upregulating cell signaling pathways and cell migration [20].

In 2019, Yang et al. [21] evaluated the preventative effect of PBM preconditioning on brain damage in a hypoxia–ischemia model using neonatal rats and determined the best PBM preconditioning time for this model. Mice received 808 nm laser transcranial PBM (12 J/cm^2), 6 h before the animals were subjected to a cerebral hypoxia protocol. The pre-treated animals showed improved cognitive performance and also showed less pronounced volume shrinkage in the brain, reduced neuron loss, and less dendritic and synaptic injury. Given the above-mentioned results and a series of experiments that measured ATP levels in the brains of intact mice, they proposed that the 6-h time point could be the optimum preconditioning time [21].

References

1. Stagliano, N.E., et al. 1999. Focal ischemic preconditioning induces rapid tolerance to middle cerebral artery occlusion in mice. *Journal of Cerebral Blood Flow & Metabolism* 19 (7): 757–761.
2. Hai-Xia, Z., D. Guan-Hua, and Z. Jun-Tian. 2003. Ischemic pre-conditioning preserves brain mitochondrial functions during the middle cerebral artery occlusion in rat. *Neurological research* 25 (5): 471–476.
3. Agrawal, T., et al. 2014. Pre-conditioning with low-level laser (light) therapy: light before the storm. *Dose-Response* 12(4): p. 14–032. Agrawal.
4. Ying, R., et al. 2008. Pretreatment with near-infrared light via light-emitting diode provides added benefit against rotenone-and MPP+-induced neurotoxicity. *Brain Research* 1243: 167–173.
5. Lim, W., et al. 2009. Inhibition of mitochondria-dependent apoptosis by 635-nm irradiation in sodium nitroprusside-treated SH-SY5Y cells. *Free Radical Biology and Medicine* 47 (6): 850–857.
6. Wong-Riley, M.T., et al. 2005. Photobiomodulation directly benefits primary neurons functionally inactivated by toxins role of cytochrome c oxidase. *Journal of Biological Chemistry* 280 (6): 4761–4771.
7. Liang, H., et al. 2006. Photobiomodulation partially rescues visual cortical neurons from cyanide-induced apoptosis. *Neuroscience* 139 (2): 639–649.
8. Yang, X., et al. 2010. Low energy laser light (632.8 nm) suppresses amyloid-β peptide-induced oxidative and inflammatory responses in astrocytes. *Neuroscience* 171(3): p. 859–868.
9. Lee, H.I., et al. 2016. Pre-conditioning with transcranial low-level light therapy reduces neuroinflammation and protects blood-brain barrier after focal cerebral ischemia in mice. *Restorative Neurology and Neuroscience* 34 (2): 201–214.
10. Lee, H.I., et al. 2017. Pretreatment with light-emitting diode therapy reduces ischemic brain injury in mice through endothelial nitric oxide synthase-dependent mechanisms. *Biochemical and Biophysical Research Communications* 486 (4): 945–950.
11. Kim, B., et al. 2018. Remote tissue conditioning is neuroprotective against MPTP insult in mice. *IBRO Reports* 4: 14–17.
12. Reinhart, F., et al. 2016. Near-infrared light (670 nm) reduces MPTP-induced parkinsonism within a broad therapeutic time window. *Experimental Brain Research* 234 (7): 1787–1794.
13. Uozumi, Y., et al. 2010. Targeted increase in cerebral blood flow by transcranial near-infrared laser irradiation. *Lasers in Surgery and Medicine* 42 (6): 566–576.
14. de Freitas, L.F., and M.R. Hamblin. 2016. Proposed mechanisms of photobiomodulation or low-level light therapy. *IEEE Journal of Selected Topics in Quantum Electronics* 22 (3): 348–364.
15. Hamblin, M.R. 2019. *Mechanisms of photobiomodulation in the brain Photobiomodulation in the Brain*, 97–110. Elsevier.
16. Chen, X., et al. 2014. Effect and mechanism of 808 nm light pretreatment of hypoxic primary neurons. *International Journal of Photoenergy*, **2014**.
17. Jiang, W., et al. 2014. Red photon treatment inhibits apoptosis via regulation of bcl-2 proteins and ROS levels, alleviating hypoxic–ischemic brain damage. *Neuroscience* 268: 66–74.
18. Mirzaii-Dizgah, I., et al. 2009. Attenuation of morphine withdrawal signs by low level laser therapy in rats. *Behavioural Brain Research* 196 (2): 268–270.
19. Mathangi, D., and R. Shyamala. 2016. Effect of LED photobiomodulation on fluorescent light induced changes in cellular ATPases and Cytochrome c oxidase activity in Wistar rat. *Lasers in Medical Science* 31 (9): 1803–1809.

20. Ganeshan, V., et al. 2019. Pre-conditioning with remote photobiomodulation modulates the brain transcriptome and protects against MPTP insult in mice. *Neuroscience* 400: 85–97.
21. Yang, L., et al. 2019. Photobiomodulation preconditioning prevents cognitive impairment in a neonatal rat model of hypoxia-ischemia. *Journal of Biophotonics* 12 (6): e201800359.

Photobiomodulation in Neuronal Cell Cultures

8

8.1   Importance of In Vitro Studies

The use of in vitro cell culture experiments is one of the most common laboratory methods employed for research into biological mechanisms. In general, the study of biological processes outside of a living organism, under controlled conditions, is known as in vitro or ex vivo studies. These types of studies are considered to be a bridge between computer-based (in silico or simulation) and experimental animal studies, and have been used in a broad range of biological sciences, including but not limited to neuroscience. Furthermore, in terms of ethical considerations, the use of in vitro studies has been assigned a priority, with fewer ethical limitations compared to animal and human clinical studies [1]. In vitro models, in addition to the possibility of employing specific relevant cell types, allow researchers to carry out more molecular, functional, and electrophysiological studies than could be carried out in animals. Moreover, since in experimental animal studies, the microenvironment of the tissue can affect the response of specific cells to interventions, the use of in vitro studies can help to overcome this complexity. However, from another point of view, this issue could be considered a disadvantage, because it may not be able to completely mimic the actual result of the intervention on real living tissue.

Due to the specific characteristics of living brain tissue, the design of in vitro studies in the field of neuroscience is complex and sometimes impossible. In addition, since most brain functions require the integration of a huge number of neurons and synaptic connections, studies on individual cell units in culture may not be helpful for these studies. In some cases, ex vivo methods such as brain slices, are not suitable for behavioral, cognitive, connectivity, and electrophysiological studies. However, brain slices can be useful in neurotoxicity, neuropharmacology, and neuromodulation studies. Recently, new in vitro techniques have been developed for modeling brain structures and functions, such as the brain blood barrier and neuron-glial communications. Moreover, microfluidic

F. Salehpour et al., *Photobiomodulation for the Brain*, Synthesis Lectures on Biomedical Engineering, https://doi.org/10.1007/978-3-031-36231-6_8

platforms are also under development to act as a model for a variety of brain structures and diseases in vitro.

8.2   In Vitro Photobiomodulation Studies

After undertaking a systematic evaluation of publications using the terms "in vitro neuronal culture and PBM", there are only limited ex vivo studies on this topic, and most of the ex vivo studies have been performed on cortical or hippocampal slices removed from rodents (Table 8.1). The majority of studies have been conducted on primary neuronal or glial traditional cell cultures, which are often isolated from the brains of laboratory animals such as rats or mice. There are also some studies on brain-derived stem cells. Many of the studies have been performed on neuronal or astroglial cell lines, which are commercially available in the market. Despite many in vivo animal studies of PBM, in which a variety of mechanisms could be evaluated, in the in vitro studies, the scope of research is limited and most studies have been focused on mechanisms like cytochrome c oxidase (CCO) activity, nitric oxide (NO), and reactive oxygen species (ROS) production, mitochondrial membrane potential (MMP), and Ca^{2+} regulation [2]. These mechanisms have been frequently discussed in the literature [3], so we will briefly discuss these mechanisms below.

8.2.1   Cytochrome C Oxidase Activity

Tissue bioenergetics is one of the most important pathways by which PBM can affect brain structure and function. There is strong in vitro evidence supporting the positive effects of PBM on the energy levels of neural cells. CCO is the terminal enzyme in the mitochondrial electron transport chain playing a key role in the regulation of bioenergetics after PBM in neuroglial structures. Light of specific wavelengths stimulates CCO and improves its catalytic activity, and increases ATP synthesis. In one study, irradiation of cultured rat cortical neurons using 670 nm light emitting diodes (LEDs) at a dose of 4 J/cm^2 increased the CCO activity [4]. Another study by the same research team showed that both 670 nm and 830 nm light could increase CCO activity in cultured rat visual cortical neurons [5]. It was shown that a similar dose of 670 nm LED could increase the activity of CCO in potassium cyanide (KCN)-intoxicated neurons from the occipital cortex or striatum [6]. Also, the application of 808 nm LEDs once a day for 3 days with 2 J/cm^2 at the culture surface could increase the CCO activity in the primary mouse cortical neurons in hypoxic conditions [7]. Overall, there is strong evidence from in vitro studies on the boosting effects of PBM on CCO activity, and activation of subsequent pathways.

Table 8.1 Summary of in vitro studies on the effects of photobiomodulation in neuronal activities from 1988 to 2019

Study/Year	Cell types	Light source	Wavelengths	Irradiation parameters	Findings
Wade et al. 1988 [27]	Slice of cerebral cortex	Tungsten halogen lamp	NR	650 W, 0.6, 1.3, or 1.9 mW/cm²	At 0.6 and 1.3 mW/cm²: enhanced K⁺-induced [³H]GABA release. At 1.9 mW/cm²: suppressed K⁺-induced [³H]GABA release
Tsai and Kao 1991 [28]	Rat brain astrocytes (RBA-1) cell line	Argon laser (Model 770, Cooper Laser Sonics, Inc.); CO₂ (Model 1060, Sharplan, Israel); He–Ne laser (Omniprobe); GaAs laser (Omniprobe)	Argon (488–512 nm); CO₂ (10,600 nm); He–Ne (632.8 nm); GaAs (904 nm)	Argon laser: 0.4–2 W, 1.1–5.4 W/cm², 0.4–22 J/cm², 0.1–20 s; CO₂ laser: 0.8 W, 2.16 W/cm², 0.5–5 s, 1.1–11 J/cm²; He–Ne laser: 1 mW, 2.7–326 mJ/cm², 1–120 s; GaAs laser: 0.4 mW, 9–380 mJ/cm², 9–350 s	For all lasers, no significant stimulatory effects were observed
Karu et al. 1996 [29]	Pyramidal hippocampal neurons; C6 glial cells	Laser, He–Ne (LG-78, Lvov, Ukraine)	632.8 nm	2 μW, 5 mW/cm²	Suppressed open-state probability of the background single channels, delayed average time of the close-state time
Wollman et al. 1996 [30]	Fetal brain cells	Laser, He–Ne	632.8 nm	0.3 mW, 8 min; once, two, or three times irradiation with 24 h interval	Using 2 and 3 times irradiation regimens: enhanced the appearance of brain cells around the treated aggregates
Iwase et al. 1996 [31]	Hippocampal brain slices	Laser, He–Ne (Model Soft Laser 632, manufactured by World Wide Laser Industry, SA Geneva)	632.8 nm	6 mW	Increased the time required for loss of excitability and increased recovery from the ischemic
Wollman and Rochkind 1988 [32]	Adult brain cells	Laser, He–Ne	632.8 nm	3.6 J/cm², 8 min, two times irradiation with 24 h interval	Caused a significant amount of sprouting of cellular processes outgrowth in microexplants

(continued)

Table 8.1 (continued)

Study/Year	Cell types	Light source	Wavelengths	Irradiation parameters	Findings
Wong-Riley et al. 2001 [4]	Cultured rat cortical neurons (TTX-induced neurotoxicity)	LEDs, GaAlAs	670 nm	50 mW/cm^2, 4 J/cm^2, 80 s, CW	Increased CCO activity in all three metabolic categories of neurons (daily irradiation for 5 days); increased CCO activity in darkly reactive cell type (a single irradiation)
Duan et al. 2003 [33]	PC12 cell (Aβ_{25-35}-induced neurotoxicity)	LEDs, self-made GaAlAs	640 nm	0.05–1 mW/cm^2, 30–60 min, single irradiation, CW	At 0.09 mW/cm^2 and 60 min diminished apoptosis and attenuated DNA fragmentation
Jou et al. 2004 [16]	Rat brain astrocytes (RBA-1) cell line	Mercury lamp; Argon laser	Mercury lamp (450–490 nm); Argon (488 nm)	Mercury lamp: 100 mW, 2 min Argon laser: 1 mW/cm^2, 1 min	Increased mitochondrial ROS and Ca^{2+} levels
Byrnes et al. 2005 [34]	Olfactory ensheathing cells	Laser, Thor DDII Laser, Thor International, LTD (Amersham, Bucks, UK)	810 nm	127 mW, 0.2, or 68 J/cm^2; with corresponding duration of 4 s or 9 min and 21 s, respectively	0.2 J/cm^2 significantly increased BDNF, collagen, and GDNF gene expression, while 68 J/cm^2 did not produce significant change; both light fluencies increased olfactory ensheathing cells proliferation on day 7 post-irradiation, but not day 3
Wong-Riley et al. 2005 [5]	Cultured rat visual cortical neurons (KCN-induced neurotoxicity)	LEDs, Quantum Devices, Inc. (Barnaveld, WI, USA)	670, 728, 770, 830, or 880 nm	50 mW/cm^2, 4 to 30 J/cm^2, 80 to 600 s, CW	670 and 830 nm increased CCO activity and ATP content back to control levels compared to 728, 880, and 770 nm (each at 4 J/cm^2) 670 nm: pre-irradiation at 30 J/cm^2 reduced cell death
Liang et al. 2006 [17]	Cultured rat visual cortical neurons (KCN-induced neurotoxicity)	LEDs, Quantum Devices, Inc. (Barnaveld, WI, USA)	670 nm	50 mW/cm^2, 30 J/cm^2, single irradiation, CW	Pre-irradiation reduced cell death (100 μM of KCN) and (300 μM of KCN); reduced number of ssDNA-positive neurons (100 μM of KCN) and (300 μM of KCN); reduced caspase-3 and Bax levels, and increased Bcl-2 levels (both 100 and 300 μM of KCN); reduced ROS production (300 μM of KCN)
Oron et al. 2007 [35]	Cultured human neuronal cells	Laser, GaAs, Photothera, Inc. (Carlsbad, CA, USA)	808 nm	600 mW, 50 mW/cm^2, 0.05 J/cm^2, 1 s	Increased ATP content at 10 min post-irradiation

(continued)

Table 8.1 (continued)

Study/Year	Cell types	Light source	Wavelengths	Irradiation parameters	Findings
Higuchi et al. 2007 [36]	PC12 cells	LEDs, Ikegami Tsusho Co., Ltd., (Tokyo, Japan); Hamamatsu Photonics K.K., (Shizuoka, Japan)	455, 470, 525, 600, 630, 880, or 945 nm	Various irradiances	525 nm: suppressed neurite outgrowth at 0.5 and 0.75 mW/cm^2; Nerve growth ratio at 0.25, 0.5, or 0.75 mW/cm^2 was lower than that in 1 and 1.8 mW/cm^2 for all wavelengths; 2.0 mW/cm^2 showed even greater suppression of neurite outgrowth almost for all wavelengths
Zhang et al. 2008 [37]	PC12 cells ($A\beta_{25-35}$-induced neurotoxicity)	Laser, HN-1000 (Guangzhou, China)	632.8 nm	5 mW, 0.52 mW/cm^2, 0.156, 0.312, 0.624, or 1.248 J/cm^2; with corresponding duration of 5, 10, 20, or 40 min, respectively	Increased cell proliferation (at 0.156, 0.312, and 0.624 J/cm^2), inhibited chromatin condensation in the cells (only at 0.156 J/cm^2); activated PKC and diminished apoptosis by decreasing levels of Bax/Bcl-xl mRNA ratio (only at 0.156 J/cm^2)
Liang et al. 2008 [6]	Cultured rat occipital cortical and striatal neurons (KCN- or MMP$^+$- or rotenone-induced neurotoxicity)	LEDs, Quantum Devices, Inc. (Barnaveld, WI, USA)	670 nm	50 mW/cm^2, 4 J/cm^2, 80 s, 1 to 4 ×/day, CW	KCN: reduced apoptosis (1 irradiation) and (2 irradiations); reduced ROS production (2, 3, and 4 irradiations); reduced NO production (2 and 3 irradiations); reduced nitrotyrosine expression (2 irradiations); highest increase in CCO activity and ATP level (2 irradiations); MPP$^+$: twice a day irradiation suppressed ROS and NO generation, increased ATP level and attenuated apoptosis in both types of neurons; Rotenone: twice a day irradiation reduced apoptosis, ROS and NO levels, and increased ATP level in both types of neurons
Ying et al. 2008 [38]	Cultured rat visual cortical and striatal neurons (Rotenone- or MPP$^+$-induced neurotoxicity)	LEDs, Quantum Devices, Inc. (Barnaveld, WI, USA)	670 nm	50 mW/cm^2, 4 J/cm^2, 80 s, CW	Rotenone: LED irradiation and pre-irradiation decreased apoptosis in both types of neurons; MPP$^+$: LED irradiation and pre-irradiation decreased apoptosis in both types of neurons; LED irradiation and pre-irradiation increased ATP content in striatal neurons

(continued)

Table 8.1 (continued)

Study/Year	Cell types	Light source	Wavelengths	Irradiation parameters	Findings
Rochkind et al. 2009 [39]	Rat embryos brain	Laser	780 nm	10, 30, 50, 110, 160, 200, 250 mW; 1, 4, or 7 min	Accelerated nerve cell sprouting and cell migration (at 50 mW for 1 or 4 min); resulted in thick elongated fibers (at 50 mW for 4 or 7 min); exhibited much larger neurons (at 50 mW for 1 min); exhibited large size neurons with a dense branched interconnected network of neuronal fibers (at 50 mW for 1 or 4 min)
Lim et al. 2009 [8]	Cultured SH-SY5Y cells (sodium nitroprusside-induced neurotoxicity)	LEDs, Biophoton Co. (Gwangju, Korea)	635 nm	5 mW/cm^2, 60 min	Inhibited mitochondrial-dependent apoptotic pathway via suppression of cytochrome c release, Bax protein down-regulation, and caspase-9 and caspase-3 dysfunctions; protected against neuronal oxidative damage via blocking the mitochondrial apoptotic pathway induced by elevated ONOO$^-$ synthesis as well as NO and ROS production
Giuliani et al. 2009 [24]	PC12 cell (H$_2$O$_2$-induced neurotoxicity)	Laser, SANYO DL3149-055A, (RGM, Genoa, Italy)	670 nm	0.005 or 0.011 mW/cm^2; 0.11, 0.22, 5.06 or 10.12 J/cm^2; 20 or 900 s, single irradiation, PW at 100-Hz with DC of 1% or 50%	Enhanced axonal protection via stimulation of NGF-induced neurite outgrowth; rescued MMP (at all fluencies); increased cell viability (at 0.11 and 0.22 J/cm^2)
Hymer et al. 2009 [40]	Cultured rat pituitary cells	LEDs, Quantum Devices Inc. (Barnaveld, WI, USA)	670 nm	50 mW/cm^2, 4 J/cm^2, 80 s	Up-regulated release of growth hormone from primary rat pituitary cell cultures (in first and second days of irradiation); up-regulated release of growth hormone from rat hemi-pituitary glands
Trimmer et al. 2009 [41]	PD cybrid cells	Laser, Acculaser, PhotoThera, Inc. (Carlsbad, CA, USA)	810 nm	50 mW/cm^2, 2 J/cm^2, 40 s, single irradiation, CW	Increased total distance traveled and velocity of mitochondria at 2 h post-irradiation
Yang et al. 2010 [15]	Primary astrocytes (Aβ$_{1-42}$-induced neurotoxicity)	Laser, He-Ne	632.8 nm	1.5 mW/cm^2, 16.2 J/cm^2, 3 h, single irradiation, CW	Decreased oxidative stress burden via suppression of superoxide anion production, NADPH oxidase; and phosphorylation of cPLA$_2$; inhibited pro-inflammatory markers including IL-1β and iNOS

(continued)

Table 8.1 (continued)

Study/Year	Cell types	Light source	Wavelengths	Irradiation parameters	Findings
Sharma et al. 2011 [9]	Cultured mouse cortical neurons	Laser, Photothera, Inc. (Carlsbad, CA, USA)	810 nm	25 mW/cm², 0.03, 0.3, 3, 10, or 30 J/cm², single irradiation, CW	Highest increase in mitochondrial ROS (at 3 and 30 J/cm²); increased intracellular NO (at 0.3 J/cm²); increased MMP (at 0.3 and 3 J/cm²); increased intracellular Ca²⁺ (at 3 J/cm²); increased intracellular ATP (at 3 J/cm²)
Saito et al. 2011 [42]	PC12 cell	Laser, LD 15, Dentek Laser Systems Production, (Vienna, Austria)	810 nm	10 W, 5 or 20 J/cm²; with corresponding duration of 1.26 or 5.04 s, respectively	Decreased cell numbers after 24 and 48 h (at 20 J/cm²); increased neurite outgrowth after 24 and 48 h (at 5 J/cm²); increased expression of neurofilament and β-tubulin proteins after 48 h (at 5 and 20 J/cm²); enhanced phospho-p38 expression after 1 to 3 h post-irradiation (only at 5 J/cm²)
Sommer et al. 2012 [43]	SH-EP and PC12 cells (Aβ₄₂-induced neurotoxicity)	Laser	670 nm	17.36 mW/cm², 1 J/cm², 1 min, single irradiation, PW at 1-Hz	Aβ₄₂-free SH-EP cells: increased ATP levels; SH-EP cells: reduced intracellular Aβ₄₂ aggregate amounts; increased cell proliferation; PC12 cells: small decrease in ATP levels in Aβ₄₂-challenged
Choi et al. 2012 [44]	Cultured rat cortical neurons (OGD-induced neurotoxicity)	LEDs, QRAY, Inc. (Seoul, Korea)	710 nm	50 mW/cm², 4 J/cm², 4 min, 1 to 4 × within 8 h at 2 h intervals for 7 days, CW	Enhanced cell protection; promoted neurite outgrowth and synaptogenesis mediated by MAPK activation
Zhang et al. 2012 [45]	PC12 cells (Aβ₂₅₋₃₅-induced neurotoxicity)	Laser, HN-1000 (Guangzhou, China)	632.8 nm	5.4 mW, 6.89 mW/cm², 2 J/cm²	Protected cells against apoptosis through activation of Akt and subsequent inhibition of YAP translocation from cytoplasm to nucleus; inhibited expression and activation of Bax through Akt/YAP/p73 pathway
Song et al. 2012 [14]	Murine microglia-like cell line BV-2 or primary microglia	Laser, HN-1000, Laser Technology Application Research Institute Co., Ltd. (Guangzhou, China)	632.8 nm	64.6 mW/cm²; 3, 5, 10, 20, 25, or 50 J/cm², with corresponding duration of 0.8, 1.33, 2.66, 5.32, 6.66, or 13.33 min, respectively	In both cell types: induced the lowest microglia-mediated neurotoxicity (at 20, 25, and 50 J/cm² with a maximize effects in 20 J/cm²); inhibited iNOS protein expression (at fluencies greater than 10 J/cm² with a maximize effects in 50 J/cm²); inhibited lipopolysaccharide-activated microglia-mediated neuroinflammation and enhanced its phagocytic activity through activation of Src/PI3K/Akt/Rac1 signaling pathway (at 20 J/cm²)

(continued)

Table 8.1 (continued)

Study/Year	Cell types	Light source	Wavelengths	Irradiation parameters	Findings
Liang et al. 2012 [46]	SH-SY5Y, PC12, and HEK293T cells; ($A\beta_{25-35}$-induced neurotoxicity)	Laser, HN-1000 (Guangzhou, China)	632.8 nm	12.74 mW/cm^2, 2 J/cm^2, single irradiation, CW	In all cell types: decreased apoptosis via Akt/GSK3b/b-catenin pathway
Fukuzaki et al. 2013 [47]	Human-derived glioblastoma cells	Laser, SUWTECH, LDC-2500 (China)	532 nm	60 mW, 845 mW/cm^2, 10.1, 20.3, or 30.4 × 10^2 J/cm^2; with corresponding duration of 20, 40 or 60 min; CW	Increased cell proliferation at 48 h post-irradiation through elevation of Akt expression mediated by suppression of PTEN production (at 20.3 and 30.4 × 10^2 J/cm^2)
von Leden et al. 2013 [10]	Murine microglia-like cell line BV-2 or primary microglia	Laser, (LabTHOR, Stuarts Draft, VA, USA)	808 nm	50 mW, 7.1 mW/cm^2; 0.2, 4, 10, or 30 J/cm^2; with corresponding duration of 28, 565, 1413, or 4239 s, respectively	Increased expression of M1 marker, CD86, in microglia (at fluencies between 4 and 30 J/cm^2); increased expression of M2 phenotype markers, CD206 and TIMP1 (at fluencies between 0.2 and 10 J/cm^2); increased NO expression in BV2 cells at 24 h post-irradiation (at 4 and 30 J/cm^2); increased ROS production in primary microglia at 2 or 24 h post-irradiation (at 10 and 30 J/cm^2); no significant effect on TNF-α, IL-1β, and IL-6 expressions in primary or BV2 microglia at 2 or 24 h post-irradiation; up-regulated modulatory cytokine and chemokine levels in microglial cultures as evidenced by increased MCP-1 expression in BV2 cells (at 0.2 J/cm^2) and increased TIMP1 (at 0.2, 4, and 10 J/cm^2); increased number of neurites (at 0.2 and 30 J/cm^2); increased the length of neurites (at 4 J/cm^2)
Huang et al. 2013 [18]	Cultured mouse cortical neurons (H$_2$O$_2$- or CoCl$_2$- or rotenone-induced neurotoxicity)	Laser, Photothera, Inc. (Carlsbad, CA, USA)	810 nm	20 mW/cm^2; 3 J/cm^2, 150 s, single irradiation, CW	Increased cell viability (at 10 and 20 M μ of H$_2$O$_2$, 0.2, 0.5, 1, and 2 mM of CoCl$_2$, and 0.2, 2, and 5 M μ of rotenone); decreased mitochondrial and cytoplasmic ROS production, and increased MMP (at 500 M μ of CoCl$_2$, 20 M μ of H$_2$O$_2$, and 200 nM of rotenone)

(continued)

Table 8.1 (continued)

Study/Year	Cell types	Light source	Wavelengths	Irradiation parameters	Findings
Meng et al. 2013 [48]	SH-SY5Y cell and mice hippocampal primary neuron ($A\beta_{25-35}$ and $A\beta_{1-42}$-induced neurotoxicity)	Laser, HN-1000, Laser Technology Application Research Institute (Guangzhou, China)	632.8 nm	12.74 mW/cm^2; 0.5, 1, 2, or 4 J/cm^2; with corresponding duration of 0.7, 1.25, 2.5, and 5 min in the dark, respectively; single irradiation, CW	At 2 J/cm^2: promoted cell survival and improved dendrite growth atrophy through up-regulation of BDNF mediated by activation of ERK/CREB signaling pathway
Li et al. 2014 [49]	Cultured rat primary neurons (OGD-induced neurotoxicity)	LEDs	660 nm	60 mW/cm^2	Promoted bone marrow mesenchymal stem cell migration toward primary neurons at 40 h post-irradiation
Chen et al. 2014 [7]	Primary mouse cortical neurons (CoCl$_2$-induced neurotoxicity)	LEDs	808 nm	25 mW/cm^2, 2 J/cm^2, 80 s, once a day for 3 days	Increased cell viability improved morphology of neurons; increased CCO activity and ATP contents
Duggett and Chazot 2014 [50]	Cath.a-differentiated cells ($A\beta_{1-42}$-induced neurotoxicity)	LEDs, Virulite Distribution Ltd (UK)	1068 nm	5 mW/cm^2, 5 sets of 3 min irradiation (with 30 min interval) for 3 days, PW at 600-Hz, with DC of 300 s μ	Decreased cell death (3.5–25 μ M of $A\beta_{42}$)
Huang et al. 2014 [11]	Primary mouse cortical neuron (glutamate, NMDA, or kainite-induced neurotoxicity)	Laser, Photothera, Inc. (Carlsbad, CA, USA)	810 nm	25 mW/cm^2, 3 J/cm^2, 2 min, single irradiation, CW	Glutamate: increased neuronal survival (at 30 μM); increased ATP levels and MMP, and decreased NO and intracellular Ca^{2+} levels; NMDA: increased neuronal survival (at 100 μM); increased ATP levels and MMP, as well as decreased intracellular Ca^{2+} levels; Kainite: increased neuronal survival (at 50 μM); increased ATP levels and MMP, as well as decreased ROS and NO and intracellular Ca^{2+} levels

(continued)

Table 8.1 (continued)

Study/Year	Cell types	Light source	Wavelengths	Irradiation parameters	Findings
Bungart et al. 2014 [51]	Primary rat cortical astrocytes (Aβ-induced neurotoxicity)	Bioluminescence Resonance Energy Transfer to Quantum Dots, Zymera (San Jose, CA, USA)	800 nm	NR	Decreased superoxide anion production; reduced inflammatory markers of IL-1β and iNOS
Burland et al. 2014 [52]	Dorsal Root Ganglion neurons	LEDs, LS E63F, Osram (Regensburg, Germany)	645 nm	20 mW, 11.3 mW/cm^2, 2.72 J/cm^2, 4 min, single irradiation, CW	Accelerated neurite growth in non-injured neurons and in axotomized neurons
Jiang et al. 2014 [19]	PC12 cells (OGD-induced neurotoxicity)	LEDs, Biological Engineering Institute of Chongqing University (Chongqing, China)	660 nm	30 mw/cm2; for various irradiation duration of 4, 6, 10, 12, or 14 h	Decreased MMP, ROS, and apoptosis rates (with 6 h irradiation); decreased protein and mRNA expression of Bax and increased Bcl-2 mRNA and protein levels (with 6 h irradiation)
Fukuzaki et al. 2015 [53]	Neural stem/progenitor cell derived	Laser, SUWTECH, LDC-2500 (China)	532 nm	60 mW, 845 mW/cm^2, 10.1, 20.3, or 30.4 × 10^2 J/cm^2; with corresponding duration of 20, 40 or 60 min; CW	Increased cell proliferation (at 30.4 × 10^2 J/cm^2); promoted migration of NSPCs through increased Akt expression
Renno et al. 2015 [54]	Olfactory ensheathing cell	Laser, Smart Laser Medilaze, Adlaser Pty Ltd. (New South Wales, Australia)	830 nm	30 mW, 10 J/cm^2, 33 s, single irradiation, CW	Decreased proliferation of olfactory ensheathing cell on glass–ceramic discs and increased proliferation of olfactory ensheathing cell on collagen scaffolds; decreased cell growth on the Biosilicate scaffolds and increased cell proliferation on collagen scaffolds

(continued)

Table 8.1 (continued)

Study/Year	Cell types	Light source	Wavelengths	Irradiation parameters	Findings
Dong et al. 2015 [20]	Cultured SH-SY5Y cells (CoCl₂-induced neurotoxicity)	LEDs, PhotoMedex (Horsham, PA, USA)	830 nm	0.1, 0.5, 1, 3, or 10 J/cm², CW	Increased cell viability and ATP production (at 3 and 10 J/cm²); decreased lactate production at 18 h post-toxin treatment (at 3 J/cm²); decreased ROS production and increased MMP; reduced cytochrome c leakage and diminished caspase-3 activation; suppressed apoptosis (at 3 J/cm²)
Giacci et al. 2015 [55]	PC12 cells (glutamate-induced neurotoxicity)	Non-coherent Xenon light	440, 550, 670, or 810 nm	0.008, 0.041, 0.19, or 0.38 J/cm²	No significant effects on the H₂O₂ with all fluencies at 670 nm
Zheng et al. 2015 [12]	Primary Dorsal Root Ganglion neurons	Laser, Newport Corporation (Irvine, CA, USA)	658 nm	26.85 mW; 2, 6, or 16 J/cm²; with corresponding duration of 3, 9 or, 24 min, respectively	Increased NO release shortly after irradiation (at 6 J/cm²); increased (at 6 J/cm²) and decreased (at 16 J/cm²) NOS levels
Wang et al. 2015 [56]	Dorsal Root Ganglion neurons	Laser, Semiconductor, Model: SB2007047, Shanghai University of TCM (Shanghai, China)	657 nm	35 mW, 18 mW/cm², 21 J/cm², CW	Attenuated extracellular ATP by up-regulating ecto-ATPase activity
Yu et al. 2015 [13]	Cultured mouse cortical neurons (OGD-induced neurotoxicity)	Laser, Photothera, Inc. (Carlsbad, CA, USA)	810 nm	25 mW/cm², 3 J/cm², 2 min, single irradiation, CW	Decreased NO production and nNOS activity (at 5 and 30 min post-irradiation); decreased NO donor SNAP-induced neuron death; promoted Akt and Bcl-2 expression (at 1 and 2 h); ameliorated Bax and BAD expression (at 1 and 2 h); suppressed caspase-3 and cleaved caspase-3 expression (at 2 h)

(continued)

Table 8.1 (continued)

Study/Year	Cell types	Light source	Wavelengths	Irradiation parameters	Findings
Choi et al. 2015 [57]	Cultured goldfish brain neurons	LEDs, Daesin LED Co. (Kyunggi, Korea); White fluorescent light	450, 530, or 630	0.9 W/m^2	450 and 530 nm: decreased GnIH and MT1 mRNA expression levels; 630 nm and white fluorescent light: decreased expression levels of Kiss1 and its receptor, GPR54
Zhu et al. 2017 [21]	Mouse neural stem cells	Laser	635 nm	750 mW: 10.95, 21.9, 43.8, or 65.7 J; with corresponding duration of 15, 30, 60, or 90 s, respectively	Increased cell proliferation at 24 h post-irradiation (at 10.95 J); inhibited cell proliferation (at 43.8 and 65.7 J); increased intracellular ROS (at 10.95 J); promoted neural stem cells differentiation
Santos et al. 2017 [22]	Cultured subventricular zone cells	Laser, Z-Laser Optoelektronik Gmbh (Freiburg, Germany)	405 nm	300 mW/cm^2, 9 or 18 J/cm^2	At 18 J/cm^2: increased mitochondrial ROS levels 30 min post-irradiation, reaching its maximum at 1 h and reverting back to basal levels at 7 h; light-induced increase in mitochondrial ROS was NADPH oxidase (Nox)-dependent, with significant up-regulation of Nox4 (at 1 h post-irradiation) as well as Nox1 and Rac1 (both at 3 h post-irradiation) levels; increased cytosolic peroxide levels at 3 h post-irradiation peaking at 7 h, and a reverting to basal levels at 36 h; activated b-catenin (at 2 h post-irradiation); induced neuronal differentiation (at 7 days post-irradiation); up-regulated RARa (at 12 h post-irradiation)
Yan et al. 2017 [26]	Dorsal Root Ganglion neurons	Laser, HN-1000, Laser Technology Application Research Institute (Guangzhou, China)	632.8 nm	12.74 mW/cm^2; 0.5, 1, 1.9, and 3.8 J/cm^2; with corresponding duration of 0.7, 1.25, 2.5, and 5 min in the dark, respectively; single irradiation, CW	Enhanced cell viability and neuritogenesis through induction of BDNF mRNA expression by increasing of Ca^{2+} influx, phosphorylated levels of CREB and ERK proteins

(continued)

Table 8.1 (continued)

Study/Year	Cell types	Light source	Wavelengths	Irradiation parameters	Findings
Gu et al. 2017 [58]	Cultured SH-SY5Y cells (MPP$^+$-induced neurotoxicity)	Laser, HN-1000, Laser Technology Application Research Institute (Guangzhou, China)	632.8 nm	10 mW, 12.74 mW/cm^2; 1, 2, or 4 J/cm^2; with corresponding duration of 1.25, 2.5, or 5 min, respectively	Increased cell viability (at 2 and 4 J/cm^2); At 2 J/cm^2: promoted cell survival and dopamine release through activation of ERK/CREB/VMAT2 pathway
Zhang et al. 2018 [59]	Cultured neuronal N2a cell line	NR	630 nm	0.5 mW/cm^2	Increased cell viability; enhanced the activities of formaldehyde dehydrogenase and catalase; reversed formaldehyde-inhibited catalase activity and H$_2$O$_2$-inhibited formaldehyde dehydrogenase activity (with 20 and 40 min irradiation)
Fan et al. 2018 [60]	BV2 microglia cells (lipopolysaccharide-induced neurotoxicity)	White LED, Intertek (London, UK)	411 to 777 nm	40 to 50 Hz	Prevented lipopolysaccharide-induced cell death; exerted neuroinflammatory action via decrease of TNF-α and IL-6 levels; inhibited ERK and p38 phosphorylation levels induced by lipopolysaccharide
Amaroli et al. 2018 [61]	Slice of cerebral cortex	Laser, AB2799 hand-piece, Doctor Smile–LAMBDA (Spa–Vicenza, Italy)	808 nm	1 W, 1 W/cm^2, 60 J/cm^2, 60 s, single irradiation, CW	Evoked a glutamate efflux; released glutamate from synaptosomes and gliosomes;
Levchenko et al. 2018 [62]	Primary rat cortex neuron	Laser, LJU0808T020; Lumix (Germany)	808 nm	50 mW/cm^2: 0.3, 3, 10, or 30 J/cm^2	Increased lipid contents in the cytoplasm (at all applied fluencies; at 2, 6, and 24 h post-irradiation); increased the number of lipid droplets (LDs) per cell; increased mitochondrial ROS levels (at all applied fluencies)

(continued)

Table 8.1 (continued)

Study/Year	Cell types	Light source	Wavelengths	Irradiation parameters	Findings
Rhee et al. 2019 [25]	Human brain cortical neuron cell line HCN-2 (ouabain-induced neurotoxicity)	LEDs, WON Technology Co., Ltd. (Korea)	660 nm	50 mW, 5.2 mW/cm^2; 0.78, 1.56, 3.12, 6.24, or 9.36 J/cm^2; with corresponding duration of 150, 300, 600, 1200, or 1800s, respectively	Increased cell survival (at 1.56, 3.12, and 6.24 J/cm^2); increased Na/K-ATPase activity (at 3.12 J/cm^2); increased relative ADP/ATP ratio (at 1.56, 3.12, and 6.24 J/cm^2); decreased expression of Src and Ras as well as Na, K-ATPase; modulated MAPK signaling through inhibition of phosphorylation of ERK and p38, but not JNK; decreased Na$^+$-dependent intracellular Ca^{2+} levels (at 15–60 min post-irradiation); increased MMP
Silveira et al. 2019 [63]	Mitochondrial suspension from adult rat brain	Laser, Ibramed Equipamentos Médicos Ltda (Amparo, Brasil)	660 nm	10, 30, or 60 J/cm^2; with corresponding duration of 20, 60, or 120 s, respectively	At 5 min post-irradiation: increased only the activity of complex IV (at all applied fluencies); At 60 min post-irradiation: increased activities of complex IV (at all applied fluencies) and complex II (at 60 J/cm^2)
Argibay et al. 2019 [64]	C17.2 immortalized mouse neural progenitor cell line	LEDs, Quantum Spectralife 830 nm LED, Quantum Devices Inc. (WI, USA)	830 nm	1, 5, or 10 mW; 0.2 to 6 J/cm^2: 6, 12, or 18 min; 2 consecutive days	Increased cellular proliferation (using 1 mW: at 0.2, 0.4, and 0.6 J/cm^2: using 5 mW: at 3 J/cm^2); no significant effects on cell viability
Heo et al. 2019 [65]	Hippocampal cell line (HT-22) and mouse organotypic hippocampal tissues	LEDs, LED4D067, Thorlabs Inc., Newton (NJ, USA)	660 nm	1 W, 20 mW/cm^2, 3 J/cm^2, PW at 2.5-Hz	Inhibited intracellular oxidative stress and cell viability; increased BDNF expression in hippocampal cells through the activation of ERK and CREB signaling pathways; increased the expression of BDNF in the mouse hippocampus; enhanced the activity of the antioxidant enzymes glutathione peroxidase (GPx), superoxide dismutase (SOD1), and glutathione reductase (GR)

(continued)

Table 8.1 (continued)

Study/Year	Cell types	Light source	Wavelengths	Irradiation parameters	Findings
Zhang et al. 2019 [66]	APP/PS1 neurons and SH-SY5YAPPswe cells	Laser, HN-1000 (Guangzhou, China)	632.8 nm	10 mW; 0.5, 1, 2, and 4 J/cm^2; with the corresponding duration of 0.7, 1.25, 2.5, and 5 min, respectively	Increased ADAM10 expression and decreased BACE1 expression in both APP/PS1 neurons and SH-SY5Y-APPswe cells; regulated ADAM10 and BACE1 expressions in a dose-dependent manner with statistically significant increases in ADAM10 levels and decreases in BACE1 levels observed at the dose of 0.5, 1, 2, and 4 J/cm^2; decreased mRNA expression of BACE1; up-regulated ADAM10 through a SIRT1-coupled RARβ-dependent transcription and down-regulates BACE1 through a SIRT1-coupled PGC-1α-dependent transcription; enhanced SIRT1 activity via the cAMP/PKA pathway; enhanced mitochondrial photoacceptor CCO activity, increases ATP and cAMP levels, and further activates PKA/SIRT1 pathway in SH-SY5Y-APPswe cells
Luisa et al. 2019 [67]	Cultured SH-SY5Y cells	Laser, K-Laser Blue series, K-laser d.o.o. (Sežana, Slovenia)	445 or 970 nm	445 nm: 0.1 W/cm^2, 3 J/cm^2, PW at 5-Hz with DC of 50%, on/off time duration 0.1 s; 970 nm: 0.1 W/cm^2, 6 J/cm^2, PW at 5-Hz	Both 445 and 970 nm: increased length and number of neuritis; 970 nm: increased levels of ROS (at 2 days post-irradiation) and ATP (at 4 days post-irradiation) as well as decreased levels of ROS (at 4 days post-irradiation); 445 nm: decreased levels of ROS (at 4 days post-irradiation); No significant effects on the expression of β-tubulin III by irradiation with both wavelengths

Abbreviations: Aβ, Amyloid beta; ADP, Adenosine diphosphate; Akt, Protein kinase B; ATP, Adenosine triphosphate; BAD, Bcl-2-associated death promoter; Bax, Bcl-2-associated X protein; Bcl-2, B-cell lymphoma-2; BDNF, Brain-derived neurotrophic; CCO, Cytochrome c oxidase; cPLA$_2$, Cytosolic phospholipase A$_2$; CREB, CAMP responsive element binding; CW, Continuous wave; DC, Duty cycle; DNA, Deoxyribonucleic acid; ERK, Extracellular signal-regulated kinase; GaAlAs, Gallium aluminum arsenide; GaAs, Gallium arsenide; GABA, Gamma-Aminobutyric acid; GDNF, Glial cell line-derived neurotrophic factor; GnIH, Gonadotropin-inhibitory hormone; GSK3b, Glycogen synthase kinase-3β gene; He–Ne, Helium–neon; IL, Interleukin; iNOS, Inducible nitric oxide; JNK, C-Jun N-terminal kinase; KCN, Potassium cyanide; Kiss1, Kisspeptin-1; LEDs, Light emitting diodes; MAPK, Mitogen-activated protein kinase; MMP, Mitochondrial membrane potential; MMP$^+$, 1-Methyl-4-phenylpyridinium ion; mRNA, Messenger ribonucleic acid; NADPH, Nicotinamide adenine dinucleotide phosphate; NGF, Nerve growth factor; NMDA, N-methyl-D-aspartate; nNOS, Neuronal nitric oxide synthase; NO, Nitric oxide; NR, Not reported; NSPCs, Neural stem/progenitor cells; OGD, Oxygen–glucose deprivation; PD, Parkinson's disease; protein kinase A, PKA; PKC, Protein kinase C; peroxisome proliferator-activated receptor-γ coactivator 1α, PGC-1α; PTEN, Phosphatase and tensin homolog deleted on chromosome ten; PW, Pulsed wave; RARα, Retinoic acid receptor alpha; ROS, Reactive oxygen species; silent information regulator 1, SIRT1; SNAP, S-nitro-N-acetylpenicillamine; ssDNA, Single-stranded DNA; TTX, Tetrodotoxin; VMAT2, Vesicular monoamine transporter 2

8.2.2   Nitric Oxide Modulation

NO is an important signaling molecule in neural cells, which modulates vascular tone and affects cerebral blood flow (CBF). Moreover, NO synthase (NOSs) enzymes such as the inducible isoform (iNOS) catalyze the production of NO and trigger subsequent pathways. It is evident that the change in regional CBF results in changes in behavioral and cognitive performance, and modulation of iNOS and NO levels in the neuro-vascular structures following laser irradiation could explain the changes in cerebral hemodynamics. In addition to catalytic effects, iNOS acts as an inflammatory mediator in neuronal structures.

Various PBM studies have examined changes in NO-related molecules in cultured cells. In an in vitro study by Liang et al. [6] using rat cortical neurons, 670 nm LEDs reduced the levels of NO after KCN-induced neurotoxicity (300 μM). On the other hand, a recent study using a single irradiation with 635 nm LEDs showed no effects on NO levels in SH-SY5Y cells [8]. Increased intracellular NO, was also reported after irradiation with 810 nm laser (0.3 J/cm^2) on cultured mouse cortical neurons [9]. According to a study by von Leden et al. [10], 808 nm laser increased NO levels in BV2 cells at 24 h post-irradiation. Huang et al. [11] reported a decrease in intracellular NO levels in glutamate and kainite intoxicated cortical neurons after irradiation with 810 nm laser. Increased NO release shortly after 658 nm laser PBM (at 6 J/cm^2) has been reported in cultured dorsal root ganglion neurons [12]. In another study, Yu et al. [13] reported decreased NO levels and NOS activity at 5 and 30 min post-irradiation using 810 nm laser (3 J/cm^2) in mouse cortical neurons. Also, the application of 632.8 nm laser to murine microglia-like cell line BV2 or primary microglial cells inhibited iNOS protein expression at fluences greater than 10 J/cm^2, with a maximum effect at 50 J/cm^2 [14]. In another study, 632.8 nm laser irradiation reduced beta-amyloid-induced iNOS up-regulation in primary astrocytes [15]. It seems that the changes in NO or other NO-related molecules caused by PBM, depends on the light parameters. Moreover, the post-irradiation measurement time is an important factor in the experimental design.

8.2.3   Reactive Oxygen Species Modulation

ROS or other types of free radicals are produced by a variety of biochemical reactions taking place inside cells. Oxidative stress has been frequently reported to be involved in the pathophysiology of neuropsychological disorders. There is strong evidence supporting the effects of PBM on the modulation of ROS levels in biological systems. However, the level of ROS inside cells changes in a time- and dose-dependent manner after PBM. This may cause some inconsistencies in research results. An in vitro study using a rat brain astrocyte cell line (RBA-1), has shown that irradiation with 450–490 nm light for 2 min increased mitochondrial ROS levels [16]. However, the application of 670 nm LEDs

(30 J/cm^2, single irradiation) to rat visual cortical neurons reduced ROS production [17]. Evaluation of laser PBM effects on rat cortical and striatal neurons showed that laser irradiation reduced toxin-induced ROS production in the cultured cells [6]. Lim et al. [8] showed the inhibitory effects of 635 nm laser irradiation on ROS production in sodium nitroprusside-intoxicated SH-SY5Y cells. Sharma et al. [9] reported an increase in ROS production after PBM in cultured mouse cortical neurons. In an in vitro study by von Leden et al. [10], an increase in ROS levels was observed in microglial cells at 2 or 24 h post-irradiation. Two other studies from Huang et al. showed decreased ROS levels after 810 nm laser irradiation in mouse cortical neurons [11, 18]. Another PBM study on PC12 cells using 660 nm laser and a study on SH-SY5Y cells using 830 nm laser have also shown the attenuating effects of laser on ROS levels [19, 20]. On the contrary, studies on neural stem cells reported an increase in mitochondrial ROS levels after irradiation with 635 and 405 nm laser [21, 22]. Recently, Luisa et al. [23] reported increased levels of ROS in SH-SY5Y cells 2 days after irradiation with 970 nm laser, and decreased levels of ROS at 4 days post-irradiation. This evidence on the up-regulation or down-regulation effects of PBM on ROS levels in neuronal cells, could allow the selection of better preconditioning or therapeutic regimens for therapeutic applications.

8.2.4 Mitochondrial Membrane Potential

The integrity of biological membranes is a critical item in the maintenance of cellular homeostasis. The MMP is an essential component for maintaining energy levels and ATP production, as well as inhibition of oxidative injury. An impairment of the MMP can affect mitochondrial permeability and lead to cell death. Apoptosis is programmed cell death initiated by a decrease in MMP leading to the release of cytochrome c and activation of caspase-3 protein. Several reports have demonstrated the positive effects of PBM on MMP in various types of cells. Giuliani et al. [24] reported the positive effects of 670 nm laser irradiation on MMP integrity in PC12 cells. Also, several in vitro studies using mouse cortical neurons have shown the protective effect of PBM on MMP integrity after toxin-induced membrane disruption [9, 11, 18]. Dong et al. [20] showed an increase in MMP after the application of 830 nm laser to SH-SY5Y cells. In another study on a human brain cortical neuronal cell line, an increase in MMP was reported after 660 nm LED irradiation [25]. The overall results show the protective effect of PBM on membrane integrity and MMP in the different types of cultured neural cells.

8.2.5 Ca^{2+} Regulation

Intercellular calcium homeostasis plays a regulatory role in the control of cellular/intracellular signaling pathways in neural cells. Controlled increases of Ca^{2+} inside cells are essential for the activation of transcription factors and signaling mediators, such as nuclear factor-κB (NF-κB) and result in long-lasting effects on cells, while calcium overload in cells triggers apoptosis and excitotoxicity-dependent pathways. The regulatory role of PBM on intracellular Ca^{2+} levels has been reported in several in vitro studies. Jou et al. [16] reported increased mitochondrial Ca^{2+} levels after PBM in the RBA-1 cell line. In other studies, increased intracellular Ca^{2+} levels were observed in mouse cortical neurons after 810 nm laser application [9, 11]. Also, Yan et al. [26] reported increased levels of Ca^{2+} after 632.8 nm laser in dorsal root ganglion neurons. Changes in Ca^{2+} levels inside cells depends on the environmental and pathophysiological conditions, and the PBM parameters should be adjusted according to these conditions.

References

1. Sadigh-Eteghad, S., et al. 2017. D-galactose-induced brain ageing model: a systematic review and meta-analysis on cognitive outcomes and oxidative stress indices. *PloS one* 12(8).
2. Salehpour, F., et al. 2018. Brain photobiomodulation therapy: A narrative review. *Molecular Neurobiology* 55 (8): 6601–6636.
3. Hamblin, M.R., and Y.-Y. Huang. 2019. *Photobiomodulation in the Brain: Low-Level Laser (Light) Therapy in Neurology and Neuroscience.* Academic Press.
4. Wong-Riley, M.T., et al. 2001. Light-emitting diode treatment reverses the effect of TTX on cytochrome oxidase in neurons. *NeuroReport* 12 (14): 3033–3037.
5. Wong-Riley, M.T., et al. 2005. Photobiomodulation directly benefits primary neurons functionally inactivated by toxins role of cytochrome c oxidase. *Journal of Biological Chemistry* 280 (6): 4761–4771.
6. Liang, H.L., et al. 2008. Near-infrared light via light-emitting diode treatment is therapeutic against rotenone-and 1-methyl-4-phenylpyridinium ion-induced neurotoxicity. *Neuroscience* 153 (4): 963–974.
7. Chen, X., et al. 2014. Effect and mechanism of 808 nm light pretreatment of hypoxic primary neurons. *International Journal of Photoenergy* 2014.
8. Lim, W., et al. 2009. Inhibition of mitochondria-dependent apoptosis by 635-nm irradiation in sodium nitroprusside-treated SH-SY5Y cells. *Free Radical Biology and Medicine* 47 (6): 850–857.
9. Sharma, S.K., et al. 2011. Dose response effects of 810 nm laser light on mouse primary cortical neurons. *Lasers in Surgery and Medicine* 43 (8): 851–859.
10. von Leden, R.E., et al. 2013. 808 nm wavelength light induces a Dose-D ependent alteration in microglial polarization and resultant microglial induced neurite growth. *Lasers in Surgery and Medicine* 45 (4): 253–263.
11. Huang, Y.Y., et al. 2014. Low-level laser therapy (810 nm) protects primary cortical neurons against excitotoxicity in vitro. *Journal of Biophotonics* 7 (8): 656–664.
12. Zheng, L.-Q., et al. 2015. Modulating nitric oxide levels in dorsal root ganglion neurons of rat with low-level laser therapy. *Optoelectronics Letters* 11 (3): 233–236.
13. Yu, Z., et al. 2015. Near infrared radiation protects against oxygen-glucose deprivation-induced neurotoxicity by down-regulating neuronal nitric oxide synthase (nNOS) activity in vitro. *Metabolic Brain Disease* 30 (3): 829–837.

14. Song, S., F. Zhou, and W.R. Chen. 2012. Low-level laser therapy regulates microglial function through Src-mediated signaling pathways: Implications for neurodegenerative diseases. *Journal of Neuroinflammation* 9 (1): 219.

15. Yang, X., et al. 2010. Low energy laser light (632.8 nm) suppresses amyloid-β peptide-induced oxidative and inflammatory responses in astrocytes. *Neuroscience* 171(3): p. 859–868.

16. Jou, M.-J., et al. 2004. Mitochondrial reactive oxygen species generation and calcium increase induced by visible light in astrocytes. In *Mitochondrial Pathogenesis*, 45–56. Springer.

17. Liang, H., et al. 2006. Photobiomodulation partially rescues visual cortical neurons from cyanide-induced apoptosis. *Neuroscience* 139 (2): 639–649.

18. Huang, Y.Y., et al. 2013. Low-level laser therapy (LLLT) reduces oxidative stress in primary cortical neurons in vitro. *Journal of Biophotonics* 6 (10): 829–838.

19. Jiang, W., et al. 2014. Red photon treatment inhibits apoptosis via regulation of bcl-2 proteins and ROS levels, alleviating hypoxic–ischemic brain damage. *Neuroscience* 268: 66–74.

20. Dong, T., et al. 2015. Low-level light in combination with metabolic modulators for effective therapy of injured brain. *Journal of Cerebral Blood Flow & Metabolism* 35 (9): 1435–1444.

21. Zhu, W., et al. 2017. 3D printing scaffold coupled with low level light therapy for neural tissue regeneration. *Biofabrication* 9 (2): 025002.

22. Santos, T., et al. 2017. Blue light potentiates neurogenesis induced by retinoic acid-loaded responsive nanoparticles. *Acta Biomaterialia* 59: 293–302.

23. Luisa, Z., et al. 2019. Photobiomodulation therapy at different wavelength impacts on retinoid acid–dependent SH-SY5Y differentiation. *Lasers in Medical Science* p. 1–6.

24. Giuliani, A., et al. 2009. Low infra red laser light irradiation on cultured neural cells: Effects on mitochondria and cell viability after oxidative stress. *BMC Complementary and Alternative Medicine* 9 (1): 8.

25. Rhee, Y.-H., et al. 2019. Effect of photobiomodulation therapy on neuronal injuries by ouabain: The regulation of Na, K-ATPase; Src; and mitogen-activated protein kinase signaling pathway. *BMC Neuroscience* 20 (1): 19.

26. Yan, X., et al. 2017. Low-level laser irradiation modulates brain-derived neurotrophic factor mRNA transcription through calcium-dependent activation of the ERK/CREB pathway. *Lasers in Medical Science* 32 (1): 169–180.

27. Wade, P.D., J. Taylor, and P. Siekevitz. 1988. Mammalian cerebral cortical tissue responds to low-intensity visible light. *Proceedings of the National Academy of Sciences* 85 (23): 9322–9326.

28. Tsai, J.-C., and M.-C. Kao. 1991. The biological effects of low power laser irradiation on cultivated rat glial and glioma cells. *Journal of Clinical Laser Medicine and Surgery* 9 (1): 35–41.

29. Karu, T., et al. 1996. He-Ne laser radiation influences single-channel ionic currents through cell membranes: A patch-clamp study. *Lasers in the Life Sciences* 7 (1): 35–48.

30. Wollman, Y., S. Rochkind, and R. Simantov. 1996. Low power laser irradiation enhances migration and neurite sprouting of cultured rat embryonal brain cells. *Neurological Research* 18 (5): 467–470.

31. Iwase, T., et al. 1996. Low power laser irradiation reduces ischemic damage in hippocampal slices in vitro. *Lasers in Surgery and Medicine: The Official Journal of the American Society for Laser Medicine and Surgery* 19 (4): 465–470.

32. Wollman, Y., and S. Rochkind. 1998. In vitro cellular processes sprouting in cortex microexplants of adult rat brains induced by low power laser irradiation. *Neurological Research* 20 (5): 470–472.

33. Duan, R., et al. 2003. Light emitting diode irradiation protect against the amyloid beta 25–35 induced apoptosis of PC12 cell in vitro. *Lasers in Surgery and Medicine: The Official Journal of the American Society for Laser Medicine and Surgery* 33 (3): 199–203.
34. Byrnes, K.R., et al. 2005. Low power laser irradiation alters gene expression of olfactory ensheathing cells in vitro. *Lasers in Surgery and Medicine: The Official Journal of the American Society for Laser Medicine and Surgery* 37 (2): 161–171.
35. Oron, U., et al. 2007. Ga-As (808 nm) laser irradiation enhances ATP production in human neuronal cells in culture. *Photomedicine and Laser Surgery* 25 (3): 180–182.
36. Higuchi, A., et al. 2007. Visible light regulates neurite outgrowth of nerve cells. *Cytotechnology* 54 (3): 181–188.
37. Zhang, L., et al. 2008. Low-power laser irradiation inhibiting Aβ25-35-induced PC12 cell apoptosis via PKC activation. *Cellular Physiology and Biochemistry* 22 (1–4): 215–222.
38. Ying, R., et al. 2008. Pretreatment with near-infrared light via light-emitting diode provides added benefit against rotenone-and MPP+-induced neurotoxicity. *Brain Research* 1243: 167–173.
39. Rochkind, S., et al. 2009. Increase of neuronal sprouting and migration using 780 nm laser phototherapy as procedure for cell therapy. *Lasers in Surgery and Medicine: The Official Journal of the American Society for Laser Medicine and Surgery* 41 (4): 277–281.
40. Hymer, W., et al. 2009. Modulation of rat pituitary growth hormone by 670 nm light. *Growth Hormone & IGF Research* 19 (3): 274–279.
41. Trimmer, P.A., et al. 2009. Reduced axonal transport in Parkinson's disease cybrid neurites is restored by light therapy. *Molecular Neurodegeneration* 4 (1): 26.
42. Saito, K., et al. 2011. Effect of diode laser on proliferation and differentiation of pc12 cells. *The Bulletin of Tokyo Dental College* 52 (2): 95–102.
43. Sommer, A.P., et al. 2012. 670 nm laser light and EGCG complementarily reduce amyloid-β aggregates in human neuroblastoma cells: Basis for treatment of Alzheimer's disease? *Photomedicine and Laser Surgery* 30 (1): 54–60.
44. Choi, D.-H., et al. 2012. Effect of 710 nm visible light irradiation on neurite outgrowth in primary rat cortical neurons following ischemic insult. *Biochemical and Biophysical Research Communications* 422 (2): 274–279.
45. Zhang, H., S. Wu, and D. Xing. 2012. Inhibition of Aβ25–35-induced cell apoptosis by Low-power-laser-irradiation (LPLI) through promoting Akt-dependent YAP cytoplasmic transloca-tion. *Cellular Signalling* 24 (1): 224–232.
46. Liang, J., L. Liu, and D. Xing. 2012. Photobiomodulation by low-power laser irradiation atten-uates Aβ-induced cell apoptosis through the Akt/GSK3β/β-catenin pathway. *Free Radical Biol-ogy and Medicine* 53 (7): 1459–1467.
47. Fukuzaki, Y., et al. 2013. 532 nm low-power laser irradiation recovers γ-secretase inhibitor-mediated cell growth suppression and promotes cell proliferation via Akt signaling. *PLoS One* 8(8).
48. Meng, C., Z. He, and D. Xing. 2013. Low-level laser therapy rescues dendrite atrophy via upreg-ulating BDNF expression: Implications for Alzheimer's disease. *Journal of Neuroscience* 33 (33): 13505–13517.
49. Li, X., et al. 2014. 660 nm red light-enhanced bone marrow mesenchymal stem cell trans-plantation for hypoxic-ischemic brain damage treatment. *Neural Regeneration Research* 9 (3): 236.
50. Duggett, N.A., and P.L. Chazot. 2014. Low-intensity light therapy (1068 nm) protects CAD neuroblastoma cells from β-amyloid-mediated cell death. *Biologie et Médecine* 6 (3): 1.

51. Bungart, B.L., et al. 2014. Nanoparticle-emitted light attenuates amyloid-β-induced superoxide and inflammation in astrocytes. *Nanomedicine: Nanotechnology, Biology and Medicine* 2014. **10**(1): p. 15–17.

52. Burland, M., et al. 2015. Neurite growth acceleration of adult Dorsal Root Ganglion neurons illuminated by low-level Light Emitting Diode light at 645 nm. *Journal of Biophotonics* 8 (6): 480–488.

53. Fukuzaki, Y., et al. 2015. 532 nm low-power laser irradiation facilitates the migration of GABAergic neural stem/progenitor cells in mouse neocortex. *PloS one*, 2015. **10**(4).

54. Renno, A.C.M., et al. 2015. Effect of 830-nm laser phototherapy on olfactory neuronal ensheathing cells grown in vitro on novel bioscaffolds. *Journal of Applied Biomaterials & Functional Materials* 13 (3): 234–240.

55. Giacci, M.K., et al. 2015. Method for the assessment of effects of a range of wavelengths and intensities of red/near-infrared light therapy on oxidative stress in vitro. *JoVE (Journal of Visualized Experiments)* 97: e52221.

56. Wang, L., et al. 2015. Modulation of extracellular ATP content of mast cells and DRG neurons by irradiation: studies on underlying mechanism of low-level-laser therapy. *Mediators of Inflammation* 2015.

57. Choi, C.Y., et al. 2015. Time-related effects of various LED light spectra on reproductive hormones in the brain of the goldfish Carassius auratus. *Biological Rhythm Research* 46 (5): 671–682.

58. Gu, X., et al. 2017. Photoactivation of ERK/CREB/VMAT2 pathway attenuates MPP+-induced neuronal injury in a cellular model of Parkinson's disease. *Cellular Signalling* 37: 103–114.

59. Zhang, J., et al. 2018. Illumination with 630 nm red light reduces oxidative stress and restores memory by photo-activating catalase and formaldehyde dehydrogenase in SAMP8 mice. *Antioxidants & Redox Signaling* 30 (11): 1432–1449.

60. Fan, S., et al. 2018. LED enhances anti-inflammatory effect of luteolin (3', 4', 5, 7-tetrahydroxyflavone) in vitro. *American Journal of Translational Research* 10 (1): 283.

61. Amaroli, A., et al. 2018. Near-infrared laser photons induce glutamate release from cerebrocortical nerve terminals. *Journal of Biophotonics* 11 (11): e201800102.

62. Levchenko, S.M., et al. 2018. Near-infrared irradiation affects lipid metabolism in neuronal cells, inducing lipid droplets formation. *ACS Chemical Neuroscience* 10 (3): 1517–1523.

63. Silveira, P.C.L., et al. 2019. Effects of photobiomodulation on mitochondria of brain, muscle, and C6 astroglioma cells. *Medical Engineering & Physics* 71: 108–113.

64. Argibay, B., et al. 2019. Light-Emitting Diode photobiomodulation after cerebral ischemia. *Frontiers in Neurology* 10: 911.

65. Heo, J.-C., et al. 2019. Photobiomodulation (660 nm) therapy reduces oxidative stress and induces BDNF expression in the hippocampus. *Scientific Reports* 9 (1): 1–8.

66. Zhang, Z., et al. 2020. Activation of PKA/SIRT1 signaling pathway by photobiomodulation therapy reduces Aβ levels in Alzheimer's disease models. *Aging Cell* 19 (1): e13054.

67. Zupin, L., et al. 2019. Photobiomodulation therapy at different wavelength impacts on retinoid acid–dependent SH-SY5Y differentiation. *Lasers in Medical Science* 1–6.

Photobiomodulation Therapy for Dementia

9.1   What is Dementia?

Dementia is an age-related neurodegenerative disorder, employed as a general term for diseases and conditions characterized by gradual and irreversible deterioration of memory and cognitive function due to the dysfunction and degeneration of specific populations of neurons within the brain structures involved in memory and cognition. There are several types of dementia, including Alzheimer's disease (AD), vascular dementia, dementia with Lewy bodies, frontotemporal dementia, and Parkinson's disease dementia [1].

There are many scales that have been extensively administered to patients in dementia treatment trials, which broadly follow guidance from the Alzheimer's Association. The Mini-Mental State Exam (MMSE) or Folstein test is a 30-point questionnaire that has been widely used in clinical and research settings to measure cognitive function. The MMSE is a paper-based test with a maximum score of 30, the lower scores indicating more severe cognitive impairment, while higher scores indicate better cognitive function. In general, MMSE scores address 5 types of cognitive function: (1) orientation (10 points); (2) short-term memory or retention (3 points); (3) attention (5 points); (4) short-term memory or recall (3 points); and (5) language (9 points). According to the Alzheimer's Association, MMSE scores are categorized as follows: severe dementia, 0–9; moderate dementia, 10–18; mild dementia, 19–24; mild cognitive impairment (MCI), 25–27; and normal; 28–30 [2]. The Montreal Cognitive Assessment (MoCA) is also a cognitive screening test originally developed to diagnose MCI patients as a possible prodromal stage predicting the development of AD. It evaluates many cognitive domains, including attention, executive function, memory, orientation, and language. MoCA scores can range from 0 to 30, with 26 and higher considered normal, 22 indicating MCI, and 16 or below indicating dementia [3]. The AD Assessment Scale (ADAS-cog) is another brief neuropsychological assessment used to measure the severity of cognitive symptoms of dementia. ADAS-cog

F. Salehpour et al., *Photobiomodulation for the Brain*, Synthesis Lectures on Biomedical Engineering, https://doi.org/10.1007/978-3-031-36231-6_9

is scored between 0 and 70, where higher scores are indicative of more cognitive dys-function [4]. The ADAS-cog includes direct evaluation of learning (word list), naming (objects), following commands, ideational praxis (mailing a letter), constructional praxis (figure copying), orientation (time, place, person), recognition memory, and remember-ing test instructions. Moreover, the Neuropsychiatric Inventory (NPI) questionnaire is a 144-point informant-based interview that assesses neuropsychiatric domains common in dementia, including frequency, severity, and effect on caregiver distress [5].

9.1.1 Mild Cognitive Impairment

MCI is considered to be the earliest stage of dementia, and is defined as a slight but noticeable and measurable decline in cognitive abilities, including memory and thinking skills, but which does not interfere with daily living. The definition of MCI has evolved over the last decade since its introduction as a new concept by Reisberg and colleagues more than 20 years ago. However, the main criteria have continued to be the same. In 2003, a Key Symposium on MCI defined the internationally agreed criteria for the con-dition, which allows clinicians to identify subjects at an intermediate cognitive state. The newly-proposed definition altered previous concepts of MCI, which restricted the con-dition specifically toward the development of AD, and concentrated only on memory impairment [6]. Accordingly, the following criteria are now applied to diagnose MCI in a susceptible population. "Cognitive concern reflecting a change in cognition reported by patient or informant or clinician (i.e., historical or observed evidence of decline over time), impairment in one or more cognitive domains, preservation of independence in functional abilities, and not demented" are the main features used to distinguish MCI from overt dementia [7]. MCI is a rather common finding in elderly individuals, where its prevalence is somewhere between 15 and 20% in persons 60 years and older [8]. MCI patients are more likely to develop AD or another type of dementia, and sometimes MCI is considered to be a transitional stage between normal aging and dementia. This clinical entity can be caused by numerous ailments, including but not limited to AD, cerebral ischemia and cerebrovascular disease, brain traumatic injury, alcohol or drug abuse, and some psychiatric disorders [9]. MCI is classified into two categories: amnestic which primarily affects memory, and non-amnestic MCI which affects thinking skills but not memory [10, 11].

At a pathophysiological level, MCI is characterized by increased levels of oxidative stress caused by the overproduction of reactive oxygen species (ROS) and impairment of oxidative defense systems. These changes could in principle be tracked in the brain of MCI subjects even before the onset of dementia symptoms [12]. Accordingly, the result-ing oxidative stress causes damage to nuclear and mitochondrial DNA, and subsequent mitochondrial dysfunction [13]. Further studies have suggested that the amyloid-beta (Aβ) peptide can play a crucial role in the initiation and progression of oxidative stress in MCI patients [14]. In addition, it has been shown that subjects, especially those with amnestic MCI, who have low $A\beta_{42}$ levels in the cerebrospinal fluid (CSF), and increased levels of total tau and phosphorylated tau are at an increased risk for progression to dementia than the others [8].

Structural brain imaging studies have shown that the entorhinal cortex and the hippocampus are the first regions affected by MCI. Patients with MCI display smaller hippocampus and entorhinal cortex volumes, white matter abnormalities in the hippocampus and thalamus, and metabolic abnormalities in the hippocampus, medial thalamus, and posterior cingulate [15, 16]. MCI patients show somewhat decreased cerebral blood flow (CBF) in the posterior cingulate gyrus. This pathological finding is also found in the medial precuneus area when compared with healthy control subjects. On the other hand, it was found that MCI patients have increased CBF in the right amygdala, left hippocampus, the rostral tip of the right caudate nucleus, the ventral putamen, and the globus pallidus as compared to healthy controls [17].

Several criteria have been used for the accurate diagnosis of MCI, among which those proposed by Ronald C. Petersen could be used in clinical settings [18, 19]. Mental assessment tests using the MMSE, the MoCA, or the Short Test of Mental Status (STMS) are also useful in the diagnosis of MCI. Nevertheless, clinicians should bear in mind that these screening tests are insufficient to diagnose MCI with certainty [18, 19].

Pharmacologic therapy using drugs, such as donepezil, galantamine, memantine, rivastigmine, and ginkgo or ginkgo biloba has shown to be only minimally effective in MCI. On the other hand, exercise has been found to have a substantial benefit in the treatment of MCI. However, before the initiation of any therapeutic strategy, efforts should be made to direct treatment toward the underlying cause [20].

9.1.2 Alzheimer's Disease

9.1.2.1 The Problem of Alzheimer's Disease

AD was first described by the German physician, Alois Alzheimer in 1907 [21]. By far, AD is the most common cause of dementia. Approximately 5.5 million Americans have Alzheimer's dementia, which will increase to 13.8 million by 2050 due to the constant aging of the population. It has been predicted that by 2050, one new person will be afflicted by AD every 33s, resulting in nearly 1 million new cases per year [22]. The global trend of AD is similar to the USA, and around 106.8 million people in the world will be afflicted by 2050. AD is a progressive disorder and its symptoms have recently been classified into 3 phases or stages: the mild (early) phase is a presymptomatic phase during which, despite normal cognitive function, there is evidence of amyloid deposition; the moderate (middle) phase or symptomatic phase is characterized by MCI along with amyloid deposition and neurodegeneration; the severe (late) phase is characterized by pronounced cognitive impairment which may interfere with daily activity [23, 24].

It has been proposed that amyloid-beta is a crucial pathological factor in AD. This has led researchers to propose that the amyloid hypothesis is the main explanation for the initiation and progression of the disease. The core components of this hypothesis include the accumulation of extracellular senile plaques containing Aβ, intracellular neurofibrillary tangles (NFTs), and synaptic degeneration [25]. Another hypothesis commonly known as the cholinergic hypothesis proposes that cholinergic degeneration (especially in the nucleus of Meynert in the basal forebrain which innervates the hippocampus and cortex and is involved in learning and memory) results in AD-related cognitive impairment and memory loss [26]. Other hypotheses propose that microglial and astrocytic nicotinic acetylcholine receptors (nAChRs) may have a role in AD pathogenesis [27]. It has also been suggested that abnormal inflammatory responses and microglial activation, mitochondrial dysfunction, progressive decreases in neural glucose metabolism, and the reduced synthesis of acetylcholine could contribute to AD pathophysiology [24, 28–30].

No matter which hypothesis turns out to be correct to explain the initiation and progression of AD, the biological changes ignite a signaling cascade, which leads to increased oxidative stress and mitochondrial dysfunction. Oxidative stress is the hallmark of the AD brain, which causes structural and functional damage to cerebral mitochondria and leads to their fragmentation [31]. Accordingly, due to the atrophy of brain regions involved in learning and memory, a range of symptoms and signs develop, such as memory impairment interfering with daily activities, challenges in problem-solving, confusion, problems in writing, reading, and speaking, problems in visual and spatial understanding, and diverse neuropsychiatric symptoms (NPS) [32].

Neuroimaging studies often show evidence of brain atrophy in the medial temporal lobe, particularly in the hippocampus, entorhinal cortex, amygdala, and parahippocampal gyrus, ventricular expansion, reduced total brain volume, as well as decreased glucose consumption and regional blood flow in the temporoparietal cortex [29, 33–35]. Imaging studies have also shown that AD patients have increased CBF in the posterior cingulate gyrus which is in line with the changes observed in MCI patients. The detected decrease in CBF in the respected area could be found in the medial precuneus as opposed to the control subjects. Decreased CBF have also been found in the left inferior parietal, lateral frontal, superior temporal, and orbitofrontal cortices when compared with both MCI patients and healthy controls. Conversely, increased CBF have been shown in the right anterior cingulate gyrus as compared with healthy controls [17].

Several biomarkers have been proposed to allow the early detection of AD. These biomarkers include: CSF assays for tau and amyloid-β isoforms; magnetic resonance imaging (MRI) examination of brain atrophy; positron emission tomography (PET) techniques to evaluate the brain metabolic ratio; PET assays for Aβ and tau deposition in various regions of the brain [36]. However, none of these non-invasive diagnostic approaches has been shown to be definitive in this regard. To establish a definite diagnosis, several criteria have been proposed. In 2007, the International Working Group (IWG) for New Research Criteria for the Diagnosis of AD proposed IWG criteria (then IWG-2) for typical AD (A plus B at any stage). This includes A: specific clinical phenotype, and B: in-vivo evidence of Alzheimer's pathology which has been discussed in detail elsewhere [37].

No cure exists for AD, however, several therapeutic strategies have been proposed to halt its progression and relieve signs and symptoms. Traditionally, cholinesterase inhibitors are prescribed for mild to moderate AD. These drugs such as Razadyne® (galantamine), Exelon® (rivastigmine), and Aricept® (donepezil) can offer some symptomatic relief for AD patients. Thus, more targeted therapies are being developed for this purpose, including vaccination against amyloid-β-peptide, gamma-secretase inhibitors, cholesterol-lowering drugs, metal chelators, and anti-inflammatory medications [38].

Some non-pharmacological treatments have also been developed or are being investigated for AD. A recent study showed that AD patients who had undergone neurofeedback therapy had stable or improved cognitive performance [39]. Other therapeutic strategies including, repetitive transcranial magnetic stimulation (rTMS) have been shown to improve cognitive performance in AD and MCI. These changes could be due to the effects of rTMS on cortical excitability and long-lasting neuroplasticity in the brain [40]. Deep brain stimulation (DBS) could also be an option for the improvement of cognitive performance in AD, however, its effects in this regard are still under investigation [41].

9.1.2.2 Animal Models of Alzheimer's Disease

Animal models are extremely valuable to investigate the pathophysiology of AD and to test new therapeutic approaches. These models can be generally classified into, spontaneous models, pharmacological, chemical, and lesion-induced rodent models, Aβ infusion rodent models, and transgenic models that develop AD. Spontaneous animal models include dogs, cats, (polar) bears, goats, sheep, wolverines, and several nonhuman primate species. However, their use is of limited value due to ethical issues, high costs, and lack of wide availability. Among these models, rodent models such as the senescence-accelerated mouse (SAM) are extremely valuable tools to investigate AD pathology [42]. Genetic models include but are not limited to, PDAPP mice, APP mice, Tg2576 mice, PSAPP mice, Tg-tau mice, and Ttpa−/−APPsw mice. Genetic models can be used to obtain information about the pathophysiology of familial or early onset AD by transfection of a mutant human amyloid precursor protein (APP) [43]. Aβ-based models can lead to neurotoxicity and damage to cholinergic basal forebrain projections; these changes result in impairment of memory function in rodents. Aβ can be introduced into the brain via either intracerebroventricular or intrahippocampal routes, and this mimics AD-like changes in the rodent brain [25]. Nevertheless, the lack of agreement between the results of preclinical AD animal studies and human clinical trials is clearly evident, and should be meticulously addressed in future studies.

9.1.3   Vascular Dementia

Vascular dementia is a heterogeneous group of disorders that are mainly caused by different types of vascular pathology resulting in dementia. The entity encompasses a wide range of disorders, including but not limited to multi-infarct dementia (cortical vascular dementia), small vessel dementia (subcortical vascular dementia), strategic infarct dementia, hypoperfusion dementia, hemorrhagic dementia, hereditary vascular dementia (CADASIL), and AD co-occurring with cardiovascular disease [44]. Studies have found that vascular dementia is the second most common cause of dementia after AD [45]. Several risk factors can predispose the brain to vascular dementia, such as older age, low education level, female gender, typical vascular risk factors, strokes in certain locations, and temporal atrophy both global and medial as seen on structural imaging [46]. A variety of symptoms and signs can be present in vascular dementia. The cognitive disturbances show much more variability compared to other types of dementia, such as AD. It has been found that the MMSE is less sensitive for establishing the diagnosis of vascular dementia compared to AD. However, the MoCA or the vascular dementia assessment scale (VADAS-cog) is more sensitive in this regard [47]. The confirmed diagnosis requires the presence of sufficient cerebrovascular disease seen on brain imaging, which is in agreement with the degree of cognitive impairment in clinical examination [48]. Genetic evaluations are reserved for patients suffering from CADASIL (cerebral autosomal dominant arteriopathy with subcortical infarcts and leukoencephalopathy).

Evaluation and treatment of the comorbidities, which contribute to vascular dementia are of high priority in disease management. Cholinesterase inhibitors and memantine are among the best-studied drugs for this purpose; however, their efficacy is modest in this regard [44]. Newer studies have shown the positive effects of cerebrolysin (a porcine brain-derived peptide) on cognition [49]. However, due to the inconsistent results, this medication should be used with caution in clinical settings [44]. Other studies have shown that a primary intervention by lowering cholesterol is of no value in the prevention of vascular dementia [50]. Environmental enrichment has been also used to decrease cognitive deficits in vascular-type dementia. Evidence has shown that environmental enrichment decreased behavioral impairment and oxidative stress in a rat model of chronic cerebral hypoperfusion [51]. It also increased brain-derived neurotrophic factor (BDNF) and N-methyl-D-aspartate (NMDA) receptor subunit 1 (NR1) levels in the hippocampus, and reversed the cognitive decline in rats with chronic cerebral hypoperfusion [52]. Therefore it seems that the management of vascular dementia should focus on identifying and managing comorbidities, because most drugs show only limited benefits [44].

9.2 Photobiomodulation Therapy for Dementia

9.2.1 Photobiomodulation Therapy for Dementia in Animal Models

Lu et al. [53] investigated the potential therapeutic benefits of NIR PBM therapy in an Aβ-induced rat model of AD. They administrated 808 nm laser light to the $Aβ_{1-42}$-treated AD Wistar rats, and found that daily treatment for 5 days could mitigate neurodegeneration in the hippocampal neurons, and enhance spatial and object recognition memory. Overall, the biochemical results also indicated that low-levels of NIR light (scalp fluence of 3 J/cm^2 per session) significantly improved mitochondrial dynamics, increased mitochondrial membrane potential, lowered oxidized mitochondrial DNA and mitophagy, suppressed apoptosis and neuroinflammation, increased mitochondrial antioxidant capacity, improved CCO activity and ATP levels, as well as inhibited Aβ-induced reactive gliosis and tau hyperphosphorylation [53]. Duggett and Chazot [54] also showed the neuroprotective effect of NIR 1068 nm LEDs with a scalp fluence of 4.5 J/cm^2 could successfully protect CAD neuroblastoma cells from $Aβ_{1-42}$-induced cell death. In another laboratory study in rats, da Luz Eltchechem and colleagues [55] also injected $Aβ_{25-35}$ peptide into the hippocampus, and treated the animals with 627 nm transcranial LEDs with a scalp fluence of 7 J/cm^2, once a day for 3 consecutive weeks. After 21 days of treatment, they found that PBM therapy profoundly reduced the level of Aβ plaques, and improved spatial memory and motor skills as measured by Morris water maze and open field tests. Additionally, a study carried out by Blivet et al. [56] using a Swiss mouse model of $Aβ_{25-35}$–induced AD showed that PBM therapy (combined 625 nm LED plus 850 nm laser, with a skin fluence of 8.4 J/cm^2) applied to either the head or the abdomen could suppress levels of hippocampal inflammatory markers, TNF-α, IL-6, and IL-1β. The treatment also markedly inhibited the activation of astrocytes and microglia.

De Taboada et al. [57] for the first time administered 810 nm laser irradiation transcranially to Aβ-PP transgenic mice. Mice in 3 different PBM groups received fluences of 1.2, 6, or 12 J/cm^2 at the cortical surface using either continuous or pulsed wave mode. Laser treatment was administered 3 times/week for a total of 6 months. The number of Aβ plaques was significantly reduced in the brain tissue by PBM therapy in a dose-dependent manner. PBM treatment also showed a dose-dependent decrease in the amyloid load, soluble AβPPα, and cerebral inflammatory markers such as IL-1β, TNF-α, and TGF-β. In addition to an increase in ATP levels, oxygen consumption, and c-fos expression, spatial learning and memory as evaluated by the Morris water maze were also improved by PBM therapy. It should be noted that PBM at 6 J/cm^2 in pulsed mode produced greater therapeutic benefits as compared to the other treatment regimens [57]. Furthermore, Purushothuman et al. [58] reported the therapeutic benefits of PBM therapy in two different transgenic mouse models of AD, namely the APPswe/PSEN1dE9 transgenic model (engineered to develop Aβ plaques) and the K369I tau transgenic model (engineered to develop neurofibrillary tangles). Transcranial PBM therapy was administrated once daily using 670 nm LEDs for 4 weeks. In the Aβ model, PBM therapy significantly decreased the size and number of Aβ plaques in the neocortex and hippocampal neurons. Similarly, in the tau model, PBM therapy decreased hyperphosphorylated tau, neurofibrillary tangles, and oxidative stress indices in the neocortex and hippocampal neurons, and significantly restored the activity of mitochondrial cytochrome c oxidase in surviving neurons. It is interesting to note that a follow-up study conducted by Purushothuman et al. [59] also extended the above-mentioned beneficial effects to the brain cerebellum region. Lately, two different studies from another research team also showed that transcranial 670 nm LED PBM therapy for 4 consecutive weeks, could protect neurons against the synaptic accumulation of toxic tau oligomers in two transgenic mouse models of human tauopathy disorders (hTau and 3xTgAD) [60, 61]. Finally, Farfara et al. [62] used 5XFAD transgenic male mice (Tg6799) to explore the potential neuroprotective benefits of systemic PBM therapy. Irradiation used an 810 nm laser applied to the middle portion of the tibia, which was remote from the brain. Two months of this systemic PBM therapy, starting at the age of 4 months, considerably improved the cognitive performance, along with a 68% decrease in amyloid plaque burden in the brain of PBM-treated mice.

The age-related decrease in the expression level of heat shock proteins (HSPs) can cause abnormal polypeptide folding, leading to the accumulation of toxic protein aggregates, and potentially trigger intrinsic and extrinsic apoptotic pathways, and eventually increase the risk of neurodegenerative diseases like AD. It has been suggested that the application of PBM could prevent the progression of some neurodegenerative diseases through the regulation of HSPs [63]. In this respect, Grillo et al. [64] reported that PBM therapy could up-regulate a group of stress-responsive and heat shock proteins in the brain tissue, resulting in a decrease in protein aggregates and apoptosis. They observed that whole-body PBM therapy using 1072 nm LEDs (1.8 J/cm^2) biweekly for 5 months markedly increased HSP60, HSP70, HSP105, and p-HSP27 protein levels. Given this, they proposed that: (1) HSP60 may contribute to the reduced Aβ deposition via aiding protein folding; (2) HSP70 could inhibit neuronal death through the c-Jun N-terminal kinase (JNK)-BID-mediated mitochondrial apoptotic pathway and apoptosis-inducing factor (AIF); (3) HSP105 could suppress the mitogen-activated protein kinase-38 (MAPK38) and JNK stress-pathways causing a decrease in pro-apoptotic proteins. PBM suppressed JNK3 by activation of ERK/MKP7 to attenuate AMPA receptor endocytosis in AD [65].

9.2.2 Photobiomodulation Therapy for Dementia in Human Studies

In recent years, researchers from Toronto, ON (Canada) have carried out a series of studies to treat dementia patients using transcranial photobiomodulation (PBM) and/or intranasal PBM therapy. In the first study [66], Lim reported that single-modality intranasal PBM using a VieLight LED applicator (810 nm, 10-Hz frequency) once a day for 1 year, significantly enhanced cognitive and memory function in two moderate dementia patients as measured by the MMSE. One of the patients (Rudy) had a total pre- and post-treatment MMSE score of 16 and 24, respectively, with cognitive subdomain scores of: orientation (pre = 4; post = 8); registration (pre = 2; post = 3); attention (not reported); short-term memory or recall (pre = 1; post = 2); and language (pre = 8; post = 8) [66, 67].

In a study by Saltmarche et al. [68], a pilot 16-week, single-blinded, randomized, placebo-controlled study ($n = 19$) with combined intranasal and transcranial PBM therapy showed significant positive changes in cognitive function, and neurobehavioral regulation, with no adverse effects in moderate dementia subjects. Transcranial PBM therapy was applied using the VieLight Neuro LED device (810 nm), along with an intranasal applicator (with a total power of 515 mW) twice weekly for 2 weeks and then once weekly for 10 weeks. Patients also self-administered treatment at home using the intranasal applicator (with a power of 13 mW). Cognitive assessment was conducted at baseline, at week 12, and at follow-up after 4 weeks of no PBM therapy. Mean (SD) baseline MMSE and ADAS-cog scores were 18.4 (9.37) and 32.1 (21.41) in the active treatment group ($n = 13$) compared with 25.8 (4.36) and 14.8 (7.91) in the placebo group ($n = 6$). The MMSE scores increased by 2.00 points for the 8 subjects in the moderate-severely impaired active treatment subgroup (baseline MMSE = 0–24) after a 12 week treatment, and their ADAS-cog scores decreased by 5.00. The only placebo subject in this subgroup dropped out before completing the trial. Slight declines in the performance of both groups were noted at follow-up after 4 weeks of no treatment. No significant differences were observed in the higher functioning group (MMSE = 25–30) between active and placebo treatment, so that the mean changes at Week 12 in MMSE and ADAS-cog were 1.80 and −2.27 in the active group ($n = 5$), compared to 1.50 and −3.67 in the placebo group ($n = 5$), respectively. In addition, qualitative feedback from subjects and their caregivers in both groups reported better sleep, fewer angry outbursts, reduced anxiety, and less wandering behavior [68].

In another study on a patient with mild dementia, Zomorrodi et al. [69], showed that a 2-week (6 days/week) home-use application of transcranial plus intranasal PBM therapy using the VieLight LED device (810 nm, 40-Hz) resulted in significant improvement of cognitive function as evidenced by an increase in the MMSE score from 21 to 24, Alzheimer's Disease Cooperative Study Activities of Daily Living Scale (ADCS-ADL) from 43 to 58, and a decrease in ADAS-cog from 35.33 to 23.34. This rapid cognitive improvement was accompanied by an overall increase in the absolute power of the electroencephalogram (EEG) signal across all brain network oscillations. The absolute power of alpha-band increased considerably from 5.1 to 12.7 μV^2, and the peak of the alpha wave oscillation shifted from 8.6 to 9.3-Hz. Moreover, the absolute power of theta and delta bands increased from 2.12 μV^2 to 3.7 μV^2 and 1.86 μV^2 to 2.9 μV^2, respectively. These data suggest that the improved overall EEG results may be associated with improved function of the default mode network (DMN) in this mild dementia patient [69].

A pilot, placebo-controlled, double-blind trial involving early to moderately severe AD patients ($n = 11$; 6 active, 3 placebo, and 2 dropouts) was carried out by Berman et al.

[70] in 2007 at the QuietMIND Foundation suburban Philadelphia clinic. Brain PBM was applied using a transcranial and intraocular light helmet originally designed by Gordon Dougal, with subsequent technical design input from Marvin Berman and James Halper. The original apparatus design involved printed circuit boards with 1060–1080 nm LEDs with 10-Hz pulse frequency. The treatment protocol included 28 consecutive daily 6-min irradiation sessions. Participants underwent Quantitative EEG (QEEG) recordings with eyes open and closed, and were administered the ADAS-cog, the Boston Naming Test, and the Trail Making Test on the first day of treatment, and within 3 days of completing the treatment plan. Surface cortical perfusion was also measured before and after each PBM therapy session using NIRS on the frontal region (Fp1 and Fp2), while a 2-min baseline assessment was recorded using a red and infrared spectroscopy headband. There were significant changes in cortical blood flow in the active treatment group compared to the placebo group, so that the placebo subjects showed an average 0.50% increase in pre- and post-measures of cortical perfusion, compared to 4.07% for the active treatment group. In addition, there was a significant pre- and post-treatment difference in components of the ADAS-cog, as well as in measures of ideational praxis, Boston Naming Test, and Trail Making Test, demonstrating an overall improvement in executive function, memory, and visual attention. With respect to the QEEG data, the abnormal delta absolute and relative power became significantly normalized over the treatment period, as did the alpha relative power [70].

In a recent case series conducted by Saltmarche et al. [71], twelve weeks of transcranial plus intranasal PBM therapy using the VieLight Neuro Alpha transcranial headset (810 nm, 10-Hz) and intranasal applicator (810 nm, 10-Hz) considerably enhanced cognitive function in the two moderate and three mild dementia patients, as evidenced by MMSE and ADAS-cog scores. This was a single-blind study to explore the effects of PBM therapy on memory and cognition. Their protocol involved weekly, in-clinic use of a combined transcranial and intranasal PBM device and daily at home-use of an intranasal device. At baseline, the mean (SD) MMSE and ADAS-cog scores were 17.4 (6.84) and 35.47 (21.00), respectively. After 6 weeks of treatment, there was a trend toward a significant improvement, with +2.40 points on the MMSE, and with −7.40 points on the ADAS-cog. After 12 weeks of treatment, the mean scores improved to 20.00 (7.10) on the MMSE, and to 28.73 (18.85) on the ADAS-cog scores. At follow-up after 4 weeks of no treatment, 3 out of 4 of the patients showed lower MMSE scores (vs. at 12 week of treatment); and 2 out of 4 showed worse ADAS-cog scores (vs. at 12 week of treatment). Only one participant continued to improve on his ADAS-cog test after 12 weeks of treatment, but the MMSE was worse at that time point. The improved test scores were also accompanied by improved quality of life feedback from patients and their caregivers [71].

Most recently, Salehpour et al. [72] reported the rapid reversal of cognitive decline and olfactory dysfunction in a 64-year-old Caucasian female with MCI, after 4 weeks of combined transcranial and intranasal PBM therapy. Along with a transcranial light helmet

(635 + 810 nm; ProNeuroLIGHT LLC, Phoenix, AZ, USA), an intranasal LED applicator (810 nm, 10-Hz; VieLight LLC, Toronto, On, Canada) was placed in the left nostril and 10.65 J/cm^2 fluence was delivered twice daily. Cognitive improvement was demonstrated by an improvement in executive function/visuospatial ability, mathematical ability, and orientation (while delayed recall did not improve) as indicated by the MoCA, so that the pre- and post-treatment MoCA scores were 18 and 24, respectively. The Working Memory Questionnaire (WMQ) is a validated, self-administered scale addressing three domains of WM, consisting of short-term storage, attention, and executive control. The total score is out of 120, where higher scores indicate more impairment [73]. The initial gross score of the patient was 53, which improved to 10 at week 4. Although the patient showed a change in WM (20 to 6), attention (15 to 2), and executive (18 to 2) domains, her subjective disability measured by 44% in WM pre-treatment score decreased to only 8% post-treatment. In addition to the cognitive performance, the patient's pre-treatment Alberta Smell Test (AST) scores were 0 (right nostril) and 0 (left), which improved to 2 (right) and 2 (left) at post-treatment. Also, the patient's pre-treatment score for the peanut butter test, a quick, non-invasive odor detection test, was 0 cm bilaterally with no ability to identify the odor. At post-treatment, the patient was able to recognize the peanut butter smell and could detect it at 18 cm (left) and 10 cm (right). It should be mentioned that in the study of Salehpour et al. [72], there was a one week delay in the use of the intranasal device. The authors justified that delay because they were examining tolerance of low dose irradiation while monitoring any adverse effects. Given their dose calculations, since the ProNeuroLIGHT prototype helmet consisted of 50 red plus 150 NIR LEDs, therefore, considering the 25 min irradiation in each treatment session, the fluence per each red LED was 112.5 J/cm^2 and each NIR LED was 46.5 J/cm^2. Hence, a total energy of 12,600 J [50 × 112.5 (J/cm^2) = 5625 J] + [150 × 46.5 (J/cm^2) = 6975 J] was delivered to the patient's head during each 25 min session of treatment. On the other hand, in the Saltmarche et al. study [71], the VieLight transcranial "Neuro" device consisted of a headset frame, holding four separate LED cluster heads which each contained three LEDs. The fluence per LED was 24.6 J/cm^2. Therefore, a total energy of 295.2 J (12 × 24.6 (J/cm^2) was delivered to the patient's head during each 20 min session of treatment. Overall, a comparison of the results for the two studies suggests that applied total energy in Salehpour et al.'s study [72] was higher than that reported by Saltmarche et al. [71] by ~42 times.

In another case report, Nawashiro et al. [74] have shown a potential neurostimulatory effect of transcranial PBM in a 77-year-old male with multi-infarct dementia. The patient was diagnosed with memory disturbance, aphasia, and decreased spontaneity. Data from head magnetic resonance imaging (MRI) exhibited multiple lacunae in the bilateral basal ganglia and a cortical infarct in the left temporoparietal region. Pre-treatment cerebral blood flow measurements also showed hypofrontality. The treatment protocol consisted of 28 consecutive daily 15-min 850 nm LEDs irradiation to the forehead. During and after transcranial PBM therapy, they performed blood-oxygen-level-dependent (BOLD) functional MRI (fMRI) in a resting-state. The patient showed increased spontaneity and enhanced speech fluency following treatment. Immediately after the treatment, striking BOLD responses were observed at an irradiance of 30 mW/cm^2 in the dorsolateral prefrontal cortex just beneath the fiber optic application site, and widespread areas of the whole brain, such as the ipsilateral parietal cortex and the contralateral caudate nucleus. In addition, significant BOLD responses were observed at an irradiance of 30 mW/cm^2 in the brain regions, such as the bilateral occipital cortices and the contralateral temporal cortex. At 9-min after completion of treatment, a restricted BOLD response was observed at an irradiance of 200 mW/cm^2 in the dorsolateral prefrontal cortex just under the fiber optic sites; however, the BOLD response in the widespread area of the whole brain was no longer observed [74].

In 2019, Chao [75] used transcranial PBM with the VieLight Neuro Gamma headset combined with the intranasal applicator to treat eight patients with dementia (4 active and 4 placebo). PBM therapy was administered at home for 3 days a week for 12 weeks. The subjects were evaluated with the ADAS-cog subscale and the NPI at baseline and at 6 and 12 weeks, and with arterial spin-labeled perfusion MRI and resting-state functional MRI at baseline and 12 weeks. The mean (SEM) baseline ADAS-cog score was 37.5 (5.5), which declined to 35.7 (4.7) at week 6, and to 32.3 (4.8) at week 12, suggesting a significant improvement in cognitive function. Likewise, the mean (SEM) baseline NPI frequency severity score was 35.0 (11.6), which decreased to 22.8 (4.0) at week 6 and to 13.5 (2.0) at week 12, suggesting a significant improvement in behavioral function. Based on neuroimaging data, PBM also increased the cerebral perfusion, and the connectivity between the posterior cingulate cortex and lateral parietal nodes within the DMN [75].

Over the past decade, Ivan V. Maksimovich has utilized a different PBM irradiation method, named intracerebral transcatheter laser therapy to deliver red laser light into the brain in a large number of dementia patients [76–78]. Under local anesthesia, the common femoral artery was catheterized and then a thin fiber optic with 25–100 μm diameter was advanced to the distal section of the anterior and middle cerebral arteries, where intracerebral transcatheter irradiation was performed using a 632.8 nm HeNe laser for 20–40 min of treatment. The treatment improved cerebral microcirculation, decreased dementia symptoms and restored cognitive function in the active laser-treatment group. Although the mechanism of action for this type of brain PBM therapy is not completely understood, some factors including increased capillary blood supply, enhanced tissue metabolism, promotion of neurogenesis, and the clearance of Aβ have been postulated to explain the mechanisms of intracerebral transcatheter PBM therapy [79] (see Tables 9.1 and 9.2).

Table 9.1 Summary of in vivo studies on the effects of photobiomodulation therapy in dementia

Study/year	Animal/model	Light source	Wavelengths	Irradiation parameters	Irradiation approach/sites	Findings
De Taboada et al., 2011 [57]	Mouse (APP transgenic Alzheimer's disease model)	Laser, Photothera, Inc. (Carlsbad, CA. USA)	808 nm	1.2, 6, or 12 J/cm^2 at cortex; fiber diameter of 3 mm, 2 min, 3×/ week for 6 months, PW at 100-Hz with 2 ms pulse duration, or CW	Transcranially; at a point in sagittal suture, 4 mm caudal to coronal suture	Improved learning and memory in MWM test (by all regimens); decreased CSF (by PW regimens), plasma (by all regimens), and brain (by all regimens) Aβ level; decreased brain inflammatory markers such as IL-1β, TNF-α, and TGF-β (by all regimens); increased ATP level and O_2 consumption (by PW at 6 J/cm^2)
Grillo et al., 2013 [64]	Mouse (TASTPM Alzheimer's disease model)	LEDs, Virulite Distribution Limited	1072 nm	5 mW/cm^2, 1.8 J/cm^2, 6 min/day for 2 days, biweekly for 5 months, PW at 600-Hz with duty cycle of 300 μs	Whole-body irradiation	Increased HSP27, 60, 70, 105 and P-HSP27, and decreased PS1, αB-crystallin, APP, Aβ$_{1-40}$, Aβ$_{1-42}$, and phosphorylated tau proteins expression; decreased small Aβ$_{1-40/42}$ and Aβ$_{1-42}$ plaque deposition in dentate gyrus and cerebral cortex
Sutalangka et al., 2013 [87]	Rat (AF64A Alzheimer's disease model)	Laser, Xinland International Limited (Xi'an, Shaanxi, China)	405 nm	52.63 J/cm^2, spot diameter of 500 μm, 10 min/day for 2 weeks, CW	Laser acupuncture; at the HT7 point or a point 2–4 mm lateral to the HT7 acupoint	Improved learning and memory in MWM test; increased catalase and SOD activities and decreased AChE activity in hippocampus

(continued)

Table 9.1 (continued)

Study/year	Animal/model	Light source	Wavelengths	Irradiation parameters	Irradiation approach/sites	Findings
Purushothuman et al., 2014 [88]	Mouse (K3 and APP/PS1 transgenic Alzheimer's disease models)	LEDs, WARP 10; Quantum Devices (Barneveld, WI, USA)	670 nm	4 J/cm², 90 s, 5×/week for 4 weeks, CW	Transcranially; holding probe 1–2 cm above the head	K3 mouse: decreased phosphorylated tau and NFTs in neocortex and hippocampus; decreased oxidative stress markers such as 8-OHDG and 4-HNE in neocortex; enhanced CCO expression patterns in neocortex and hippocampus. APP/PS1 mouse: decreased number, size, and burden of Aβ plaques in neocortex and hippocampus
Farfara et al., 2015 [62]	Mouse (5XFAD transgenic Alzheimer's disease model)	NR	NR	400 mW, 1 J/cm², spot diameter of 0.3 cm, 6 × (at 10-day intervals, for 2 months), CW	Remote tissue irradiation; at the middle portion of the medial part of the tibia	Improved memory in object recognition and fear-conditioning tests; decreased Aβ burden in hippocampus
Purushothuman et al., 2015 [59]	Mouse (K3 and APP/PS1 transgenic Alzheimer's disease models)	LEDs, WARP 10, Quantum Devices (Barneveld, WI, USA)	670 nm	4 J/cm², 90 s, 5×/week for 4 weeks, CW	Transcranially; holding probe 1–2 cm above the head	APP/PS1 mouse: reduced number, size and deposition of Aβ plaques in cerebellar cortex. K3 mouse: decreased formation of NFTs, phosphorylated tau, oxidative stress, and increased CCO expression in cerebellar cortex
Lu et al., 2017 [53]	Rat (Aβ₁₋₄₂ Alzheimer's disease model)	Laser, 808M100, Dragon Lasers (Changchun, China)	808 nm	25 mW/cm², 3 J/cm² at cerebral cortex, spot area of 1 cm², 2 min/day for 5 consecutive days, CW	Transcranially; at the 3 mm posterior to eye and 2 mm anterior to ear	Suppressed neuronal degeneration; suppressed expression of mitochondrial fission and preserved mitochondrial fusion proteins; recovered changes in mitochondrial dynamics; improved mitochondrial function through reduction of Bax/Bcl-2 ratio and increase of MMP, CCO activity and ATP levels; inhibited G6PDH and NADPH oxidase activities; enhanced total antioxidant capacity; inhibited glial activation, proinflammatory cytokines production and tau hyperphosphorylation; attenuated cytosolic level of cytochrome c, caspase-9 and-3 activities; improved spatial learning and memory in Barnes maze task and long-term recognition memory in NOR test
Da Luz Eltchechem et al., 2017 [55]	Rat (Aβ₂₅₋₃₅ Alzheimer's disease model)	LEDs, AlGaInP, RL5-R12008	627 nm	70 mW, 7 J/cm², 100 s/day for 21 days	Transcranially; holding probe at 1 cm from the frontal region of scalp	Improved motor skills in OFT at days 14 and 21; improved spatial memory in MWM test at day 14; reduced amount of Aβ

(continued)

Table 9.1 (continued)

Study/year	Animal/model	Light source	Wavelengths	Irradiation parameters	Irradiation approach/sites	Findings
Comerota et al., 2017 [89]	Mouse (Tg2576 transgenic Alzheimer's disease model)	LEDs, LEDs, WARP 10; Quantum Devices (Barneveld, WI, USA)	670 nm	4 J/cm², 90 s, 5×/week for 4 weeks, CW	Transcranially; holding probe 1–2 cm above the head	Decreased susceptibility of synapses to toxic Aβ oligomers binding; decreased Aβ oligomer level in total homogenate and at the synapses; reduces Aβ oligomer-induced deficits in long-term potentiation; increased synaptic mitochondrial membrane potential
Han et al., 2018 [90]	Mouse (APP and PSEN1 transgenic Alzheimer's disease models)	LEDs	1040–1090 nm	12 mW/cm², 6 min per day for 40 days	Whole-body irradiation	Improved learning and memory in MWM test; decreased number of small-size Aβ plaques (<20 μm) in hippocampus and cerebral cortex
Blivet et al., 2018 [56]	Mouse (Aβ$_{25-35}$ Alzheimer's disease model)	Combined laser and LEDs, RGn500, REGEnLIFE (France)	Laser: 850 nm LEDs: 850 and 625 nm	28 mW/cm², 8.4 J/cm², 10 min, PW at 10-Hz with duty cycle of 50%; once a day, twice a day, or for 7 says	Transcranially; holding probe 1 cm above the head Remote tissue irradiation; to the abdomen	Improved short-term memory as assessed by Y-maze; improved long-term memory as assessed by passive avoidance task; decreased oxidative stress (lipid peroxidation levels); decreased inflammatory markers of IL-1β, TNF-α, and IL-6; modified astrocytes and microglial activities; decreased Bax/Bcl-2 ratio; inhibited Aβ augmentation; decreased pTau-Thr181 levels
Cho et al., 2018 [91]	Mouse (5XFAD transgenic Alzheimer's disease model)	LEDs, Color Seven Co. (Seoul, South Korea)	610 nm	1.7 mW/cm², 2.0 J/cm², 20 min, 3×/week for 14 weeks	Transcranially; at the two points (the midpoint of the parietal bone and the posterior midline of the seventh cervical vertebra)	Improved learning and memory in MWM test; decreased anxiety levels in elevated plus maze test; decreased Aβ plaque pathology and alleviated neuronal degeneration in the cortex; decreased microglial activation; decreased Aβ accumulation by increasing levels of the Aβ-degrading enzyme, insulin-degrading enzyme (IDE), but not neprilysin (NEP)
Comerota et al., 2019 [60]	Mouse (hTau and 3xTgAD transgenic Alzheimer's disease models)	LEDs, LEDs, WARP 10; Quantum Devices (Barneveld, WI, USA)	670 nm	4 J/cm², 90 s, 5×/week for 4 weeks, CW	Transcranially; holding probe 1 cm above the head	In both hTau and 3xTgAD mice: decreased levels of endogenous total and oligomeric tau in both synaptosomes and total protein extracts from the hippocampus and cortex; improved deteriorating memory function in NOR test

(continued)

Table 9.1 (continued)

Study/year	Animal/model	Light source	Wavelengths	Irradiation parameters	Irradiation approach/sites	Findings
Zhang et al., 2019 [92]	Mouse (double APPswe/PSENdE9 transgenic Alzheimer's disease model)	Laser, HN-1000; Laser Technology Application Research Institute (Guangzhou, China)	632.8 nm	92 mW, 2 J/ cm^2 at hippocampus level, 10 min, once a day for 30 days, CW	Transcranially	Decreased Aβ levels and amyloid plaque burden; improved learning and memory in MWM test; increased the level of sAPPα and decreased the level of sAPPβ fragments most likely through increasing ADAM10 and decreasing BACE1 protein levels in the cortex and hippocampus
Zhang et al., 2019 [93]	Mouse (SAMP8 model of age-related dementia)	LEDs	630 nm	0.5 mW/cm^2, 30 min, 5×/ week for 4 weeks	Transcranially	Decreased early-stage and late-stage memory decline by improving spatial learning and memory in MWM test; decreased SSAO activity; enhanced FDH activity and reduced brain formaldehyde levels; increased catalase activity and reduced brain H$_2$O$_2$ levels; enhanced brain ChAT activity and improved brain Ach concentrations
Yue et al., 2019 [94]	Mouse (APP/PS1 transgenic Alzheimer's disease model)	LEDs	630 nm	0.5 mW/cm^2, 40 min, 5×/ week for 4 weeks	Transcranially + remote irradiation to the abdomen	Improved learning and memory in MWM test; destroyed Aβ assembly; activated formaldehyde dehydrogenase to degrade formaldehyde and attenuated formaldehyde facilitated Aβ aggregation; smashed Aβ deposition in the extracellular space; recovered the flow of interstitial fluid
Zinchenko et al., 2019 [95]	Mouse (Aβ$_{1-42}$ Alzheimer's disease model)	Laser, (LD-1267-FBG-350 (Innolume, Dortmund, Germany)	1267 nm	50, 100, 150, or 200 mW/ cm^2 with corresponding fluencies of 51, 102, 153, or 204 J/cm^2, respectively; 17 min	Transcranially; at area of the frontal cortex	Improved cognition sand memory in NOR test; improved neurobehavioral status on the NSS scale; decreased Aβ plaques deposition

Abbreviations: 4-HNE, 4-Hydroxynonenal: 8-OHDG, 8-Oxo-2′-deoxyguanosine; Aβ, Amyloid-beta; AChE, Acetylcholinesterase; APP, amyloid precursor protein; ATP, adenosine triphosphate; Bax, Bcl-2-associated X protein; CCO, cytochrome c oxidase; ChAT, choline acetyltransferase; CSF, cerebrospinal fluid; CW, continuous wave; FDH, Formaldehyde dehydrogenase; G6PDH, Glucose-6-phosphate dehydrogenase; HSP27, heat-shock protein-27; IL, interleukin; LEDs, light-emitting diodes; MMP, mitochondrial membrane potential; MWM, Morris water maze; NADPH, nicotinamide adenine dinucleotide phosphate; NFTs, neurofibrillary tangles; NOR, novel object recognition; NSS, neurologic Severity Score; OFT, open field test; PW, pulsed wave; SOD, superoxide dismutase; SSAO, semicarbazide-sensitive aminoxidase; TGF-β, transforming growth factor-β; TNF-α, tumor necrosis factor-α

Table 9.2 Summary of clinical studies on the effects of photobiomodulation therapy in Alzheimer's disease

Study/year	Subjects (n)	Light source	Wavelengths	Irradiation parameters	Irradiation approach/sites	Findings
Arakelyan, 2005 [80]	Alzheimer's disease (25)	Laser	632.8 nm	4 mW, every day for 5–9 min, 6 courses of 15 procedures during 18 months	Intravenously	Improved cognitive function evaluated by Alzheimer's Disease Assessment Scale-cognitive (ADAS-cog)
Maksimovich 2012 [81]	Pre-clinical stage or increased AD risk (9); mild dementia and cognitive impairment (24); moderate dementia and persistent cognitive impairment (31); severe dementia and cognitive impairment (17)	Laser, coupled with fiber-optic light guided instrument	Visible region of spectrum	20 mW, fiber diameter of 50–100 m μ, 20–40 min, CW, or PW, or combined modes	Transcatheterly; threading a fiber optic through a catheter in femoral artery (advancing fiber optic to distal site of anterior and middle cerebral arteries)	Restored capillary blood flow accompanied by reduction of arteriovenous shunts and improvement of venous return; increased the tissue mass of the temporal lobes of the brain; improved memory, intellectual capacity, and the condition of higher mental functions and of social adjustment; improved the ability to memorize; increased mental stress tolerance
Maksimovich 2015 [77]	Alzheimer's disease (89)	Laser, coupled with fiber-optic light guided instrument	Visible region of spectrum	20 mW, fiber diameter of 25–100 m μ, 20–40 min, CW, or PW, or combined modes	Transcatheterly; threading a fiber optic through a catheter in femoral artery (advancing fiber optic to distal site of anterior and middle cerebral arteries)	Improved cerebral microcirculation and cognitive recovery; decreased permanent dementia

(continued)

Table 9.2 (continued)

Study/year	Subjects (n)	Light source	Wavelengths	Irradiation parameters	Irradiation approach/sites	Findings
Stephan et al., (2017) [82]	Vascular dementia (2); early stage dementia (3); advanced dementia (1)	Laser, Theralase TLC 900 (Toronto, Canada)	905 + 660 nm	60 mW, 2.5 min, PW with pulse duration of 200 nanosec, an average of 4 times irradiation over an 8-day period	Transcranially: various areas including prefrontal cortex, temporal lobe, hippocampus, and circle of Willis	Patients showed varying degrees of improvement in cognitive function and personality, leading to enhanced quality of life and decreased caregiver burden
Berman et al., 2017 [83]	Dementia (11)	LEDs, 15 arrays of 70 LEDs/array (total of 1100 LEDs set)	1060–1080 nm	6 min/day for 28 consecutive days, PW at 10-Hz with DC of 50%	Transcranially: whole head irradiation with LED helmet	Improved executive functioning including clock drawing, immediate recall, praxis memory, visual attention and task switching; improved EEG amplitude and connectivity measures
Saltmarche et al., 2017 [84]	Dementia (5)	LEDs, "810" and "Neuro" devices, Vielight, Inc. (Toronto, Canada)	810 nm	14.2 or 41 + 23 mW/cm² per LED, 10.65 or 24.6 + 13.8 J/cm², 25 or 20 min, PW at 10-Hz, for 12 weeks	Transcranially: multiple areas, bilateral mesial prefrontal cortex, precuneus/posterior cingulate cortex, angular gyrus (correspond to Fpz, Cz, T3 and T4 EEG points) Intranasally: left nose	Improved function and sleep quality; decreased angry outbursts, anxiety, and wandering
Chao (2019) [75]	Dementia (8)	LEDs, Neuro Gamma device, Vielight, Inc. (Toronto, Canada)	810 nm	Transcranial: 75 and 100 mW, 75 and 100 mW/cm², 20 min, PW at 40-Hz with duty cycle of 50% Intranasal: 25 mW, 25 mW/cm², 20 min, PW at 40-Hz with duty cycle of 50%	Transcranially: multiple areas, bilateral mesial prefrontal cortex, precuneus/posterior cingulate cortex, angular gyrus (correspond to Fpz, Cz, T3 and T4 EEG points) Intranasally: left nose	Improved cognitive function evaluated by Alzheimer's Disease Assessment Scale-cognitive (ADAS-cog): increased cerebral perfusion; increased connectivity between the posterior cingulate cortex and lateral parietal nodes within the default-mode network

(continued)

Table 9.2 (continued)

Study/year	Subjects (n)	Light source	Wavelengths	Irradiation parameters	Irradiation approach/sites	Findings
Salehpour et al., (2019) [85]	Mild cognitive impairment (1)	LEDs, a prototype transcranial helmet (total of 200 diodes) and a body pad (total of 70 diodes) both from ProNeuroLIGHT LLC (Phoenix, AZ, USA); intranasal device from Vielight, Inc. (Toronto, Canada)	Transcranial: 810 + 635 nm Intranasal: 810 nm	Transcranial: 810 + 635 nm, 3.72 and 9 mW, 31 and 75 mW/cm², 46.5 and 112.5 J/cm², 25 min, CW Intranasal: LED, 810 nm, 14.2 mW, 14.2 mW/cm², 10.65 J/cm², 25 min, PW at 10-Hz with duty cycle of 50% Body pad: 810 + 635 nm, 3.72 and 9 mW, 31 and 75 mW/cm², 46.5 and 112.5 J/cm², 25 min, CW	Transcranially; whole head irradiation with LED helmet Intranasally; left nose Body pad: various areas on the lower back	Improved cognitive function by improvements in executive function/visuospatial ability, mathematical ability, and orientation as evaluated by the Montreal Cognitive Assessment (MoCA); improved working memory, attention, and executive domains as measured by the Working Memory Questionnaire (WMQ); reversed olfactory impairment evaluated by the Alberta Smell Test and peanut butter odor detection test; improved quality-of-life; reduced caregiver stress
Nawashiro et al., (2019) [86]	Multi-infarct dementia (1)	LEDs, total of 23 arrays	850 nm	13 mW, 11.4 mW/cm², 15 min	Transcranially; forehead	Increased spontaneity and improved speech fluency; induced striking blood oxygenation level–dependent (BOLD) responses in the dorsolateral prefrontal cortex and widespread areas of the whole brain including the ipsilateral parietal cortex and the contralateral caudate nucleus

(continued)

Table 9.2 (continued)

Study/year	Subjects (n)	Light source	Wavelengths	Irradiation parameters	Irradiation approach/sites	Findings
Maksimovich 2019 [78]	Binswanger's disease or subcortical vascular dementia (27)	Laser, ULF-01, Anod, (Bryansk, Russia)	632.8 nm	25–45 mW, 29–106 J, fiber diameter of 25–100 μm, 20–40 min, 1–2 mm beam spot diameter in the vessel, CW	Intracerebral transcatheter; threading a fiber optic through a catheter in femoral artery (advancing fiber optic to distal site of anterior and middle cerebral arteries)	Immediate results: increased angiogenesis, and collateral and capillary revascularization Early period (1–6 months) results: improved mental and motor functions; improved blood flow and volume pulse blood filling in the cerebral hemispheres; decreased general involutive changes in the brain, narrowing of the subarachnoid space Long-term period (from 1 till 8 years) results: restored mental and motor functions; decreased general brain involutive changes and Sylvian fissure narrowing; decreased nonocclusive hydrocephalus, gliosis, and leukoaraiosis; induced neurogenesis and regenerative cerebral changes

Abbreviations: AD, Alzheimer's disease; CW, continuous wave; EEG, electroencephalography; LEDs, light-emitting diodes; PW, pulsed wave

References

1. Burns, A., and P. Robert. 2009. *The national dementia strategy in England*. British Medical Journal Publishing Group.
2. Folstein, M.F., S.E. Folstein, and P.R. McHugh. 1975. "Mini-mental state": A practical method for grading the cognitive state of patients for the clinician. *Journal of Psychiatric Research* 12 (3): 189–198.
3. Nasreddine, Z.S., et al. 2005. The Montreal Cognitive Assessment, MoCA: A brief screening tool for mild cognitive impairment. *Journal of the American Geriatrics Society* 53 (4): 695–699.
4. Skinner, J., et al. 2012. The Alzheimer's disease assessment scale-cognitive-plus (ADAS-Cog-Plus): An expansion of the ADAS-Cog to improve responsiveness in MCI. *Brain Imaging and Behavior* 6 (4): 489–501.
5. Cummings, J.L. 1997. The neuropsychiatric inventory: Assessing psychopathology in dementia patients. *Neurology* 48(5 Suppl 6): 10S–16S.
6. Petersen, R.C., et al. 2014. Mild cognitive impairment: A concept in evolution. *Journal of Internal Medicine* 275 (3): 214–228.
7. Albert, M.S., et al. 2011. The diagnosis of mild cognitive impairment due to Alzheimer's disease: Recommendations from the National Institute on Aging-Alzheimer's Association workgroups on diagnostic guidelines for Alzheimer's disease. *Alzheimer's & Dementia* 7 (3): 270–279.
8. Petersen, R.C. 2016. Mild cognitive impairment. *CONTINUUM: Lifelong Learning in Neurology* 22(2 Dementia): 404.
9. Jia, J., et al. 2014. The prevalence of mild cognitive impairment and its etiological subtypes in elderly Chinese. *Alzheimer's & Dementia* 10 (4): 439–447.
10. Morris, J.C., et al. 2001. Mild cognitive impairment represents early-stage Alzheimer disease. *Archives of Neurology* 58 (3): 397–405.
11. Petersen, R.C., et al. 2001. Current concepts in mild cognitive impairment. *Archives of Neurology* 58 (12): 1985–1992.
12. Praticò, D., et al. 2002. Increase of brain oxidative stress in mild cognitive impairment: A possible predictor of Alzheimer disease. *Archives of Neurology* 59 (6): 972–976.
13. Wang, J., W.R. Markesbery, and M.A. Lovell. 2006. Increased oxidative damage in nuclear and mitochondrial DNA in mild cognitive impairment. *Journal of Neurochemistry* 96 (3): 825–832.
14. Butterfield, D.A., et al. 2007. Roles of amyloid β-peptide-associated oxidative stress and brain protein modifications in the pathogenesis of Alzheimer's disease and mild cognitive impairment. *Free Radical Biology and Medicine* 43 (5): 658–677.
15. Yin, C., et al. 2013. Brain imaging of mild cognitive impairment and Alzheimer's disease. *Neural Regeneration Research* 8 (5): 435–444.
16. Ries, M.L., et al. 2008. Magnetic resonance imaging characterization of brain structure and function in mild cognitive impairment: A review. *Journal of the American Geriatrics Society* 56 (5): 920–934.
17. Dai, W., et al. 2009. Mild cognitive impairment and alzheimer disease: Patterns of altered cerebral blood flow at MR imaging. *Radiology* 250 (3): 856–866.
18. Fujiwara, Y., et al. 2010. Brief screening tool for mild cognitive impairment in older Japanese: Validation of the Japanese version of the Montreal Cognitive Assessment. *Geriatrics & Gerontology International* 10 (3): 225–232.
19. Kokmen, E., J.M. Naessens, and K.P. Offord. 1987. *A short test of mental status: Description and preliminary results*. In *Mayo clinic proceedings*. Elsevier.

20. Ströhle, A., et al. 2015. Drug and exercise treatment of Alzheimer disease and mild cognitive impairment: A systematic review and meta-analysis of effects on cognition in randomized controlled trials. *The American Journal of Geriatric Psychiatry* 23 (12): 1234–1249.
21. Alzheimer, A. 1907. Uber eine eigenartige Erkrankung der Hirnrinde. *Allgemeine Zeitschrift für Psychiatrie und Psychisch-gerichtliche Medizin* 18: 177–179.
22. Association, A.S. 2017. 2017 Alzheimer's disease facts and figures. *Alzheimer's & Dementia* 13(4): 325–373.
23. Jack, C.R., Jr., et al. 2013. Tracking pathophysiological processes in Alzheimer's disease: An updated hypothetical model of dynamic biomarkers. *The Lancet Neurology* 12 (2): 207–216.
24. Raskin, J., et al. 2015. Neurobiology of Alzheimer's disease: Integrated molecular, physiological, anatomical, biomarker, and cognitive dimensions. *Current Alzheimer Research* 12 (8): 712–722.
25. Sadigh-Eteghad, S., et al. 2015. Amyloid-beta: A crucial factor in Alzheimer's disease. *Medical Principles and Practice* 24 (1): 1–10.
26. Majdi, A., et al. 2017. Revisiting nicotine's role in the ageing brain and cognitive impairment. *Reviews in the Neurosciences* 28 (7): 767–781.
27. Sadigh-Eteghad, S., et al. 2016. Astrocytic and microglial nicotinic acetylcholine receptors: An overlooked issue in Alzheimer's disease. *Journal of Neural Transmission* 123 (12): 1359–1367.
28. Reddy, P.H. 2011. Abnormal tau, mitochondrial dysfunction, impaired axonal transport of mitochondria, and synaptic deprivation in Alzheimer's disease. *Brain Research* 1415: 136–148.
29. Wong, P.C., et al. 2012. Neurobiology of Alzheimer's disease. In *Basic neurochemistry*, 8th ed. 815–828.
30. Braak, H., and E. Braak. 1991. Neuropathological staging of Alzheimer-related changes. *Acta Neuropathologica* 82 (4): 239–259.
31. Wang, X., et al. 2014. Oxidative stress and mitochondrial dysfunction in Alzheimer's disease. *Biochimica et Biophysica Acta (BBA)-Molecular Basis of Disease* 1842(8): 1240–1247.
32. Lyketsos, C.G., et al. 2011. *Neuropsychiatric symptoms in Alzheimer's disease*. Elsevier.
33. Herholz, K., et al. 2002. Direct comparison of spatially normalized PET and SPECT scans in Alzheimer's disease. *Journal of Nuclear Medicine* 43 (1): 21–26.
34. Nestor, P.J., P. Scheltens, and J.R. Hodges. 2004. Advances in the early detection of Alzheimer's disease. *Nature Medicine* 10 (7): S34.
35. Ferreira, L.K., and G.F. Busatto. 2011. Neuroimaging in Alzheimer's disease: Current role in clinical practice and potential future applications. *Clinics (Sao Paulo, Brazil)* 66(Suppl 1): p. 19–24.
36. Hampel, H., et al. 2010. Biomarkers for Alzheimer's disease: Academic, industry and regulatory perspectives. *Nature Reviews Drug Discovery* 9 (7): 560.
37. Dubois, B., et al. 2014. Advancing research diagnostic criteria for Alzheimer's disease: The IWG-2 criteria. *The Lancet Neurology* 13 (6): 614–629.
38. Schelterns, P., and H. Feldman. 2003. Treatment of Alzheimer's disease; current status and new perspectives. *The Lancet Neurology* 2 (9): 539–547.
39. Luijmes, R.E., S. Pouwels, and J. Boonman. 2016. The effectiveness of neurofeedback on cognitive functioning in patients with Alzheimer's disease: Preliminary results. *Neurophysiologie Clinique/Clinical Neurophysiology* 46 (3): 179–187.
40. Nardone, R., et al. 2014. Transcranial magnetic stimulation (TMS)/repetitive TMS in mild cognitive impairment and Alzheimer's disease. *Acta Neurologica Scandinavica* 129 (6): 351–366.
41. Hardenacke, K., et al. 2013. Deep brain stimulation as a tool for improving cognitive functioning in Alzheimer's dementia: A systematic review. *Frontiers in Psychiatry* 4: 159.
42. Philipson, O., et al. 2010. Animal models of amyloid-β-related pathologies in Alzheimer's disease. *The FEBS Journal* 277 (6): 1389–1409.

43. Laurijssens, B., F. Aujard, and A. Rahman. 2013. Animal models of Alzheimer's disease and drug development. *Drug Discovery Today: Technologies* 10 (3): e319–e327.
44. T O'Brien, J., and A. Thomas. 2015. Vascular dementia. *The Lancet* 386(10004): 1698–1706.
45. Jorm, A.F., and D. Jolley. 1998. The incidence of dementia: A meta-analysis. *Neurology* 51 (3): 728–733.
46. Pendlebury, S.T., and P.M. Rothwell. 2009. Prevalence, incidence, and factors associated with pre-stroke and post-stroke dementia: A systematic review and meta-analysis. *The Lancet Neurology* 8 (11): 1006–1018.
47. Ylikoski, R., et al. 2007. Comparison of the Alzheimer's disease assessment scale cognitive subscale and the vascular dementia assessment scale in differentiating elderly individuals with different degrees of white matter changes. *Dementia and Geriatric Cognitive Disorders* 24 (2): 73–81.
48. Román, G.C., et al. 1993. Vascular dementia: Diagnostic criteria for research studies: Report of the NINDS-AIREN international workshop. *Neurology* 43 (2): 250–250.
49. Guekht, A.B., et al. 2011. Cerebrolysin in vascular dementia: Improvement of clinical outcome in a randomized, double-blind, placebo-controlled multicenter trial. *Journal of Stroke and Cerebrovascular Diseases* 20 (4): 310–318.
50. Rea, T.D., et al. 2005. Statin use and the risk of incident dementia: The Cardiovascular Health Study. *Archives of Neurology* 62 (7): 1047–1051.
51. Cechetti, F., et al. 2012. Environmental enrichment prevents behavioral deficits and oxidative stress caused by chronic cerebral hypoperfusion in the rat. *Life Sciences* 91 (1–2): 29–36.
52. Sun, H., et al. 2010. Environmental enrichment influences BDNF and NR1 levels in the hippocampus and restores cognitive impairment in chronic cerebral hypoperfused rats. *Current Neurovascular Research* 7 (4): 268–280.
53. Lu, Y., et al. 2017. Low-level laser therapy for beta amyloid toxicity in rat hippocampus. *Neurobiology of Aging* 49: 165–182.
54. Duggett, N.A., and P.L. Chazot. 2014. Low-intensity light therapy (1068 nm) protects CAD neuroblastoma cells from β-amyloid-mediated cell death. *Biologie et Médecine* 1 (103): 2.
55. da Luz Eltchechem, C., et al. 2017. Transcranial LED therapy on amyloid-β toxin 25–35 in the hippocampal region of rats. *Lasers in Medical Science* 32(4): 749–756.
56. Blivet, G., et al. 2018. Neuroprotective effect of a new photobiomodulation technique against Aβ25–35 peptide–induced toxicity in mice: Novel hypothesis for therapeutic approach of Alzheimer's disease suggested. *Alzheimer's & Dementia: Translational Research & Clinical Interventions* 4: 54–63.
57. De Taboada, L., et al. 2011. Transcranial laser therapy attenuates amyloid-β peptide neuropathology in amyloid-β protein precursor transgenic mice. *Journal of Alzheimer's Disease* 23 (3): 521–535.
58. Purushothuman, S., et al. 2014. Photobiomodulation with near infrared light mitigates Alzheimer's disease-related pathology in cerebral cortex–evidence from two transgenic mouse models. *Alzheimer's Research & Therapy* 6 (1): 1–13.
59. Purushothuman, S., et al. 2015. Near infrared light mitigates cerebellar pathology in transgenic mouse models of dementia. *Neuroscience Letters* 591: 155–159.
60. Comerota, M.M., et al. 2019. Near infrared light treatment reduces synaptic levels of toxic tau oligomers in two transgenic mouse models of human Tauopathies. *Molecular Neurobiology* 56 (5): 3341–3355.
61. Comerota, M.M., B. Krishnan, and G. Taglialatela. 2017. Near infrared light decreases synaptic vulnerability to amyloid beta oligomers. *Scientific reports* 7 (1): 1–11.
62. Farfara, D., et al. 2015. Low-level laser therapy ameliorates disease progression in a mouse model of Alzheimer's disease. *Journal of Molecular Neuroscience* 55 (2): 430–436.

63. Bathini, M., C.R. Raghushaker, and K.K. Mahato. 2020. The molecular mechanisms of action of photobiomodulation against neurodegenerative diseases: A systematic review. *Cellular and Molecular Neurobiology*, 1–17.
64. Grillo, S., et al. 2013. Non-invasive infra-red therapy (1072 nm) reduces β-amyloid protein levels in the brain of an Alzheimer's disease mouse model, TASTPM. *Journal of Photochemistry and Photobiology B: Biology* 123: 13–22.
65. Shen, Q., et al. 2021. Photobiomodulation suppresses JNK3 by activation of ERK/MKP7 to attenuate AMPA receptor endocytosis in Alzheimer's disease. *Aging Cell* 20 (1): e13289.
66. Lim, L. 2014. Intranasal photobiomodulation improves cognitive and memory performance of Alzheimer's disease patients in case studies. In *NAALT/WALT2014*, Arlington, Virginia.
67. Berman, M.H., M.R. Hamblin, and P. Chazot. 2017. Photobiomodulation and other light stimulation procedures. In *Rhythmic stimulation procedures in neuromodulation*, 97–129. Elsevier.
68. Saltmarche, A.E., et al. 2016. Significant improvement in memory and quality of life after transcranial and intranasal photobiomodulation: A randomized, controlled, single-blind pilot study with dementia. *Alzheimer's & Dementia: The Journal of the Alzheimer's Association* 12 (7): P155–P156.
69. Zomorrodi, R., et al. 2017. Complementary EEG evidence for a significantly improved Alzheimer's disease case after photobiomodulation treatment. In *26th annual scientific conference, Canadian academy of geriatric psychiatry Toronto*.
70. Berman, M.H., et al. 2017. Photobiomodulation with near infrared light helmet in a pilot, placebo controlled clinical trial in dementia patients testing memory and cognition. *Journal of Neurology and Neuroscience* 8(1).
71. Saltmarche, A.E., et al. 2017. Significant improvement in cognition in mild to moderately severe dementia cases treated with transcranial plus intranasal photobiomodulation: Case series report. *Photomedicine and Laser Surgery* 35 (8): 432–441.
72. Salehpour, F., M.R. Hamblin, and J.O. DiDuro. 2019. Rapid reversal of cognitive decline, olfactory dysfunction, and quality of life using multi-modality photobiomodulation therapy: Case report. *Photobiomodulation, Photomedicine, and Laser Surgery* 37 (3): 159–167.
73. Vallat-Azouvi, C., P. Pradat-Diehl, and P. Azouvi. 2012. The working memory questionnaire: A scale to assess everyday life problems related to deficits of working memory in brain injured patients. *Neuropsychological Rehabilitation* 22(4): 634–649.
74. Nawashiro, H., et al. 2019. Time courses of BOLD responses during transcranial near-infrared laser irradiation. *Brain Stimulation: Basic, Translational, and Clinical Research in Neuromodulation* 12 (3): 778–780.
75. Chao, L.L. 2019. Effects of home photobiomodulation treatments on cognitive and behavioral function, cerebral perfusion, and resting-state functional connectivity in patients with dementia: A pilot trial. *Photobiomodulation, Photomedicine, and Laser Surgery* 37 (3): 133–141.
76. Maksimovich, I.V. 2012. Endovascular application of low-energy laser in the treatment of dyscirculatory angiopathy of Alzheimer's type. *Journal of Behavioral and Brain Science* 2 (1): 67–81.
77. Maksimovich, I.V. 2015. Dementia and cognitive impairment reduction after laser transcatheter treatment of Alzheimer's disease. *World Journal of Neuroscience* 5 (03): 189.
78. Maksimovich, I.V. 2019. Intracerebral transcatheter laser photobiomodulation therapy in the treatment of Binswanger's disease and vascular parkinsonism: research and clinical experience. *Photobiomodulation, Photomedicine, and Laser Surgery* 37 (10): 606–614.
79. Hamblin, M.R. 2019. Photobiomodulation for Alzheimer's disease: Has the light dawned? In *Photonics*. Multidisciplinary Digital Publishing Institute.
80. Arakelyan, H.S. 2005. Treatment of Alzheimer's disease with a combination of laser, magnetic field and chromo light. *Laser Therapy* 14 (1): 19–28.

81. Maksimovich, I.V. 2012. Endovascular application of low-energy laser in the treatment of dyscirculatory angiopathy of Alzheimer's type. *Journal of Behavioral and Brain Science* 2 (01): 67.
82. Stephan, W., R.A. Din, L.J. Banas, J. Thomas, C. Kochert, R.J. Lamartiniere, et al. 2017. Management of post-traumatic stress (PTSD) dementia and other neuro-degenerative disease with photo-medicine: Clinical experience and case studies. *Open Journal of Psychiatry* 7(04): 386.
83. Berman, M.H., J.P. Halper, T.W. Nichols, H. Jarrett, A. Lundy, and J.H. Huang. 2017. Photobiomodulation with near infrared light helmet in a pilot, placebo controlled clinical trial in dementia patients testing memory and cognition. *Journal of Neurology and Neuroscience* 8(1).
84. Saltmarche, A.E., M.A. Naeser, K.F. Ho, M.R. Hamblin, and L. Lim. 2017. Significant improvement in cognition in mild to moderately severe dementia cases treated with transcranial plus intranasal photobiomodulation: Case series report. *Photomedicine and Laser Surgery* 35(8): 432–441.
85. Salehpour, F., M.R. Hamblin, and J.O. DiDuro. 2019. Rapid reversal of cognitive decline, olfactory dysfunction, and quality of life using multi-modality photobiomodulation therapy: Case report. *Photobiomodulation, Photomedicine, and Laser Surgery* 37(3): 159–167.
86. Nawashiro, H., S. Kawauchi, Y. Tsunoi, and S. Sato. 2019. Time courses of BOLD responses during transcranial near-infrared laser irradiation. *Brain Stimulation: Basic, Translational, and Clinical Research in Neuromodulation* 12(3): 778–780.
87. Sutalangka, C., et al. 2013. Laser acupuncture improves memory impairment in an animal model of Alzheimer's disease. *Journal of Acupuncture and Meridian Studies* 6 (5): 247–251.
88. Purushothuman, S., et al. 2014. Photobiomodulation with near infrared light mitigates Alzheimer's disease-related pathology in cerebral cortex–evidence from two transgenic mouse models. *Alzheimer's Research & Therapy* 6 (1): 2.
89. Comerota, M.M., B. Krishnan, and G. Taglialatela. 2017. Near infrared light decreases synaptic vulnerability to amyloid beta oligomers. *Scientific Reports* 7 (1): 15012.
90. Han, M., et al. 2018. Near infra-red light treatment of Alzheimer's disease. *Journal of Innovative Optical Health Sciences* 11 (01): 1750012.
91. Cho, G.M., et al. 2018. Photobiomodulation using a low-level light-emitting diode improves cognitive dysfunction in the 5XFAD mouse model of Alzheimer's disease. *The Journals of Gerontology: Series A*.
92. Zhang, Z., et al. 2019. *Activation of PKA/SIRT1 signaling pathway by photobiomodulation therapy reduces Aβ levels in Alzheimer's disease models. Aging Cell*
93. Zhang, J., et al. 2019. Illumination with 630 nm Red light reduces oxidative stress and restores memory by photo-activating catalase and formaldehyde dehydrogenase in SAMP8 mice. *Antioxidants & Redox Signaling* 30 (11): 1432–1449.
94. Yue, X., et al. 2019. New insight into Alzheimer's disease: Light reverses Aβ-obstructed interstitial fluid flow and ameliorates memory decline in APP/PS1 mice. *Alzheimer's & Dementia: Translational Research & Clinical Interventions* 5: 671–684.
95. Zinchenko, E., et al. 2019. Pilot study of transcranial photobiomodulation of lymphatic clearance of beta-amyloid from the mouse brain: Breakthrough strategies for non-pharmacologic therapy of Alzheimer's disease. *Biomedical Optics Express* 10 (8): 4003–4017.

Photobiomodulation Therapy for Parkinson's Disease

10

10.1 Parkinson's Disease

Parkinson's disease (PD) is the second most common neurodegenerative disorder after Alzheimer's disease (AD), affecting around 2–3% of the population aged over 65 years [1]. The incidence ranges somewhere between 14 per 100,000 people of all ages, and 160 per 100 000 in people aged over 65 in high-income countries [2]. Intriguingly, men have a higher lifetime risk (2%) to develop PD, compared to women (1.3%) in subjects aged over 40 in the United States [3]. PD prevalence seems to be lower in Africa than in Europe and the Americas. However, the risk is similar in Asia to Europe and the Americas [4–6]. The black population are more likely affected by PD than their white counterparts [7]. The highest incidence is among people of Hispanic origin, while Asian people are somewhere between Hispanics and blacks [8].

PD is thought to be a direct result of dopaminergic nigrostriatal denervation in the mesencephalic substantia nigra (SN) [9]. This, in turn, leads to deficiency of striatal dopamine. From a molecular perspective, intracellular inclusion bodies of misfolded α-synuclein aggregations (known as Lewy bodies) are a neuropathological hallmark of PD. Apart from Lewy bodies, other thread-like aggregations are seen, which are called Lewy neurites, and are found in the involved nerve cells [10]. These inclusions are not only found in dopaminergic neurons of the SN, but they are also found in other parts of the central and peripheral autonomic (but not sensory and motor) nervous system later in the course of the disease [1]. These regions include the dorsal motor nucleus of the vagus (DMV) in the medulla oblongata [11]. Other areas that are involved in PD are the medullary reticular formation, the raphe nuclei, the locus coeruleus, the pedunculopontine nuclei, the ventral tegmental area, and the retrorubral area [12].

The main cause of PD has not yet been completely determined. Several risk factors are responsible for PD. Dairy products, possibly due to their urate-lowering effects, have

F. Salehpour et al., *Photobiomodulation for the Brain*, Synthesis Lectures on Biomedical Engineering, https://doi.org/10.1007/978-3-031-36231-6_10

been shown to increase the risk of PD [13]. Evidence also suggests that exposure to pesticides such as 1-methyl,-4-phenyl-1,2,3,6-tetrahydropyridine (MPTP), increases the risk of developing PD. However, the risk linked to specific compounds continues to be unclear [14]. This could be due to the effects of pesticides on mitochondrial complex I (such as rotenone), or their oxidative stress-inducing effects [15]. Other studies have shown a positive link between methamphetamine use and PD [16, 17]. Among cancers, melanoma is associated with an increased risk of PD [18]. Traumatic brain injury causing disruption of the blood–brain barrier, neuroinflammation, mitochondrial dysfunction, excitotoxicity, and accumulation of α-synuclein can all contribute to the pathophysiology of PD [19]. A substantial increase in PD risk has been found in patients with type 2 diabetes. However, no association was found between body mass index (BMI) and the risk of PD [20, 21]. A slightly lower risk of PD has been reported in alcohol drinkers compared to non-drinkers [22]. On the other hand, various factors such as tobacco smoking, coffee and caffeine, green or black tea, urate, physical activity, non-steroidal anti-inflammatory drugs (NSAIDs), calcium channel blockers, statins, flavonoids, and various dietary profiles have been shown to exert protective effects against PD [13].

From the histopathological view, PD progression can be classified into six distinct stages. Stages 1 and 2 are the pre-symptomatic stages of PD during which, the aggregation of α-synuclein is limited to areas such as the olfactory bulb, medulla oblongata, pontine tegmentum, and anterior olfactory nucleus. Subsequently, during the 3rd and 4th stages of the disease, SN and other nuclei located in the mid-and forebrain become involved. It is at this stage that the clinical manifestations of PD become obvious, and the patients become symptomatic. During the end stages of the disease, i.e., stages 5 and 6, the PD pathology spreads to the neocortex, and a wide range of nonmotor manifestations can develop [23].

PD can present with a range of motor or non-motor symptoms. However, these manifestations do not appear until at least 80% of the dopaminergic cells are lost in the SN [24]. The main motor symptoms of PD are bradykinesia, rest tremor or postural instability, and rigidity, which appear early in the course of the disease [9]. However, before making the diagnosis, other conditions, which cause similar symptoms should be excluded. These conditions include (but are not limited to) repeated strokes, repeated head injury, confirmed encephalitis, and oculogyric crises [25]. During step three of the diagnosis, findings should be sought in the affected subject, such as unilateral initiation of the symptoms, favorable therapeutic response to levodopa, and induction of dyskinesia by dopaminergic therapeutic drugs [10]. Other motor symptoms of PD are hypomimia or an expressionless face, micrographia or a decrease in the amplitudes of handwriting, the pill rolling form of limb tremor, disturbance in speech, dysphagia, dystonia, and a shuffling gait [26–28]. Besides these motor signs and symptoms, other nonmotor symptoms can also be found in both the earlier and later stages of PD. Premotor symptoms may include apathy, sleep deprivation, daytime sleepiness, anhedonia, mood disorders, loss of smell and taste, cognitive problems, fatigue, and pain. Due to the involvement of the autonomic nervous system, autonomic symptoms such as constipation, postural hypotension, difficulties in rectal evacuation, postprandial fullness, urinary frequency, urgency and incontinence, nocturia, erectile dysfunction, and hyperhidrosis may also occur [29]. As the disease progresses the nonmotor symptoms become more complex and troublesome to the patients [10]. Unfortunately, PD patients also suffer from a wide variety of neuropsychiatric symptoms such as visual hallucinations or illusions, and risk-taking behavior such as gambling, anxiety, and depression [30–32].

The diagnosis of PD is mainly clinical. Accordingly, clinical guidelines have been developed for this purpose. The criteria of the UK Parkinson's Disease Society Brain Bank is a set of guidelines with a diagnostic accuracy of up to 90%. These criteria assert that bradykinesia should be present, along with at least one of the following symptoms, rigidity, resting tremor (4–6 Hz), and postural instability. However, as described earlier, steps two and three should follow step 1 of the diagnosis procedure [33]. Apart from clinical diagnosis, magnetic resonance imaging (MRI) should be used to rule out other possibilities. Novel technologies such as positron emission tomography (PET) or single-photon emission computed tomography (SPECT) with a radiolabeled dopamine transporter ligand, transcranial Doppler ultrasonography, morphometric and functional MRI, and perfusion imaging can be used to distinguish idiopathic PD from other parkinsonian syndromes [34]. Recently, the role of biomarkers in the diagnosis and treatment of PD has been highlighted. However, a single biomarker is unlikely to be sufficient when used alone, and a combination of genetic (e.g., glucocerebrosidase [GBA], PTEN-induced kinase 1 [PINK1], SNCA, Parkin, DJ-1, and leucine-rich repeat kinase 2 [LRRK2]), and laboratory (e.g., biopsy, saliva, plasma, cerebrospinal fluid) biomarkers combined with neuroimaging, could be used to diagnose PD more accurately [35].

Pharmaceutical therapy using dopaminergic agents is the cornerstone of PD treatment. Levodopa, dopamine agonists, and monoamine oxidase B inhibitors are some of the drugs, which are currently used to treat PD patients. No proven superiority exists for any class of these drugs. However, several factors such as symptom severity, patient preference, and cost are decisive in the choice of medication. Dopamine agonists have the advantage of having less serious side effects compared to levodopa. However, their efficacy decreases over time (around ten years). On the other hand, the monoamine oxidase B inhibitor rasagiline has shown some neuroprotective properties. Nevertheless, treatment with this drug is expensive, and the cost is not covered by several insurance agencies [36, 37]. Unfortunately, during treatment with levodopa, some patients develop levodopa-induced dyskinesia (LID) which further complicates PD treatment. Dopamine agonists are better alternatives in this regard, however, they are associated with an increased risk of neuropsychiatric adverse effects. Other medications such as amantadine, anticholinergic drugs such as trihexyphenidyl, or beta-blockers may be started alongside levodopa to restrict levodopa-associated motor fluctuations or LIDs [36]. A variety of other medications have also been used to treat nonmotor symptoms, which are beyond the scope of this chapter.

Some non-pharmacological therapies are also under investigation for PD. These interventions include (but are not limited to) exercise, computer-based cognitive training, cognitive gaming, dietary modifications, transcranial magnetic stimulation, transcranial direct current stimulation, speech and language therapy, and complementary interventions [38]. Deep brain stimulation (DBS) is an approved therapeutic option for advanced PD [39]. A six-month trial of DBS to the subthalamic nucleus in patients under 75 years of age with severe motor complications of PD, showed that DBS was more effective than drugs alone [40]. Another five-year open follow-up trial also showed that bilateral stimulation of the subthalamic nucleus using DBS significantly improved mobility, and decreased dyskinesia in advanced PD [41]. Evidence suggests that similar promising improvements in motor function can be achieved after either pallidal or subthalamic stimulation [42].

10.2   Animal Models of Parkinson's Disease

Animal models are extremely useful to study the pathophysiology of PD and the effects of various treatments. These models are widely used for the above-mentioned purposes. Some models are created by the local or systemic injection of neurotoxins, such as 6-hydroxy-DOPA (6-OH-DOPA) or MPTP that destroy the dopaminergic nigrostriatal pathway and reproduce many of the pathological and behavioral characteristics of PD. This approach can be used in animals such as rodents or primates. Transgenic models can also be used to study the pathophysiology of PD [43, 44]. Newer rodent models are also under investigation, as our understanding of the disease deepens. Newer agents producing PD-like conditions in experimental animals, include lactacystin, epoximycin, or inflammogens like lipopolysaccharide (LPS), and proteasome inhibitors such as PSI [44]. However, each model has its own advantages and disadvantages and none of the present models are able to replicate all the features of PD.

10.3   Photobiomodulation Therapy for Parkinson's Disease

10.3.1 Photobiomodulation Therapy for Parkinson's Disease in Animal Models

The first evidence to report that low-levels of red or NIR light could exert a neuroprotective effect after a Parkinsonian insult was provided by an in vitro study. In 2008, it was reported that twice a day 670 nm LED irradiation could decrease the number of cortical and striatal neurons undergoing apoptosis caused by exposure to a parkinsonian toxin, either 1-methyl-4-phenylpyridinium (MPP^+) or rotenone [45]. In the same year, it was also shown that 670 nm LED treatment (4 J/cm^2) twice a day for 2 days during MPP^+ (250 μM) or rotenone (200 nM) exposure, partially but significantly, protected cortical and striatal neurons from dying, along with an increase in ATP levels [46]. In cultures of human neuroblastoma cells engineered to overexpress α-synuclein, 810 nm laser irradiation (2 J/cm^2) was also shown to significantly increase mitochondrial function and decrease oxidative stress after MPP$^+$ exposure [47].

Following on from the in vitro studies, the neuroprotective effect of PBMT in various animal models of PD was investigated in vivo by several research groups [48]. So far, twenty-two studies have reported experiments in rodents, five articles have reported studies in primates (*Macaca fascicularis*, the macaque monkey), and only one study was performed in a Pink1 mutant PD fruit-fly model. Of twenty-two articles in rodents, sixteen evaluated the effects of PBM therapy in mice, of which thirteen were on albino BALB/c mice and three were on the C57BL/6 black mice. Besides, six rodent studies were carried out on rats, of which five were on the Sprague–Dawley strain and one was on the albino Wistar strain. Note that in one animal study, experimental procedures were performed on three different animal species, namely BALB/c mice, Wistar rats, and macaque monkeys; and also in one study, two different types of irradiation protocols, namely, transcranial or remote-tissue irradiation were applied. To induce the PD model, MPTP has been injected in both mice and primates. However, other models used 6-OH-DOPA in rats, and rotenone in Drosophila Pink1 mutants. In these animal studies, the neuroprotective effects of PBM therapy were assessed in different brain regions such as the SNc, subthalamic nucleus (STN), zona incerta (ZI), zona incerta-hypothalamus (ZI-Hyp), caudate-putamen (CPu), striatum, and periaqueductal gray matter (PaG). The light irradiation approaches included the transcranial method (using hand-held device), intracranial method (via implantation of an optical fiber into the region of interest inside the brain), systemic PBM method (via remote-tissue irradiation), and laser acupuncture method have all been investigated by researchers. Despite a variety of irradiation methods being used in these animal models, the effect of neuroprotection was fairly similar. Whole-body PBM therapy was also performed in one study using the Pink1 Drosophila mutant PD model. PBM therapy was administered with red/far-red wavelengths (627 nm [one study], 630 nm [one study], 670 nm [twenty-one studies], and 675 nm [two studies]), while, four studies used NIR wavelengths (808 nm) and only in one study blue wavelength (405 nm) was delivered to an acupuncture point. It must be emphasized that although most of the animal studies have used red or far-red wavelengths (630 nm, 670 nm, or 675 nm), this does not necessarily mean that red wavelengths are better than NIR wavelengths (808 nm). The operation mode applied in all of the animal studies was continuous wave [48].

In MPTP-treated albino BALB/c mice [49–57] and 6OHDA-lesioned Wistar rats [58, 59], PBM therapy using far-red (670 nm) and NIR (810 nm) wavelengths have been shown to preserve dopaminergic cells from death. In these animal models of PD, PBM therapy saved both the expression of the dopaminergic phenotype—as measured by tyrosine hydroxylase (TH) expression (functional neuroprotection)—and the neuronal cells themselves—as measured by Nissl staining (true neuroprotection) [57]. It is interesting to mention that the neuroprotective outcomes were similar whether the PBM therapy was applied at the same time or before/after the induction of PD, suggesting that PBM therapy can both protect normal cells against an insult and can also rescue damaged cells following an insult [50, 55, 60]. It should be also stated that many of these animal studies were acute models, with survival periods of up to 7 days [49–57], while others used more chronic models, with survival times of several weeks [50].

Initial research produced promising findings in the above-mentioned toxin-induced rodent PD models, and led Australian researchers to test an MPTP-induced primate PD model, to evaluate the neuroprotective benefits of intracranial PBM therapy [61–63]. In the sub-acute PD model with a survival period of up to three weeks, intracranial 670 nm PBM-treated monkeys showed a greater number of surviving nigral neurons, both TH1 and Nissl-stained [61]. Furthermore, there were more nerve terminals in the striatum in the PBM-treated monkeys compared to those who received sham treatment [61]. Subsequent research also revealed a dose–response neuroprotective effect of PBM therapy, so that a lower dose (25–35 J: [61]) showed a stronger beneficial effect than a higher dose (125 J: [62]). It must be emphasized that there was no evidence of toxicity from the intracranial PBM device, although the fiber-optic was implanted within the brain in direct contact with deep neurons [61–63].

Besides, there have been investigations of the neuroprotective effects in transgenic rodent models of PD [64, 65]. In the K369I transgenic mouse model of frontotemporal dementia, which also exhibits progressive neurodegeneration of dopaminergic cells in the SNc with subsequent parkinsonian signs [66], 670 nm LED PBM therapy has been shown to reduce oxidative stress and hyperphosphorylated tau, and increase dopaminergic cell survival in the SNc region [65]. In an α-synuclein genetic rat model of PD [64], daily 808 nm laser PBM therapy for 4 weeks also resulted in more dopaminergic neurons (TH1) in the SNc and nerve terminations in the striatum. There is also evidence that 808 laser irradiation can rescue mitochondria defects and improve flight, along with promoting mitochondrial complex IV dependent respiration in a pink1 transgenic fruit-fly model [67].

10.3.2 Photobiomodulation Therapy for Parkinson's Disease in Human Clinical Studies

Although there have been several experimental reports on the effectiveness of PBM in animal models of PD [48], there have been few clinical studies in human patients. Up to now, the majority of the clinical PBMT research has shown neurotherapeutic benefits of transcranial PBM therapy in conditions such as traumatic brain injury (TBI), stroke, and depression, in which the target regions were the cortical area of the brain. However, PD pathogenesis is related to abnormalities in the SNc, a midbrain nucleus that is located at a depth of 80–100 mm from the cortex. Transmission studies have demonstrated that light in the far-red to NIR spectrum may not reach the human brain more deeply than

20–30 mm [68], which would be a clear limitation to the use of transcranial PBM therapy for human PD patients. Nevertheless, there are still some promising clinical findings in the literature which will be described in this section.

The effects of in vivo 632.8 nm laser PBM on the neurological status and the activity of some enzymes were studied in 70 patients with PD [69]. PBM significantly enhanced neurological status as evaluated by the Fan–Elton scale (UPDRS), and this improvement was accompanied by modulation of monoamine oxidase B (MAO-B) and Cu/Zn-SOD activity [69].

In a non-controlled, non-randomized study [70], eight PD patients (18–80 years old) received PBM therapy (PL5000, Erchonia Medical Inc.) daily for 2 weeks. PBM was applied to the brain stem, bilateral occipital, temporal, parietal, and frontal regions, as well as irradiation along the sagittal suture. The severity of the patient's symptoms such as balance, gait, freezing, cognitive function, rolling in bed, and difficulties with speech was analyzed using a Visual Analog Scale (VAS). Compared with the baseline VAS rating, an average mean reduction of 1.87 and 2.22 was observed for gait and cognitive function, respectively. In addition, difficulty with speech and occurrence of freezing were significantly lower at the study endpoint with a mean reduction of 2.22 and 1.28, respectively [70].

In an incidental case study [71], intraoral PBM resulted in a reduction of the clinical signs in one PD patient who had visited a clinic for his dental problems. The 660 nm laser probe was directed to the posterior region of the cranium/upper neck and the irradiation was applied for ~2–3 min where it was proposed that primary or scattered light had a chance of irradiating the PD-related cerebral structures via the deep oral cavity, producing beneficial effects [71].

In a randomized controlled trial study, Santos et al. [72] reported preliminary results on the effects of transcranial PBM in 35 PD patients (17 active and 18 placebo). Participants received 9 min of PBM therapy (670 nm LED), twice weekly on non-consecutive days for 9 weeks. Results from the active treatment group showed gait improvements in fast rhythm of the ten-meter walk test (TMWT) with an increase of 0.33 m/s on average. In addition, although there were no significant changes in the motor part of the Movement Disorders Society-Unified Parkinson's Disease Rating Scale (MDS-UPDRS) and the Short Parkinson's Evaluation Scale/Scales for Outcomes in Parkinson's Disease (SPES/SCOPA), there was a trend toward improvement in the timed up and go (TUG) test at the study endpoint [72].

In 2019, John Mitrofanis and his research team reported the effects of PBM on eight movement disorder patients using "PBM buckets" in two excellent papers [73, 74]. Their reports were the first to document the effect of transcranial PBM on PD patients over a long period, at 6 ($n = 1$), 8 ($n = 1$), 12 ($n = 2$), 14 ($n = 2$), and 24 ($n = 2$) months. The first patient was a 63-year-old male, diagnosed with PD 2.5 years previously. Nine months after diagnosis, the patient started 10 min of transcranial PBM therapy twice daily, using a bucket helmet lined internally with strips of 670 and 810 nm LEDs. Four weeks after the first PBM therapy, there was an evident reduction in his tremor. After eight weeks, there was a noticeable improvement in his PD symptoms. After 2 years of use, out of the patient's 12 initial signs and symptoms (resting tremor, akinesia, gait change, impaired fine motor skills, poor facial movement, trouble sleeping and swallowing, persistent cough, fatigue, low self-esteem, depression, and writing function), eleven of them were improved (90%) after PBM therapy, whereas one remained the same (10%) and none deteriorated. Of note, over this extensive 2-year period, there was no clear decline in his writing, it still being very legible as measured by ImageJ software [73]. The second patient was a 61-year-old male, who was diagnosed with PD six years previously. Three and half years after diagnosis, the patient started 10 min sessions of transcranial PBM therapy twice daily, using a bucket helmet with 670 and 810 nm LEDs. Four weeks after the first PBM session, the patient resumed his usual activities and he was more confident, socially interactive, and could think more clearly. Over the next few months of therapy, there was an obvious improvement in his sleep, speech, overall energy, and gait, together with his expression becoming more animated. Analysis of his writing over 2 years, showed that his writing stabilized and did not deteriorate. Overall, out of the patient's eleven initial signs and symptoms (resting tremor, impaired fine motor skills, poor facial movement, gait change, fatigue, apathy, difficulty maintaining thoughts, low self-esteem, hesitant speech, trouble sleeping, and writing function), seven of them improved (90%) after PBM therapy, whereas one stayed the same (10%) and none deteriorated [73]. The third patient was a 73-year-old male, diagnosed with PD 14 months previously. Soon after diagnosis, the patient commenced 15 min of PBM therapy daily, using a 670 and 810 nm LED bucket. After 14 months of PBM therapy, although there was no clear improvement in the patient's two initial signs and symptoms, he had not developed any other Parkinsonian signs or symptoms. In addition, although no improvement was evident in his writing, there was no deterioration over this extensive 14 month period [73]. The fourth patient was a 75-year-old male, diagnosed with PD five years previously. Three years after diagnosis, the patient started 30 min of transcranial PBM therapy daily using a bucket helmet with 670 nm LED for the first 3 months, and then 850 nm LEDs were added as well. After six weeks of PBM therapy, the patient started to improve his sense of smell and also his tremor became less prominent. After five months, there was a continued improvement in the sense of smell, tremors, and akinesia. The patient's right-sided rigidity showed little sign of improvement, however, although there were no improvements in the writing, there was no deterioration over this extensive 14 month period. Overall, of the patient's five initial signs and symptoms, three showed improvement (60%) after PBM therapy, whereas two stayed the same (40%) and none deteriorated [74]. The fifth patient was a 64-year-old male, diagnosed with PD 12 years previously. Some eleven years after diagnosis, the patient started 20 min of transcranial PBM therapy daily using a bucket helmet with 670 nm LEDs for the first 4 months, and then 850 nm LEDs were added. After eight months of PBM therapy, the patient and his wife reported some subtle but distinct changes in his daily number of freezing episodes. The time of effectiveness of medication was improved from 75 to 90 min with PBM. His speech was louder and was more rapid than before. The patient's gait, anxiety, and social interaction improved. Although there were no improvements in his writing, there was no deterioration over this extensive 12-month period. Overall, of the patient's 14 initial signs and symptoms i(slow gait, muscle spasms and stiffness, trouble swallowing, soft voice, bladder urgency, itchy feet, difficulty sleeping, stress, impaired social interaction, poor tolerance, low confidence, poor writing), 7 improved (50%) after PBM therapy, whereas 7 stayed the same (50%) and none deteriorated [73]. The sixth patient was a 64-year-old male, diagnosed with PD four years previously. Two years after diagnosis, the patient started 30 min of PBM therapy daily, using a 670 LED bucket. After three months of PBM therapy, the patient and his wife reported faster times for his daily run. This was followed at 8 months by improvement in his walking, being much quicker and with more arm movement. At this time, the patient added 810 nm LEDs to his helmet and used the two wavelengths sequentially for 15 min daily. The patient also reported improvements in his excessive sweating, muscle cramps, and stiffness. Although no improvement was evident in his writing, there was no deterioration over this period. Overall, of the patient's eight initial signs and symptoms (resting tremor, gait change, muscle cramps, stiffness, constipation, profuse sweating on exertion, difficulty swallowing, and poor writing), four improved (55%) after PBM, while three stayed the same (35%) and one deteriorated [73]. The seventh patient was a 68-year-old male, diagnosed with progressive supranuclear palsy three years previously. Two years after diagnosis, the patient started transcranial PBM therapy, using a bucket helmet lined internally with strips of 670 and 940 nm LEDs. 5 months after the first application, the 940 nm LEDs were replaced by 810 nm LEDs. During this period, the patient also used an intranasal 660 nm LED applicator. About a month after the first PBM therapy, improvement was obvious in many of the patient's signs and symptoms, his speech became clearer and more understandable, and his coughing was not as frequent or explosive. The analysis of the patient's writing also indicated an 80–85% increase in the perimeter of distance and a 60% increase in the area of words over the 6-month period. Overall, of the patient's eight initial signs and symptoms (impaired vertical gaze, impaired speech, impaired fine motor skills, difficulty in maintaining balance, persistent cough, emotional lability, frequent bouts of anger and mood change, impaired sleeping, and poor writing), five showed enhancement (65%) after PBM therapy, whereas three stayed the same (35%) and none deteriorated [74]. The eighth patient was a 75-year-old male, diagnosed with PD 14 years previously. Thirteen years after diagnosis, the patient started 20 min of only intranasal 660 nm LED PBM therapy once a day for 8 months. Soon after the first use of the nasal applicator, the patient and his wife reported a better mood, improved facial movement, and more energy. The patient's urinary frequency and constipation resolved, his sleep was much improved, and there was less disturbed dreaming. Also, although no improvement was evident in his writing, there was no deterioration over these extended 8 months period. Overall, of the patient's 16 major signs and symptoms (tremor, cogwheel rigidity, impaired facial movement and gait, diminished sense of smell, fatigue, anxiety, slow thinking, mild urinary frequency and constipation, memory impairment, depression, troubled sleep, restless leg, disturbed dreaming, and poor writing) 7 improved (45%) after PBM therapy, whereas 9 stayed the same (55%) and none deteriorated [73].

Chinese researchers have also investigated the potential neurotherapeutic effects of intranasal PBM therapy in PD patients. In the first report, Li et al. [75] reported that intranasal laser PBM therapy (633 nm, 3.5–5.5 mW, 30 min) once daily for 10 days, significantly decreased serum cholecystokinin-octapeptide levels. Furthermore, PD symptoms in 60% of patients (26 out of 43) were improved as indicated by the Webster Scale scores. Similarly, using the same treatment parameters described in [75], other researchers [76] reported that PBM could significantly reduce PD symptoms in 89% of the patients (31 out of 36), in which 28% of them showed substantial improvement and 58% had a gradual improvement. In another study [77], intranasal laser PBM therapy (633 nm, 3.5–4.5 mW, 30 min) once daily for 20 days, significantly improved PD symptoms in 66% of patients (31 out of 47), along with a decrease in malondialdehyde and increase of superoxide dismutase and melatonin levels.

In 2019, Ivan Maksimovich [78] for the first time reported the beneficial effects of intracerebral transcatheter laser PBM therapy in the treatment of ischemic and neurodegenerative lesions of cerebral white matter in sixty-two patients with vascular Parkinsonism (an after effect of a stroke) (37 active and 25 placebo). Intracerebral transcatheter PBM therapy was performed using continuous 632.8 nm laser with an irradiation time of 1200–2400 s. Results from the active treatment group showed an immediate improvement according to multiple-gated acquisition (MUGA) as well as significant angiogenesis, and collateral and capillary revascularization in 35 (94.60%) patients. At 1–6 months after intracerebral transcatheter laser PBM therapy, 35 (94.60%) patients showed a significant improvement in mental and motor function; 37 (100%) patients showed improvement in blood flow and volume pulse blood filling in the cerebral hemisphere as measured by scintigraphy (SG) and rheoencephalography (REG); and 37 (100%) patients showed a tendency for a decrease in general involutive changes in the brain, and the narrowing of the subarachnoid space as measured by CT and MRI. Furthermore, at 12 to 24 months after intracerebral transcatheter laser PBM therapy, 35 (94.60%) patients showed the maximum positive clinical benefit with restoration of cognitive and motor function; 35 (94.60%) patients showed that the positive benefits persisted throughout the entire observation period as measured by SG and REG. According to the CT and MRI data, 34 (91.89%) patients showed a decrease in the general involutive changes in the brain, 32 (86.49%) patients showed a narrowing of the Sylvian fissure, 14 (37.84%) patients showed decreased nonocclusive hydrocephalus, and 30 (81.08%) patients showed decreased gliosis (see Tables 10.1 and 10.2).

Table 10.1 Summary of in vivo studies on the effects of photobiomodulation therapy in Parkinson's disease

Study/year	Animal/model	Light source	Wavelengths	Irradiation parameters	Irradiation approach/sites	Findings
Shaw et al., 2010 [49]	Mouse (MPTP-induced Parkinson's disease model)	LEDs, WARP 10, Quantum Devices (Barneveld, WI, USA)	670 nm	40 mW/cm^2 at scalp, 5.3 mW/cm^2 inside skull, 0.47 J/cm^2 per irradiation (total of 4 irradiations over 30 h), 90 s, irradiation area of 10 cm^2	Transcranially; holding probe at 1 cm from the head	Increased TH$^+$ terminals in the caudate-putamen complex; no effect on the overall volume of the SNc and ZI-Hyp; increased TH$^+$ cells in the SNc and ZI-Hyp regions; no effect on the morphology of TH$^+$ cells in both the SNc and ZI-Hyp; increased number of TH$^+$ cells in the SNc (in both 50 and 100 mg/kg MPTP doses); no effect on the number of TH$^+$ cells in the ZI-Hyp (in 50 and 100 mg/kg MPTP doses)

(continued)

Table 10.1 (continued)

Study/year	Animal/model	Light source	Wavelengths	Irradiation parameters	Irradiation approach/sites	Findings
Peoples et al., 2012 [79]	Mouse (MPTP-induced Parkinson's disease model)	LEDs, WARP 10, Quantum Devices (Barneveld, WI, USA)	670 nm	5 J/cm^2 over 10 sessions, 90 s/session	Transcranially; holding probe at 1–2 cm from the head	For both simultaneous and post-treatment series: increased TH$^+$ cell number in the SNc, but not in the PaG and ZI-Hyp regions
Shaw et al., 2012 [80]	Mouse (MPTP-induced Parkinson's disease model)	LEDs, WARP 10, Quantum Devices (Barneveld, WI, USA)	670 nm	0.5 J/cm^2, 90 s	Transcranially; holding probe at 1–2 cm from the head	Decreased Fos$^+$ cell numbers in STN and ZI after acute (~ 1 day) and chronic (5 weeks) MPTP insult
Peoples et al., 2012 [60]	Mouse (MPTP-induced Parkinson's disease model)	LEDs, WARP 10, Quantum Devices (Barneveld, WI, USA)	670 nm	40 mW/cm^2 at scalp, 0.5 J/cm^2, 90 s	Transcranially; Just above the mouse's head and in full view of their eyes	For all series and regimens: no effect on the retinal areas. For all groups except simultaneous series with acute regimen: increased TH$^+$ cell number in the retinae

(continued)

Table 10.1 (continued)

Study/year	Animal/model	Light source	Wavelengths	Irradiation parameters	Irradiation approach/sites	Findings
Moro et al., 2013 [81]	Balb/c and C57BL/6 mouse (MPTP-induced Parkinson's disease model)	LEDs, WARP 10, Quantum Devices (Barneveld, WI, USA)	670 nm	0.47 J/cm² per session (total of 4 sessions over 30 h), 90 s, 4 simultaneous irradiations over 30 h	Transcranially; holding probe at 1–2 cm from the head	For BALB/c mouse: increased TH⁺ cell number in the SNc; improved locomotor activities via increase of velocity and high mobility, and decrease of immobility. For C57BL/6 mouse: no effect on the TH⁺ cell number in the SNc; no effect on the locomotor activities
Purushothuman et al., 2013 [65]	Mouse (K369I tau transgenic Parkinson's disease model)	LEDs, WARP 10, Quantum Devices (Barneveld, WI, USA)	670 nm	40 mW/cm² at scalp, 0.5 J/cm², 90 s, 20 irradiations over 4 weeks	Transcranially; holding probe at 1–2 cm from the head	Decreased markers of oxidative stress, over expression of hyperphosphorylated tau, and increased TH⁺ cell number in the SNc
Vos et al., 2013 [67]	Drosophila Pink1 null mutants (rotenone-induced Parkinson's disease model)	Laser	808 nm	25 mW/cm², 2.5 J/cm², 100 s, on session (single dose)	Whole-body	Improved CCO-dependent oxygen consumption and ATP production; rescued major systemic and mitochondrial defects

(continued)

Table 10.1 (continued)

Study/year	Animal/model	Light source	Wavelengths	Irradiation parameters	Irradiation approach/sites	Findings
Wattanathorn and Sutalangka, 2014 [82]	Rat (6OHDA-induced Parkinson's disease model)	Laser	405 nm	100 mW, 10 min, once daily for 14 days	Laser acupuncture; at HT7 acupoint	Improved spatial memory in Morris water maze test; attenuated the decreased neuron density in CA3 and dentate gyrus, but not CA1 and CA2 regions; decreased activity of monoamine oxidase-B and acetylcholinesterase in the hippocampus; mitigated the decreased GSH-Px activity and the elevation of MDA level
Johnstone et al., 2014 [53]	Mouse (MPTP-induced Parkinson's disease model)	LEDs, WARP 10, Quantum Devices (Barneveld, WI, USA)	670 nm	40 mW/cm^2, 4 J/cm^2, 90 s	– Transcranial irradiation to the head – Remote irradiation to the dorsum	At both head or body irradiations: increased TH$^+$ cell numbers in SNc (at 50 mg/kg MPTP); increased glial cell numbers in SNc (at 75 mg/kg MPTP)

(continued)

Table 10.1 (continued)

Study/year	Animal/model	Light source	Wavelengths	Irradiation parameters	Irradiation approach/sites	Findings
Moro et al., 2014 [52]	Mouse (MPTP-induced Parkinson's disease model)	LEDs, (SMT 670, Epitex) coupled with an optical fiber (FT300EMT, Thorlabs)	670 nm	1.5 mW/cm^2 and 0.13 J/cm^2 (pulse irradiation); 14.5 mW/cm^2 or 7516.8 J/cm^2 (continuous irradiation), fiber diameter of 300 $^-$m, continuous irradiation for 6 days	Intracranially; implant site: lateral ventricle	For pulse irradiation group: significantly increased TH$^+$ cell number in the SNc For continuous irradiation group: non-significantly increased TH$^+$ cell number in the SNc
Reinhart et al., 2015 [54]	Mouse (MPTP-induced Parkinson's disease model)	LEDs, LED EPITEX SMT810, EPITEX Inc. (Kyoto, Japan)	810 nm	160 μW, 57.6 mJ (at skull for total of 4 irradiations), 90 s/ session, 4 simultaneous irradiations over 30 h	Transcranially; full head irradiation	Improved locomotor activity at different time points including at immediately after first MPTP injection in OFT, at after second PBM, at after fourth PBM, and 6 days after the last MPTP injection; increased TH$^+$ cell number in the SNc

(continued)

Table 10.1 (continued)

Study/year	Animal/model	Light source	Wavelengths	Irradiation parameters	Irradiation approach/sites	Findings
Darlot et al., 2016 [61]	Monkey (MPTP-induced Parkinson's disease model)	Laser, coupled with an optical fiber (HCP-MO200T)	670 nm	10 mW; 25 or 35 J over 5 or 7 days, respectively; with 5 s ON/60 s OFF	Intracranially; implant site: 4 mm rostral to posterior commissure, 3 mm below anterior commissure-posterior commissure line	Improved PD signs and locomotor activities including movement and velocity; increased number of nigral Nissl-stained and TH^+ cells and striatal TH^+ terminals; no observable behavioral and tissue deficits
Oueslati et al., 2015 [64]	Rat (AAV-based genetic Parkinson's disease model)	Laser, RLTMDL_808-2W, Roithner Lasertechnik GmbH (Vienna, Austria)	808 nm	2.5 or 5 mW/cm^2 at midbrain, 2 irradiation spots of about 1 cm^2, 100 s/day for 4 weeks	Transcranially; irradiation of both sides of the head	Improved motor performance in cylinder test (at 2.5 or 5 mW/cm^2); decreased dopaminergic neuronal loss in the substantia nigra and preserved dopaminergic fibers in the ipsilateral striatum (at 5 mW/cm^2)

(continued)

Table 10.1 (continued)

Study/year	Animal/model	Light source	Wavelengths	Irradiation parameters	Irradiation approach/sites	Findings
Moro et al., 2016 [62]	Monkey (MPTP-induced Parkinson's disease model)	Laser, coupled with an optical fiber (HCP-MO200T)	670 nm	10 mW, 125 J for 25 days continuous irradiation, with 5 s ON/60 s OFF, continuous irradiation for 25 days	Intracranially; implant site: a region close to midline in the midbrain, encompassing the ventral tegmental area	Improved clinical scores as indicated by locomotive traces; increased TH^+ cell number in the SNc; no effect on the striatal TH^+ terminal density
Salgado et al., 2016 [83]	Rat (6OHDA-induced Parkinson's disease model)	LEDs or laser	– LEDs, 627 nm – Laser, 630 nm	LEDs: 70 mW, 70 mW/cm², 4 J/cm², 57 s, once a day for 7 days Laser: 45 mW, 45 mW/cm², 4 J/cm², 88 s, once a day for 7	Transcranially	For laser and LEDs sources: increased locomotive traces in open field test; decreased TNF-α levels; no effect on the IL-4, IL-6, and IL-10 levels For LEDs source: increased IFN-γ levels For laser source: increased IL-2 levels

(continued)

Table 10.1 (continued)

Study/year	Animal/model	Light source	Wavelengths	Irradiation parameters	Irradiation approach/sites	Findings
Reinhart et al., 2016 [58]	Rat (6OHDA-induced Parkinson's disease model)	LEDs, (SMT 670; Epitex) coupled with an optical fiber (FT300EMT; Thorlabs)	670 nm	333 nW or 0.16 mW, 634 mJ or 304 J, fiber diameter of 300 μm, for 23 consecutive days, pulse irradiation ($2\times$/day for 90 s) or continuous irradiation for 23 days	Intracranially; implanted in region near the SNc, incorporating the red nucleus and ventral tegmental area, toward the midline	Decreased rotational behavior at days 14 (at CW of 304 J) and 21 (at CW of 304 J, and PW of 634 mJ); increased TH^+ cell numbers in SNc (at PW of 634 mJ)
Reinhart et al., 2016 [55]	Mouse (MPTP-induced Parkinson's disease model)	LEDs, WARP 10, Quantum Devices (Barneveld, WI, USA)	670 nm	5.3 mW/cm², ~0.5 J/cm² at midbrain, 90 s, $2\times$/day for 2, 4, or 6 days	Transcranially; holding probe at 1–2 cm from the head	Improved locomotor activity in OFT (at all regimens); increased TH^+ cell numbers in SNc (at all regimens)
El Massri et al., 2016 [84]	Monkey (MPTP-induced Parkinson's disease model)	Laser, coupled with an optical fiber (HCP-MO200T)	670 nm	10 mW; 25 or 35 J over 7 days; with 5 s ON/60 s OFF, continuous irradiation for 5 or 7 days	Intracranially; implanted in 1 to 2 mm to the left side of the midline in the midbrain	Decreased number of $GFAP^+$ astrocytes and astrocyte cell body size in the SNc and striatum; decreased microglia cell body size in the SNc and striatum

(continued)

Table 10.1 (continued)

Study/year	Animal/model	Light source	Wavelengths	Irradiation parameters	Irradiation approach/sites	Findings
El Massri et al., 2016 [57]	Mouse (MPTP-induced Parkinson's disease model)	LEDs, WARP 10, Quantum Devices (Barneveld, WI, USA)	670 nm	40 mW/cm^2, 4 J/ cm^2, 90 s, once a day for 2, 4, or 8 days	Transcranially; holding probe at 1 cm from the head	Increased TH$^+$ cell number in the SNc; decreased number of GFAP$^+$ cells in the CPu
El Massri et al., 2017 [85]	Mouse, monkey (MPTP-induced Parkinson's disease model) and rat (6-OHDA-induced Parkinson's disease model)	Laser, coupled with an optical fiber	670 nm	0.16 mW for mouse and rat, and 10 mW for monkey, fiber diameter of 300 μm, continuous irradiation for 2 (mouse), 23 (rat) and 5 (monkey) days	Intracranially; mouse: implanted in lateral ventricle rat and monkey: implanted in midline region of the midbrain	Mouse: no effect Rat: no effect Monkey: increased TH$^+$ cell number and terminal density in the striatum; increased GDNF expression in the striatum
Reinhart et al., 2017 [56]	Mouse (MPTP-induced Parkinson's disease model)	LEDs, Epitex devices (models 670–66-60 and 810–66-60; epoxy lens infrared illuminators)	670 and/or 810 nm	15 or 30 mW, total dosage of ~11 or 22 J, 45 or 90 s, 2×/ day for 2 days	Transcranially; full head irradiation	Improved locomotor activity in OFT and increased TH$^+$ cell numbers in SNc (at all regimens, especially in concurrent and sequential irradiation regimens)
El Massri et al., 2018 [86]	Monkey (MPTP-induced Parkinson's disease model)	Laser, coupled with an optical fiber (HCP-MO200T)	670 nm	10 mW, 25 J, with 5 s ON/60 s OFF, continuous irradiation for 5 days	Intracranially; implanted in 1 to 2 mm to the left side of the midline in the midbrain	No effect on the number and somal sizes of encephalopsin $^+$ cells in the striatum

(continued)

Table 10.1 (continued)

Study/year	Animal/model	Light source	Wavelengths	Irradiation parameters	Irradiation approach/sites	Findings
Kim et al., 2018 [87]	Mouse (MPTP-induced Parkinson's disease model)	LEDs, WARP 10, Quantum Devices (Barneveld, WI, USA)	670 nm	50 mW/cm^2 (at skin), 9 J/cm^2 (at skin), 3 min, twice (24 h apart)	Remotely; irradiation to the dorsum	Increased TH$^+$ cell number in the SNc; no effect on the density of TH$^+$ terminations in the dorsal CPu
O'Brien and Austin, 2019 [88]	Rat (lipopolysaccharide-induced Parkinson's disease model)	LEDs, WARP 10, Quantum Devices (Barneveld, WI, USA)	675 nm	500 mW, 40 mW/cm^2 (at scalp), 3.6 J/cm^2 (at scalp), 88 s, 13 irradiations (once 2 h following the completion of the lipopolysaccharide injection + twice daily for 6 days	Transcranially; holding probe at 1 cm from the head	With 10 μg lipopolysaccharide: increased TH$^+$ cell number in the SNc; no effect on the IBA1$^+$ cell densities in the SNc With 20 μg lipopolysaccharide: no significant effect on the motor behavior in the cylinder, rotarod, and adjusted stepping tests

(continued)

Table 10.1 (continued)

Study/year	Animal/model	Light source	Wavelengths	Irradiation parameters	Irradiation approach/sites	Findings
Miguel et al., 2019 [89]	Mouse (MPTP-induced Parkinson's disease model)	LEDs, WARP 10, Quantum Devices (Barneveld, WI, USA)	675 nm	50 mW/cm^2 (at scalp), 9 J/cm^2 (at scalp), 180 s, once a day for 7 days	Transcranially	Decreased vascular leakage in the SNc and CPu
Ganeshan et al., 2019 [90]	Mouse (MPTP-induced Parkinson's disease model)	LEDs, WARP 10, Quantum Devices (Barneveld, WI, USA)	670 nm	50 mW/cm^2 (at skin), 4.5 J/cm^2 (at skin), 90 s, once a day for 2, 5, or 10 days prior to injection of MPTP	Remotely; irradiation to the dorsum and hind limbs	In PBM (2 and 5 days groups): increased Fos$^+$ cell number in the CPu In PBM (10 days) group: increased TH$^+$ cell number in the SNc; increased Fos$^+$ cell number in the CPu; upregulated cell signaling and migration (including CXCR4$^+$ stem cell and adipocytokine signaling), oxidative stress response pathways and modulated blood–brain barrier

Abbreviations: 6OHDA, 6-hydroxydopamine; α-syn, α-synuclein; CPu, caudate putamen; CW, continuous wave; DMS, delayed match-to-sample; GDNF, glial cell line–derived neurotrophic factor; GFAP$^+$, glial fibrillary acidic protein positive; GSH-Px, Glutathione peroxidase; IFN-γ, interferon-γ; IL, interleukin; LEDs, light-emitting diodes; MDA, Malondialdehyde; MPTP, methyl-4-phenyl-1,2,3,6-tetrahydropyridine; PaG, periaqueductal grey matter; PBM, photobiomodulation; PD, Parkinson's disease; PVT, psychomotor vigilance task; SNc, substantia nigra, STN, subthalamic nucleus; TNF-α, tumor necrosis factor-α; WT, wild type; ZI, zona incerta; ZI-Hyp, zona incerta-hypothalamus

Table 10.2 Summary of clinical studies on the effects of photobiomodulation therapy in Parkinson's disease

Study/year	Subjects (n)	Light source	Wavelengths	Irradiation parameters	Irradiation approach/sites	Findings
Vitreshchak et al., 2003 [91]	Parkinson's disease (70)	Laser, ALOK-1 He–Ne generator	632.8 nm	1 mW, 0.5 W/cm^2, 20 min daily for 5 days	Intravenously	Improved neurological status as assessed by Fan–Elton scale; decreased Mn-SOD activity in platelets
Maloney et al., 2010 [92]	Parkinson's disease (8)	Laser, PL5000, Erchonia Medical Inc. (Melbourne, FL, USA)	NR	Daily for 2 weeks	Transcranially; multiple areas (bilateral occipital, parietal, temporal, frontal lobes and along sagittal sutures)	Improved balance, gait, freezing, cognitive function, rolling in bed, and difficulties in speech assessed by Visual Analog Scale
Santos et al., 2019 [93]	Parkinson's disease (35)	LEDs	670 nm	6 min, 18 irradiations (twice weekly) over 9 weeks	Transcranially; right and left temples	Improved gait and gait speed measured by ten-meter walk test (TMWT)
Hamilton et al., 2019 [94]	Parkinson's disease (6)	LEDs (total of 270 diodes), Eliza, C & D Hamilton (Australia)	Transcranial: 670, 810, and 850 nm Intranasal: 660 nm	Transcranial: 6.96 W (670 nm), 26.4 W (810 nm), and 6 W (850 nm); 600–900 s, 1–2 daily irradiations for various duration of 14–24 months, CW Intranasal: 20 min/day for 8 months	Transcranially; whole head irradiation with LEDs bucket Intranasally; left nose	Improved overall signs and symptoms in majority of patients; improved movement, mood, and confidence; improved writing skill as indicated by improvement in area and perimeter of words

(continued)

Table 10.2 (continued)

Study/year	Subjects (n)	Light source	Wavelengths	Irradiation parameters	Irradiation approach/sites	Findings
Hamilton et al., 2019 [95]	Progressive supranuclear palsy (1); Parkinson's disease (3)	LEDs, Light Ahead Inc. (Rancho Palos Verdes, CA, USA)	Transcranial: 670, 810, 850, and 940 nm Intranasal: 660 nm	Transcranial: 1–2 daily irradiations for various duration of 14–24 months Intranasal: 10 min	Transcranially; whole head irradiation with LEDs helmet Intranasally; left nose	Improved tremor, akinesia, and gait in two patients; improved social interactions and confidence levels in two other patients; improved ability of sensing smell and sleeping patterns; improved writing skill as indicated by improvement in area and perimeter of words

(continued)

Table 10.2 (continued)

Study/year	Subjects (n)	Light source	Wavelengths	Irradiation parameters	Irradiation approach/sites	Findings
Maksimovich, 2019 [78]	Vascular parkinsonism (62)	Laser, ULF-01, Anod, (Bryansk, Russia)	632.8 nm	25–45 mW, 29–106 J, fiber diameter of 25–100 m µ, 20–40 min, 1–2 mm beam spot diameter in the vessel, CW	Intracerebral transcatheter; threading a fiber optic through a catheter in femoral artery (advancing fiber optic to distal site of anterior and middle cerebral arteries)	Immediate results: increased angiogenesis, and collateral and capillary revascularization Early period (1–6 months) results: improved mental and motor functions; improved blood flow and volume pulse blood filling in the cerebral hemispheres; decreased general involutive changes in the brain, narrowing of the subarachnoid space Long-term period (from 1 till 8 years) results: restored mental and motor functions; decreased general brain involutive changes and Sylvian fissure narrowing; decreased nonocclusive hydrocephalus, gliosis, and leukoaraiosis; induced neurogenesis and regenerative cerebral changes

Abbreviations: CW, continuous wave; LEDs, light-emitting diodes; Mn–SOD, manganese superoxide dismutase

References

1. Poewe, W., et al. 2017. Parkinson disease. *Nature Reviews Disease Primers* 3: 17013.
2. Hirtz, D., et al. 2007. How common are the "common" neurologic disorders? *Neurology* 68 (5): 326–337.
3. Elbaz, A., et al. 2002. Risk tables for parkinsonism and Parkinson's disease. *Journal of Clinical Epidemiology* 55 (1): 25–31.
4. Winkler, A.S., et al. 2010. Parkinsonism in a population of northern Tanzania: A community-based door-to-door study in combination with a prospective hospital-based evaluation. *Journal of Neurology* 257 (5): 799–805.
5. Okubadejo, N.U., et al. 2006. Parkinson's disease in Africa: A systematic review of epidemiologic and genetic studies. *Movement Disorders: Official Journal of the Movement Disorder Society* 21 (12): 2150–2156.
6. Zhang, Z.-X., et al. 2005. Parkinson's disease in China: Prevalence in Beijing, Xian, and Shanghai. *The Lancet* 365 (9459): 595–597.
7. Mayeux, R., et al. 1995. The frequency of idiopathic Parkinson's disease by age, ethnic group, and sex in northern Manhattan, 1988–1993. *American Journal of Epidemiology* 142 (8): 820–827.
8. Van Den Eeden, S.K., et al. 2003. Incidence of Parkinson's disease: Variation by age, gender, and race/ethnicity. *American journal of epidemiology* 157 (11): 1015–1022.
9. Magrinelli, F., et al. 2016. Pathophysiology of motor dysfunction in Parkinson's disease as the rationale for drug treatment and rehabilitation. *Parkinson's Disease* 2016.
10. Sveinbjornsdottir, S. 2016. The clinical symptoms of Parkinson's disease. *Journal of Neurochemistry* 139: 318–324.
11. Del Tredici, K., and H. Braak. 2012. Lewy pathology and neurodegeneration in premotor Parkinson's disease. *Movement Disorders* 27 (5): 597–607.
12. Surmeier, D.J., and D. Sulzer. 2013. The pathology roadmap in Parkinson disease. *Prion* 7 (1): 85–91.
13. Ascherio, A., and M.A. Schwarzschild. 2016. The epidemiology of Parkinson's disease: Risk factors and prevention. *The Lancet Neurology* 15 (12): 1257–1272.
14. Langston, J.W., et al. 1983. Chronic Parkinsonism in humans due to a product of meperidine-analog synthesis. *Science* 219 (4587): 979–980.
15. Tanner, C.M., et al. 2011. Rotenone, paraquat, and Parkinson's disease. *Environmental Health Perspectives* 119 (6): 866–872.
16. Callaghan, R.C., et al. 2010. Incidence of Parkinson's disease among hospital patients with methamphetamine-use disorders. *Movement Disorders* 25 (14): 2333–2339.
17. Curtin, K., et al. 2015. Methamphetamine/amphetamine abuse and risk of Parkinson's disease in Utah: A population-based assessment. *Drug and Alcohol Dependence* 146: 30–38.
18. Liu, R., et al. 2011. Meta-analysis of the relationship between Parkinson disease and melanoma. *Neurology* 76 (23): 2002–2009.
19. Marras, C., et al. 2014. Systematic review of the risk of Parkinson's disease after mild traumatic brain injury: Results of the International Collaboration on Mild Traumatic Brain Injury Prognosis. *Archives of Physical Medicine and Rehabilitation* 95 (3): S238–S244.
20. Driver, J.A., et al. 2008. Prospective cohort study of type 2 diabetes and the risk of Parkinson's disease. *Diabetes Care* 31 (10): 2003–2005.
21. Savica, R., et al. 2012. Metabolic markers or conditions preceding Parkinson's disease: A case-control study. *Movement Disorders* 27 (8): 974–979.

22. Yamamoto, T., Y. Moriwaki, and S. Takahashi. 2005. Effect of ethanol on metabolism of purine bases (hypoxanthine, xanthine, and uric acid). *Clinica Chimica Acta* 356 (1–2): 35–57.
23. Aarsland, D., et al. 2007. The effect of age of onset of PD on risk of dementia. *Journal of Neurology* 254 (1): 38–45.
24. Chung, K.K., et al. 2001. Parkin ubiquitinates the α-synuclein–interacting protein, synphilin-1: Implications for Lewy-body formation in Parkinson disease. *Nature Medicine* 7 (10): 1144.
25. Hughes, A.J., et al. 1992. Accuracy of clinical diagnosis of idiopathic Parkinson's disease: A clinico-pathological study of 100 cases. *Journal of Neurology, Neurosurgery & Psychiatry* 55 (3): 181–184.
26. Jankovic, J., and M. Stacy. 2007. Medical management of levodopa-associated motor complications in patients with Parkinson's disease. *CNS Drugs* 21 (8): 677–692.
27. Perez-Lloret, S., et al. 2012. Oro-buccal symptoms (dysphagia, dysarthria, and sialorrhea) in patients with Parkinson's disease: Preliminary analysis from the French COPARK cohort. *European Journal of Neurology* 19(1): 28–37.
28. Tolosa, E., and Y. Compta. 2006. Dystonia in Parkinson's disease. *Journal of Neurology* 253(7): vii7–vii13.
29. Pont-Sunyer, C., et al. 2015. The onset of nonmotor symptoms in Parkinson's disease (The ONSET PD Study). *Movement Disorders* 30(2): 229–237.
30. Onofrj, M., A. Thomas, and L. Bonanni. 2007. New approaches to understanding hallucinations in Parkinson's disease: Phenomenology and possible origins. *Expert Review of Neurotherapeutics* 7 (12): 1731–1750.
31. Evans, A.H., and A.J. Lees. 2004. Dopamine dysregulation syndrome in Parkinson's disease. *Current Opinion in Neurology* 17 (4): 393–398.
32. Reijnders, J.S., et al. 2008. A systematic review of prevalence studies of depression in Parkinson's disease. *Movement Disorders* 23 (2): 183–189.
33. Vingerhoets, F.J., et al. 1997. Which clinical sign of Parkinson's disease best reflects the nigrostriatal lesion? *Annals of Neurology: Official Journal of the American Neurological Association and the Child Neurology Society* 41 (1): 58–64.
34. Rizek, P., N. Kumar, and M.S. Jog. 2016. An update on the diagnosis and treatment of Parkinson disease. *CMAJ* 188 (16): 1157–1165.
35. Delenclos, M., et al. 2016. Biomarkers in Parkinson's disease: Advances and strategies. *Parkinsonism & Related Disorders* 22: S106–S110.
36. Connolly, B.S., and A.E. Lang. 2014. Pharmacological treatment of Parkinson disease: A review. *JAMA* 311 (16): 1670–1683.
37. Grimes, D., et al. 2012. Canadian guidelines on Parkinson's disease. *The Canadian Journal of Neurological Sciences (Le Journal Canadien des Sciences Neurologiques)* 39(4 Suppl 4): S1.
38. Van de Weijer, S., et al. 2018. Promising non-pharmacological therapies in PD: Targeting late stage disease and the role of computer based cognitive training. *Parkinsonism & Related Disorders* 46: S42–S46.
39. Little, S., et al. 2016. Bilateral adaptive deep brain stimulation is effective in Parkinson's disease. *Journal of Neurology, Neurosurgery and Psychiatry* 87 (7): 717–721.
40. Deuschl, G., et al. 2006. A randomized trial of deep-brain stimulation for Parkinson's disease. *New England Journal of Medicine* 355 (9): 896–908.
41. Krack, P., et al. 2003. Five-year follow-up of bilateral stimulation of the subthalamic nucleus in advanced Parkinson's disease. *New England Journal of Medicine* 349 (20): 1925–1934.
42. Follett, K.A., et al. 2010. Pallidal versus subthalamic deep-brain stimulation for Parkinson's disease. *New England Journal of Medicine* 362 (22): 2077–2091.
43. Blandini, F., and M.T. Armentero. 2012. Animal models of Parkinson's disease. *The FEBS Journal* 279 (7): 1156–1166.

44. Duty, S., and P. Jenner. 2011. Animal models of Parkinson's disease: A source of novel treatments and clues to the cause of the disease. *British Journal of Pharmacology* 164 (4): 1357–1391.

45. Liang, H.L., et al. 2008. Near-infrared light via light-emitting diode treatment is therapeutic against rotenone-and 1-methyl-4-phenylpyridinium ion-induced neurotoxicity. *Neuroscience* 153 (4): 963–974.

46. Ying, R., et al. 2008. Pretreatment with near-infrared light via light-emitting diode provides added benefit against rotenone-and MPP+-induced neurotoxicity. *Brain Research* 1243: 167–173.

47. Trimmer, P.A., et al. 2009. Reduced axonal transport in Parkinson's disease cybrid neurites is restored by light therapy. *Molecular Neurodegeneration* 4 (1): 1–11.

48. Salehpour, F., and M.R. Hamblin. 2020. Photobiomodulation for Parkinson's disease in animal models: A systematic review. *Biomolecules* 10 (4): 610.

49. Shaw, V.E., et al. 2010. Neuroprotection of midbrain dopaminergic cells in MPTP-treated mice after near-infrared light treatment. *Journal of Comparative Neurology* 518 (1): 25–40.

50. Peoples, C., et al. 2012. Survival of dopaminergic amacrine cells after near-infrared light treatment in MPTP-treated mice. *International Scholarly Research Notices* 2012.

51. Moro, C., et al. 2013. Photobiomodulation preserves behaviour and midbrain dopaminergic cells from MPTP toxicity: Evidence from two mouse strains. *BMC Neuroscience* 14 (1): 1–9.

52. Moro, C., et al. 2014. Photobiomodulation inside the brain: A novel method of applying near-infrared light intracranially and its impact on dopaminergic cell survival in MPTP-treated mice. *Journal of Neurosurgery* 120 (3): 670–683.

53. Johnstone, D., et al. 2014. Indirect application of near infrared light induces neuroprotection in a mouse model of parkinsonism–an abscopal neuroprotective effect. *Neuroscience* 274: 93–101.

54. Reinhart, F., et al. 2015. 810 nm near-infrared light offers neuroprotection and improves locomotor activity in MPTP-treated mice. *Neuroscience Research* 92: 86–90.

55. Reinhart, F., et al. 2016. Near-infrared light (670 nm) reduces MPTP-induced parkinsonism within a broad therapeutic time window. *Experimental Brain Research* 234 (7): 1787–1794.

56. Reinhart, F., et al. 2017. The behavioural and neuroprotective outcomes when 670 nm and 810 nm near infrared light are applied together in MPTP-treated mice. *Neuroscience Research* 117: 42–47.

57. El Massri, N., et al. 2016. The effect of different doses of near infrared light on dopaminergic cell survival and gliosis in MPTP-treated mice. *International Journal of Neuroscience* 126 (1): 76–87.

58. Reinhart, F., et al. 2016. Intracranial application of near-infrared light in a hemi-parkinsonian rat model: The impact on behavior and cell survival. *Journal of Neurosurgery* 124 (6): 1829–1841.

59. Reinhart, F., et al. 2015. Evidence for improved behaviour and neuroprotection after intracranial application of near infrared light in a hemi-parkinsonian rat model. *Journal of Neurosurgery* 27: 1–13.

60. Peoples, C., et al. 2012. Photobiomodulation enhances nigral dopaminergic cell survival in a chronic MPTP mouse model of Parkinson's disease. *Parkinsonism & Related Disorders* 18 (5): 469–476.

61. Darlot, F., et al. 2016. Near-infrared light is neuroprotective in a monkey model of P arkinson disease. *Annals of Neurology* 79 (1): 59–75.

62. Moro, C., et al. 2016. Effects of a higher dose of near-infrared light on clinical signs and neuroprotection in a monkey model of Parkinson's disease. *Brain Research* 1648: 19–26.

63. Moro, C., et al. 2017. No evidence for toxicity after long-term photobiomodulation in normal non-human primates. *Experimental Brain Research* 235 (10): 3081–3092.

64. Oueslati, A., et al. 2015. Photobiomodulation suppresses alpha-synuclein-induced toxicity in an AAV-based rat genetic model of Parkinson's disease. *PLoS ONE* 10 (10): e0140880.

65. Purushothuman, S., et al. 2013. The impact of near-infrared light on dopaminergic cell survival in a transgenic mouse model of parkinsonism. *Brain Research* 1535: 61–70.

66. Ittner, L.M., et al. 2008. Parkinsonism and impaired axonal transport in a mouse model of frontotemporal dementia. *Proceedings of the National Academy of Sciences* 105 (41): 15997–16002.

67. Vos, M., et al. 2013. Near-infrared 808 nm light boosts complex IV-dependent respiration and rescues a Parkinson-related pink1 model. *PLoS ONE* 8 (11): e78562.

68. Salehpour, F., et al. 2019. Penetration profiles of visible and near-infrared lasers and light-emitting diode light through the head tissues in animal and human species: A review of literature. *Photobiomodulation, Photomedicine, and Laser Surgery* 37 (10): 581–595.

69. Vitreshchak, T., et al. 2003. Laser modification of the blood in vitro and in vivo in patients with Parkinson's disease. *Bulletin of Experimental Biology and Medicine* 135 (5): 430–432.

70. Maloney, R., S. Shanks, and J. Maloney. 2010. The application of low-level laser therapy for the symptomatic care of late stage Parkinson's disease: A non-controlled, non-randomized study: # 185. In *Lasers in surgery and medicine*, vol. 42.

71. Burchman, M. 2011. Using photobiomodulation on a severe Parkinson's patient to enable extractions, root canal treatment, and partial denture fabrication. *J Laser Dent* 19: 297–300.

72. Santos, L., et al. 2019. Photobiomodulation in Parkinson's disease: A randomized controlled trial. *Brain Stimulation: Basic, Translational, and Clinical Research in Neuromodulation* 12 (3): 810–812.

73. Hamilton, C.L., et al. 2019. "Buckets": Early observations on the use of red and infrared light helmets in Parkinson's disease patients. *Photobiomodulation, Photomedicine, and Laser Surgery* 37 (10): 615–622.

74. Hamilton, C., et al. 2019. Transcranial photobiomodulation therapy: Observations from four movement disorder patients. In *Photobiomodulation in the brain*, 463–472. Elsevier.

75. Li, Q., et al. 1999. The effect of endonasal low energy He-Ne laser treatment of Parkinson's disease on CCK-8 content in blood. *Chinese Journal of Neurology* 32 (6): 364.

76. Zhao, G., K. Guo, and J. Dan. 2003. Case analysis of Parkinson's disease treated by endonasal low energy He-Ne laser. *Acta Academiae Medicinae Qingdao Universitatis* 39: 398.

77. Xu, C., et al. 2003. The effects of endonasal low energy He-Ne laser therapy on antioxydation of Parkinson's disease. *International Journal of Medicine & Pharmacy* 20 (11): 816–817.

78. Maksimovich, I.V. 2019. Intracerebral transcatheter laser photobiomodulation therapy in the treatment of Binswanger's disease and vascular parkinsonism: Research and clinical experience. *Photobiomodulation, Photomedicine, and Laser Surgery* 37 (10): 606–614.

79. Peoples, C., et al. 2012. Survival of dopaminergic amacrine cells after near-infrared light treatment in MPTP-treated mice. *ISRN Neurology* 2012.

80. Shaw, V.E., et al. 2012. Patterns of cell activity in the subthalamic region associated with the neuroprotective action of near-infrared light treatment in MPTP-treated mice. *Parkinson's Disease* 2012.

81. Moro, C., et al. 2013. Photobiomodulation preserves behaviour and midbrain dopaminergic cells from MPTP toxicity: Evidence from two mouse strains. *BMC Neuroscience* 14 (1): 40.

82. Wattanathorn, J., and C. Sutalangka. 2014. Laser acupuncture at HT7 acupoint improves cognitive deficit, neuronal loss, oxidative stress, and functions of cholinergic and dopaminergic systems in animal model of Parkinson's disease. *Evidence-Based Complementary and Alternative Medicine* 2014.

83. Salgado, A.S., et al. 2016. Effects of light emitting diode and low-intensity light on the immunological process in a model of Parkinson's disease. *Medical Research Archives* 4(8).

84. El Massri, N., et al. 2016. Near-infrared light treatment reduces astrogliosis in MPTP-treated monkeys. *Experimental Brain Research* 234 (11): 3225–3232.

85. El Massri, N., et al. 2017. Photobiomodulation-induced changes in a monkey model of Parkinson's disease: Changes in tyrosine hydroxylase cells and GDNF expression in the striatum. *Experimental Brain Research* 235 (6): 1861–1874.

86. El Massri, N., et al. 2018. Evidence for encephalopsin immunoreactivity in interneurones and striosomes of the monkey striatum. *Experimental Brain Research* 236 (4): 955–961.

87. Kim, B., et al. 2018. Remote tissue conditioning is neuroprotective against MPTP insult in mice. *IBRO Reports* 4: 14–17.

88. O'Brien, J.A., and P.J. Austin. 2019. Effect of photobiomodulation in rescuing lipopolysaccharide-induced dopaminergic cell loss in the male Sprague-Dawley rat. *Biomolecules* 9 (8): 381.

89. Miguel, M.S., et al. 2019. Photobiomodulation mitigates cerebrovascular leakage induced by the parkinsonian neurotoxin MPTP. *Biomolecules* 9 (10): 564.

90. Ganeshan, V., et al. 2019. Pre-conditioning with remote photobiomodulation modulates the brain transcriptome and protects against MPTP insult in mice. *Neuroscience* 400: 85–97.

91. Vitreshchak, T., V. Mikhailov, M. Piradov, V. Poleshchuk, S. Stvolinskii, and A. Boldyrev. 2003. Laser modification of the blood in vitro and in vivo in patients with Parkinson's disease. *Bulletin of Experimental Biology and Medicine* 135(5): 430–432.

92. Maloney, R., S. Shanks, and J. Maloney. 2010. The application of low-level laser therapy for the symptomatic care of late stage Parkinson's disease: A non-controlled, non-randomized study. *Lasers in Surgery and Medicine* 42.

93. Santos, L., S. Olmo-Aguado, P.L. Valenzuela, K. Winge, E. Iglesias-Soler, J. Argüelles-Luis, et al. 2019. Photobiomodulation in Parkinson's disease: A randomized controlled trial. *Brain Stimulation* 12(3): 810.

94. Hamilton, C.L., H. El Khoury, D. Hamilton, F. Nicklason, and J. Mitrofanis. 2019. The "Buckets": Early observations on the use of red and infrared light helmets in Parkinson's disease patients. *Photobiomodulation, Photomedicine, and Laser Surgery*.

95. Hamilton, C. 2019. D Hamilton, F Nicklason, and J Mitrofanis, Transcranial photobiomodulation therapy: Observations from four movement disorder patients. In *Photobiomodulation in the brain*, 463–472. Elsevier.

Photobiomodulation Therapy for Stroke

11

11.1 Stroke

11.1.1 Acute and Chronic Stroke

Stroke is a neurological deficit caused by a vascular insult leading to an acute focal injury in the central nervous system (CNS), with three distinct categories, ischemic stroke (IS), intracerebral hemorrhage (ICH), and subarachnoid hemorrhage (SAH). Stroke is the second leading cause of death worldwide [1]. Ischemic stroke is the most common type constituting about 80% of all stroke patients.

ICH comprises around 10% of all strokes, however, its rate is higher in Asian countries being twice that in Western counterparts [2]. From an anatomical point of view, ischemic stroke is caused by the occlusion of a cerebral artery either by an embolus moving from the heart, or by thrombus formation in the atherosclerotic wall of the blood vessel [3]. ICH mostly results from a rupture in an artery in the brain leading to bleeding inside the brain parenchyma [4]. The most significant risk factors for all stroke subtypes are concurrent or previous hypertension, smoking, unhealthy diet, sedentary lifestyle, diabetes mellitus, excessive alcohol intake, psychosocial stress, and depression [5]. However, hypertension is the single most common risk factor for the development of ICH [2].

Pathophysiologically, several changes occur during and after stroke, leading to permanent damage to the brain tissue. Oxidative stress directly affects brain vessels, and plays a crucial role in post-stoke ischemic injury. It also destroys the blood–brain barrier, opens potassium channels, and leads to vasodilation [6]. Oxidative stress also promotes cellular DNA damage and thus induces neuronal and glial cell death [7]. Oxidative changes in the brain during stroke reduce the production of ATP and other energy-related metabolites, leading to secondary mitochondrial dysfunction. The secondary release of several mitochondrial apoptogenic proteins has been found to exacerbate ischemic and post-ischemic brain damage, especially in neurons [8]. Post-stroke oxidative stress and excitotoxicity also activates astrocytes and microglia and increases the production and release of inflammatory cytokines, such as tumor necrosis factor-α (TNF-α), interleukins (IL), IL-1β, IL-6, IL-20, IL-10, and transforming growth factor (TGF)-β in the brain, all of which can induce neuronal cell death [9].

Stroke has a variety of signs and symptoms which are strongly dependent on the affected blood vessels and the region of the brain suffering damage, and can range from vision loss or visual field deficit, muscular weakness, to ataxia or aphasia [10]. Several tools have been developed to assess the severity of the damage caused by stroke, and to predict subsequent functional impairment and prognosis. The National Institutes of Health Stroke Scale (NIHSS) is a powerful tool used by healthcare providers to measure the functional impairment resulting from a stroke. It was used in thrombolysis stroke trials to include or exclude subjects to receive treatment [11]. Moreover, the Barthel Index (BI) and the modified Rankin Scale (mRS) are widely applied in clinical trials to evaluate ischemic stroke patients. The BI assesses the functioning of the patient in 10 daily activities and is a reliable stroke disability scale. On the other hand, the mRS assesses independence instead of functioning in specific tasks [12].

Neurovascular imaging is mandatory for the diagnosis of stroke. The gold standard for the diagnosis of ischemic stroke is a non-contrast computed tomography (CT) scan because it is readily available and can be rapidly performed. Non-contrast CT is also of diagnostic value in ICH, and if it is interpreted by an expert it can provide > 95% accuracy [13]. However, this method is not sensitive in cases of minor stroke, and a scan can neither confirm nor exclude the diagnosis. In these cases, magnetic resonance imaging (MRI), which has a higher spatial resolution, is recommended [14].

An infusion of intravenous recombinant tissue plasminogen activator (rtPA) is still the best therapeutic option for stroke patients suffering from moderate to severe neurological deficits, and who are referred to a stroke facility within 4.5 h from the onset of symptoms. However, reocclusion can occur following rtPA therapy, and is a major concern occurring in as many as 25–34% of patients [3]. In cases of acute ischemic stroke and proximal occlusion of a cerebral artery, endovascular reperfusion therapy can be performed, which may improve outcomes in the affected patients [15]. Neuroprotective agents are another novel treatment, which are currently being investigated for the treatment of ischemic stroke. Clinical trials for this approach are ongoing [16].

Stroke can become a chronic disease with acute exacerbations. Chronic stroke is manifested by cognitive deficits, gait disorder typically a vascular gait dyspraxia, swallowing disorders, and last but not least, depression. These changes appear gradually and may be suddenly exacerbated by other stimuli. Obtaining a thorough medical history and neuroimaging procedures such as CT and MRI are mandatory for diagnosis. Rehabilitation therapy is the mainstay of treatment in these patients. Pharmacologic therapy using antiplatelet drugs, antihypertensive medications, and memantine can also be helpful in selected patients [17, 18].

11.1.2 Stroke-Induced Aphasia

Stroke-induced aphasia is a type of language impairment caused by damage to the areas of the brain responsible for speaking, reading, and writing, leading to difficulty in performing each of the mentioned functions. Aphasia is not a disease, but a symptom of damage to the brain. It has been shown that aphasia changes the course of recovery, increases mortality, and increases the incidence of depression. A large multicenter Canadian study found that as many as 35% of stroke patients suffered from aphasia at the time of discharge. It also reported that the incidence of aphasia increased from admission (30%) to discharge (35%), possibly due to a milder presentation at the beginning [19]. Advanced age has been found to be a major risk factor for developing post-stroke aphasia. However, no significant difference was found between males and females [20]. Evidence has shown that Broca-type aphasia is the single most common type of aphasia in both genders. It has also been found that patients afflicted by ischemic stroke and ICH, are most likely to suffer from Broca-type aphasia or global and anomic aphasia, except for females with ICH. This order is unlikely to be affected by age group. Aphasia is traditionally divided into Broca's, Wernicke's, conductive, transcortical sensory, transcortical motor, transcortical combined, global, and anomic subtypes. This categorization is based on the Benson Classification, which assumes that three distinct speech functions are influenced by stroke, namely fluency, comprehension, and naming abilities [21]. Several therapeutic strategies are used to improve stroke-induced aphasia. Language therapy, such as constraint-induced aphasia therapy (CIAT) or constraint-induced language therapy (CILT) has been effectively used to treat stroke-induced aphasia. CIAT is a form of intensive language therapy, which promotes language application in communicative situations [22]. CILT is based on three principles namely, massed practice, behavioral and communicative relevance, and focusing, which are provided to the patient through language action games. It has been shown that the method can provide a significant improvement for patients suffering from aphasia [23]. Transcranial magnetic stimulation (TMS) and transcranial direct-current stimulation (tDCS) are other effective non-invasive procedures that stimulate or improve synaptic plasticity in the affected region of the brain, with minimal adverse effects [24]. On the other hand, pharmacologic therapies which modulate a neurotransmitter imbalance in the damaged brain region can improve language via changes in synaptic plasticity and long-term potentiation (LTP) [25].

11.1.3 Animal Models of Stroke

Animal models are a useful tool to examine the biological mechanisms, validate clinical hypotheses, and assess the efficacy of treatments [26]. Several animal models have been used to study the pathophysiology of stroke. In every one of these models, there are advantages and disadvantages, making each one suitable for specific investigations

[27]. Two- and four-vessel occlusion (2- and 4-VO) models are frequently used for stroke induction, and allow better histological evaluation on brain samples. However, their use is associated with unwanted hypotension, and can be complicated by the effects of anesthesia on the brain [28]. On the other hand, embolic models are good imitators of human stroke, but can produce variable infarct size, high mortality rates, and can cause ICH [29]. Photothrombosis models can be performed easily, are less invasive, and are a good way to predict the exact lesion location. Nevertheless, their use is associated with brain edema and blood–brain barrier (BBB) breakdown. Also, these models are not suitable for the assessment of the effects of neuroprotective agents. Photothrombosis is a process during which a cortical infarct is induced by the systemic administration of a photoactive dye (such as Rose Bengal). The injection is accompanied by irradiation with a specific wavelength laser beam onto the specific region of the brain where the stroke is desired to be produced. The combination of dye injection with laser irradiation causes focal endothelial damage, platelet activation, and aggregation, and induces an ischemic lesion. Due to the lack of penumbra, the photothrombotic model is not suitable for drug studies on the penumbra. However, newer models are able to produce a penumbra-like lesion in the affected area. This model is also unsuitable for the evaluation of treatments, which work on collateral perfusion, as it causes arterial damage. Photothrombotic models are commonly used in rats, however, mouse models are also available [30, 31].

Other stroke models include endothelin-1 induced stroke, macrosphere embolization, craniotomy models, cardiac arrest, decapitation, and autologous blood or collagenase-induced hemorrhage [27, 32]. The stroke models can also be divided into two separate categories including, focal and global cerebral ischemia. Focal ischemia is subdivided into models needing craniotomy and those models not requiring craniotomy [27, 33]. Global cerebral ischemia occurs when the cerebral perfusion decreases in most parts of the brain. However, focal cerebral ischemia is induced by a decrease in the blood flow to a specific area of the brain. It is worth noting that both types result in oxygen and glucose deprivation within the brain. Nevertheless, these changes occur to a different extent. Global ischemia is divided into complete and incomplete types. Mechanisms for complete ischemia include cardiac arrest, aortic dissection, neck cuff, and cephalic artery occlusion. On the other hand, incomplete ischemia includes hemorrhage, intracranial hypertension, 2-VO, and 4-VO stroke models. Pathophysiologically, focal cerebral ischemia is also divided into multifocal and non-multifocal types. The former is then subdivided into embolic models, and localized models, including middle cerebral artery occlusion (MCAO) + carotid, proximal MCAO, photochemical thrombosis, and application of an intraluminal filament [34].

11.2 Photobiomodulation Therapy for Stroke

11.2.1 Photobiomodulation Therapy for Stroke in Animal Models

The development of PBM therapy for stroke originated with some research performed by Uri Oron and colleagues. In their early investigations, they had shown that PBM therapy could have notable effects on a heart attack or myocardial infarction in several animal models. They used two different animal models of heart attack, namely occlusion of the coronary artery in both rats and dogs [35]. Oron et al. used three different irradiances of an 810 nm laser administered to the infarcted area of the heart through an open chest procedure immediately following the heart attack, and showed a biphasic dose–response in the rats. The best reduction in infarct area by 60% was observed with 6 mW/cm^2, with a lesser reduction observed with 2.5 or 20 mW/cm^2. A similar study on dogs also showed a slight but significant reduction in infarct area of 4%. Following these promising findings, three additional studies on this remarkable discovery were carried out by their laboratory [36–38]. Following this discovery, the company PhotoThera Inc. (Carlsbad, CA, USA) was established two years later by Jackson Streeter and colleagues to commercialize this approach. They decided to focus on acute ischemic stroke because they realized that red/NIR photons could reach the brain after non-invasive transcranial irradiation, whereas light delivery to the heart typically requires invasive surgical intervention. The pathological similarities between the biological processes occurring after ischemic stroke and myocardial infarction, both resulting in the growth of a necrotic lesion in the brain or the heart were noted [39].

Nevertheless, the first actual study on transcranial PBM therapy for ischemic stroke was carried out by Mason Leung and colleagues [40]. These workers argued that ischemia/reperfusion injury is partly mediated by NO, while TGF-β1 is neuroprotective in the stroke model. MCAO was carried out in Sprague–Dawley rats for 1 h. Transcranial 660 nm laser (average power of 8.8 mW, fluence for each minute of 2.64 J/cm^2, pulse frequency of 10 kHz, with different irradiation times of 1, 5, or 10 min) was then applied directly to the affected area of the brain through a burr hole drilled in the skull immediately after the end of the MCAO. Results showed that the activity of three isoforms of nitric oxide synthase (NOS) was significantly suppressed after PBM therapy, whereas expression of TGF-β1 was increased. However, no neurological performance parameters were evaluated in this study.

De Taboada et al. [41] published the first study using transcranial PBM to treat MCAO-induced stroke in rats accompanied by neurological testing. Stroke was induced in 169 rats in four groups: control non-PBM and three transcranial PBM-treated groups where an 808 nm laser was applied to the ipsilateral, contralateral, and both sides of the head to the site of induced stroke. The laser irradiance at the tip of the fiber optic was 7.5 mW/cm^2. A 2-min irradiation period delivered an estimated fluence of 0.9 J/cm^2 to the cortical surface. Rats were tested for neurological function using a modified neurological score (MNS) 24 h after MCAO. In all three PBM-treated groups, a marked improvement in neurological deficits was obvious at 14, 21, and 28 days post-stroke, compared to the non-PBM group. The improvement at 4 weeks post-stroke in the PBM-treated animals was about twice that of the non-PBM rats.

Another study by Oron et al. [42] explored the timing of the transcranial PBM treatment (either 4 or 24 h post-stroke). In experiment 1, they carried out a craniotomy and MCA ligation, and in experiment 2, they used a non-invasive filament insertion via the carotid artery to induce a stroke. An 808 nm laser (7.5 mW/cm^2 for 2 min irradiation providing an estimated fluence of 0.9 J/cm^2 per site) was delivered via a custom-designed fiber-coupled optical assembly (4 mm diameter beam) placed onto the shaved scalp at two sites on the head on the contralateral hemisphere to the stroke [43]. According to their previous data, laser irradiation of the contralateral or both hemispheres made no difference to the functional outcome in the animals [42]. In experiment 1, the laser operation mode was continuous wave (CW), whereas in experiment 2, both CW and pulsed wave (PW) mode 70 Hz were compared. Although PBM therapy at 4 h did not show any significant effects, the improvement at 24 h was significant, with CW mode being slightly better than PW mode. In addition, the number of neuroprogenitor cells in the ipsilateral subventricular zone was increased by CW mode PBM therapy. Markers of migrating neuroprogenitor cells including TUJ1 (neuron-specific class III beta-tubulin) and DCX (doublecortin) were also increased in the CW mode group.

A decade later, Lee et al. [44] performed a study to examine whether pre-conditioning with transcranial PBM therapy could be neuroprotective against cerebral ischemia–reperfusion injury caused by MCAO in mice. Transcranial LED irradiation was conducted using a 610 nm probe (1.7 W/cm^2, 2.0 J/cm^2 for 20 min), twice a day for 2 days prior to MCAO. Twenty-four hours after MCAO, the animals underwent behavioral evaluation and were then sacrificed. LED-treated animals showed a significantly smaller infarct and edema volume, higher cerebral blood flow (CBF), and fewer neurobehavioral abnormalities in the wire grip test as compared to untreated injured animals. There were also higher levels of endothelial NOS (eNOS) phosphorylation in the brain of PBM pre-treated mice. The elevated phospho-eNOS was suppressed by LY294002 [a phosphoinositide 3-kinase (PI3K) inhibitor], suggesting that the beneficial effects of PBM therapy on the ischemic brain could be related to increased eNOS phosphorylation through the PI3K/Akt signaling pathway. Furthermore, no significant reductions in infarct size or edema were seen in PBM pre-treated eNOS-deficient animals.

In addition to the MCAO model discussed above, Lapchak et al. performed an interesting series of studies to assess the therapeutic effects of transcranial PBM in the rabbit small clot embolic stroke model (SCEM) [45]. In the first study [46], they compared the effects of different doses of PBM therapy applied at different times after embolization. They compared different times between clot injection and low dose 808 nm laser therapy. They also compared a low dose of PBM (7.5 mW/cm^2 for 2 min providing 0.9 J/cm^2) with a high dose (25 mW/cm^2 for 10 min providing 15 J/cm^2). Their results suggested a "therapeutic window" using a low dose ranging from 0 to 6 h post-embolization (with 6 h being best), but PBM after a delay of 24 h did not show any benefit. The high dose of 15 J/cm^2 was also effective when applied at 1 or 6 h post-stroke. Nevertheless, 15 J/cm^2 did not affect the physiological parameters that were evaluated [46].

In a follow-up study, Lapchak et al. [47] sought to compare CW mode with PW transcranial PBM therapy at either 6 or 12 h after embolization. They examined three different treatment regimens: (1) CW with an irradiance of 7.5 mW/cm^2; (2) PW1 using a pulse duration of 300 μs pulse at a frequency of 1 kHz; or (3) PW2 using a pulse duration of 2 ms at a frequency 100 Hz. Behavioral analysis was carried out 48 h following embolization, allowing for the determination of the effective stroke dose (P50). According to the quantal dose–response analysis, both PW1 and PW2 modes were significantly better than CW mode at 6-h post-embolization. At the 12-h post-embolization, neither the CW nor either PW modes resulted in significant behavioral improvement.

Subsequently, Lapchak and De Taboada [48] went on to measure cortical ATP levels after transcranial PBM therapy in the rabbit SCEM model. Five minutes after embolization, animals were irradiated with 2 min of 808 nm transcranial laser in either CW or PW mode. Three hours following embolization, the cortical tissue was excised and processed to measure ATP levels. Embolization decreased cortical ATP levels in the ischemic cortex by 45% compared to intact animals. This decrease was attenuated by CW laser light, which resulted in a 41% increase in cortical ATP levels compared to the sham-embolized group. The absolute increase in ATP levels was 22.5% compared to intact rabbits. Using PW irradiation regimens, which delivered 5 times (PW1) or 35 times (PW2) higher fluence than CW mode, they measured 157% and 221% increases in the cortical ATP levels, respectively, compared to the sham-embolized group.

Huisa et al. [49] used the rabbit SCEM model to compare zero (sham), one, two, and three transcranial laser PBM sessions performed in the 5 h following embolization. The single group received PBM at 3 h (7.5–10.8 mW/cm^2), the double treatment group received PBM at 3 and 5 h, and the triple treatment group received PBM at 2, 3, and 4 h (7.5–20 mW/cm^2) post-embolization. The neurobehavioral analysis was carried out 24 h after embolization using a dichotomized behavioral score. The evaluation of the effective clot amount (milligrams) that produced a neurological deficit in 50% of the animals (P 50) was used to compare PBM therapy with sham. The triple treatment produced a 91% increase in the clot amount compared with the single treatment, and a 245% increase when compared to the sham group. Similarly, Meyer et al. [50] used the rabbit SCEM model to study a dose-escalation regimen of laser PBM therapy. They compared CW with PW (10 or 100 Hz), cortical irradiance values ranging between 7.5 and 333 mW/cm^2, and a single treatment at 2 h with a triple treatment at 2, 3, and 4 h. A significant behavioral improvement was observed with the triple 100 Hz PW regimen at 111 mW/cm^2 irradiance compared with the other tested regimens.

Lapchak et al. [51] used the rabbit large clot embolic stroke model (LCEM) to test whether tissue plasminogen activator (tPA, the only FDA-approved treatment for stroke) could be safely combined with transcranial laser PBM therapy. Transcranial 808 nm laser (10 mW/cm^2 for 2 min providing 1.2 J/cm^2) was applied at 90 min post-embolization. Transcranial PBM did not significantly alter hemorrhage incidence after embolization, but there was a trend toward a modest decrease in hemorrhage volume (by 65%) in the PBM-treated animals compared with controls. However, intravenous administration of tPA at 60 min post-embolization, using an optimized dosing regimen, significantly increased hemorrhage incidence by 160%. The tPA-induced increase in hemorrhage incidence was not significantly affected by PBM, even though there was a 30% reduction in hemorrhage incidence in the combination-treated animals. There was no effect of PBM on hemorrhage volume measured in tPA-treated animals, and no effect of any treatment on the 24 h survival rate. In another similar study, Lapchak and Boitano [52] also asked whether there was any synergic benefit with a combination of PBM and tPA. Using the rabbit SCEM model, they studied the effects of CW laser PBM therapy (7.5 mW/cm^2) alone or in combination with IV tPA (3.3 mg/kg) administered 60 min post-embolization. PBM and tPA used alone significantly increased the P50 values by 95% and 56%, respectively. The combination of PBM and tPA also increased P50 by 136% compared to intact animals. Embolization decreased cortical ATP levels by 39%; decreases that were attenuated by either PBM or tPA treatment, whereas the PBM and tPA combination further improved cortical ATP levels to 22% above intact animals.

Moreover, two research teams have tested transcranial PBM therapy in photothrombotic ischemic stroke models. A team in South Korea led by Hwa Kyoung Shin tested the effects of 610 nm LED therapy on photothrombotic-induced ischemic brain damage in mice [53]. They used a skin-adherent LED probe (1.7 mW/cm^2, 2.0 J/cm^2, for 20 min) twice a day for 3 days starting at 4 h post-ischemia. The LED probes were simultaneously affixed onto two sites on the head (the right midpoint of the parietal bone and the posterior midline of the seventh cervical vertebra). LED PBM therapy significantly decreased neuroinflammatory responses (e.g., microglial activation and neutrophil infiltration) in the ischemic cortex. PBM also reduced neuronal cell death and decreased NLRP3 inflammasomes, along with a decrease in pro-inflammatory cytokines, such as IL-1β and IL-18 in the ischemic brain. Moreover, PBM-treated mice had lower TLR-2 levels, as well as less MAPK signaling and NF-κB activation. Their results suggested that by suppressing inflammasomes, PBM therapy could attenuate neuroinflammatory reactions and preserve neurons following ischemic stroke. In a follow-up study using the same stroke model, Shin and colleagues [54] also reported the long-term functional outcome in mice following cerebral ischemia, and the optimal timing of transcranial LED PBM therapy to produce the best functional recovery. Animals were allocated to a sham-operation control, ischemic (sham-PBM treatment), or one of three different real-PBM groups depending on the time elapsed post-stroke before therapy was started [immediate (acute), 4 days (sub-acute), or 10 days (delayed)]. Each treatment group received once-daily treatment for 7 days. Behavioral outcomes were measured 21 and 28 days post-ischemia, and histopathological analysis was carried out at 28 days. The acute and sub-acute PBM-treated groups showed a significant improvement in motor function up to 28 days post-ischemia, even though no decrease in brain lesion size was seen. They observed proliferating cells in the ischemic brain region of the sub-acute PBM-treated group. Furthermore, brain-derived neurotrophic factor (BDNF) was significantly increased in the sub-acute PBM-treated group. Taken together, their results suggested that PBM therapy applied during the sub-acute stage had a beneficial effect on long-term functional outcomes, most likely through stimulating neuronal and astrocyte proliferation, blood vessel restoration, as well as increased BDNF expression.

Quanguang Zhang's team in Augusta, Georgia, USA, used a rat stroke model and investigated both behavioral deficit and neurogenesis markers [55]. Transcranial PBM therapy was delivered to the infarct injury area daily from day 1 to day 7, using an 808 nm laser (350 mW/cm^2, total scalp dose of 294 J, 2 min). Rats received intraperitoneal injections of BrdU twice daily (50 mg/kg) from day 2 to 8 post-stroke, and samples were collected on day 14. PBM therapy significantly attenuated behavioral deficits and decreased infarct volume. Also, PBM markedly improved neurogenesis as well as synaptogenesis, as indicated by immunostaining of BrdU, Ki67, MAP2, DCX, spinophilin, and synaptophysin. The neuroprotective effects of PBM were accompanied by strong inhibition of reactive gliosis and pro-inflammatory cytokines. On the other hand, PBM significantly increased the release of anti-inflammatory cytokines, cytochrome c oxidase activity, and ATP levels in the peri-infarct regions. Interestingly, PBM also effectively switched the M1 microglial phenotype to an anti-inflammatory M2 phenotype.

11.2.2 Photobiomodulation Therapy for Stroke in Human Studies

To date, besides the three major double-blind randomized clinical trial studies [56–59], occasional reports have shown the possible neuroprotective effect of PBM therapy in both acute and chronic stroke patients. These studies used intravascular laser [60, 61], laser acupuncture [62], and multiple area [63] irradiation methods.

The first study in 1995 by Naeser et al. [62] reported that laser acupuncture might be an effective strategy for neurorehabilitation of stroke patients with paralysis. Seven patients, five with hemiplegia, displaying severely reduced or absent voluntary finger movement, and two patients with hand paresis, were included in the study. The PBM consisted of the direct irradiation of 780 nm laser onto various acupuncture points, 23 sessions per week for 34 months. Laser acupuncture was conducted for 20 s/point (51 J/cm^2) on the shallower points on the hands and face, and for 40 s/point (103 J/cm^2) on the deeper points on the legs and arms. The results showed that five of seven of the patients (71%) exhibited significant improvement following treatments, including increases in the range of motion, grip strength, and hand dexterity tests. In particular, patients with arm/leg paralysis showed an improvement in knee flexion, knee extension, and/or shoulder abduction, whereas the patients with hand paralysis showed an improvement in finger and hand strength. According to the CT scan data, all the patients who showed improvement had a lesion in < 50% of the motor pathway area (representing mild-moderate paralysis), while those with no significant improvement had a lesion in > 50% of the motor pathway area (representing severe paralysis) [62].

Xiaoa and co-workers [60] used single photon emission computed tomography (SPECT) brain perfusion imaging to investigate changes in regional CBF and brain function in stroke patients treated with intravascular laser PBM therapy. Thirty-five patients with cerebral infarction were enrolled in the study. Of those, 8 patients were less than one week, 18 patients were from one week to one month, and 9 patients were longer than one month after the stroke event. The baseline CT and MRI data showed the sites of stroke: 13 in the basal ganglia, 6 in the internal capsule, 5 in the cerebral cortex, and 11 in other sites. Seventeen of 35 patients were treated with 632.8 nm intravascular laser PBM in addition to routine drug therapy. The PBM protocol consisted of 60 min irradiation, once a day for 10 consecutive days. The results from this group showed an improvement in total regional CBF and cerebral infarction at the study endpoint as compared to the baseline, especially for the affected side of the brain. Although the improvement in the focal area was significant, the opposite side exhibited no significant change. Moreover, the brain blood flow was obviously higher in the stroke side than in the opposite side. The remaining eighteen patients underwent SPECT, then were treated with 30 min of intravascular 632.8 nm laser PBM, followed by SPECT again. The results from this group also showed an increase in regional CBF both in the treated area and in the entire brain, as well as an improvement in brain function [60].

In 2012, Boonswang et al. [63] reported a case study of a 29-year-old woman diagnosed with a brainstem stroke. She suffered acute infarctions involving the medulla, superior aspect of the cervical spinal cord, and bilateral cerebellar hemispheres. The PBM protocol consisted of combined 660 and 850 nm LED irradiation on a total of 32 locations, including the cerebral cortices, brainstem, cervical spine (8 locations), and the body musculature and lymphatics (24 locations), while the patient sat in her wheelchair. After 8 weeks of treatment, the patient's mood became much less melancholic and she was less anxious and fearful as well as more alert and aware of her environment. The patient's double vision was eliminated. She was cured of her right sixth nerve palsy and her dizziness as well. The patient's right and left upper extremity strength improved and her right-hand spasticity decreased. Her right and left lower extremity strength also improved, so that she could stand with minimal or no assistance and was able to ambulate using a walker. It is interesting to note that the improvements at 8 weeks post-treatment were maintained for the next 12 months [63].

In 2017, Yang et al. [61] reported a case study of a 77-year-old man who had suffered a stroke. The patient was diagnosed with progressive weakness of left limbs, and underwent repeated falls, choking, and slow responses, based on clinical examination. MRI showed an infarction in the right anterior cerebral artery. CT of the brain also revealed low density over the right anterior cerebral artery. The PBM protocol consisted of 632.8 nm intravascular laser irradiation for 60 min over 10 consecutive days. The patient underwent 3 different SPECT procedures, before and after PBM therapy, as well as at 63 days post-stroke. Comparison of the 2 regional perfusion SPECT images revealed that the blood perfusion in the right cerebellar hemisphere became more similar to the left cerebellar hemisphere in the second SPECT after PBM therapy. The SPECT data also showed that the crossed cerebellar diaschisis (CCD) was decreased in the patient after treatment. Furthermore, the patient appeared to be more energetic, and his muscle power and functional ability were improved as measured by the Barthel index.

Three clinical trials called "Neurothera Effectiveness and Safety Trials" (NEST-1 [56], NEST-2 [57, 59], and NEST-3 [58]) have been conducted in acute stroke patients to evaluate the safety and effectiveness of 808 nm transcranial laser PBM therapy.

The NEST-1 clinical trial was the first prospective, intention-to-treat, international multicenter, double-blinded study involving 120 ischemic stroke patients (79 active and 41 placebo) [56]. Patients with baseline stroke severity measured by the National Institutes of Health Stroke Scale (NIHSS) scores of 7–22 were enrolled. The PBM therapy consisted of the application of a transcranial 808 nm laser to 20 specified locations on the shaved scalp for 2 min at each site. The mean time between stroke and PBM therapy was 16 h (range 2–24 h). The results from the active treatment group showed more patients with improved NIHSS scores (70%) as compared to the placebo group (51%). Similar results were found for the Barthel Index (BI) and Glasgow Stroke Outcome Scale. In addition, the mortality rates did not differ statistically significant (8.9% active versus 9.8% placebo), nor serious adverse effects (25.3% active versus 36.6% placebo) [56].

The NEST-2 clinical trial was a prospective, double-blind, randomized, sham-controlled, multicenter study involving 660 acute stroke patients (331 active, 327 placebo, and 2 dropouts) [57]. The patients were followed for 90 days post-stroke event. The primary endpoint for this trial was a simple binary division that defined success as a modified Rankin Scale (mRS) score of 0–2, and failure as a mRS score of 3–6 at 90 days post-treatment, or at the last evaluation. PBM therapy parameters were the same as in NEST-1 [56]. PBM therapy was administered within 24 h from the stroke onset (range 3–24 h). One hundred twenty patients (36.3%) in the active treatment group showed improved mRS scores versus 101 patients (30.9%) in the placebo group. Comparable results were also observed for the other outcome measures. Further statistical analysis of patients with a baseline NIHSS score of < 16 showed a favorable outcome at 90 days on the primary endpoint. Moreover, the mortality rates did not differ statistically significant (17.5% active versus 17.4% placebo), nor serious adverse events (37.8% active versus 41.8% placebo). Overall the study confirmed safety, although it did not meet the formal statistical significance criteria for effectiveness [57].

In a subsequent evaluation [59], researchers used brain scans (CT or MRI, if clinically indicated) to analyze additional measures in the NEST-2 study patients [57]. These measures included the cortical infarct volume in patients with cortical involvement, and the cortical Alberta Stroke Program Early CT Score (cASPECTS) components (M1–M6, anterior, posterior) on a 0- to 8-point modified scale. The brain scans were performed on a total of 640 subjects on day 5 ($\pm$2). The total ASPECTS score was correlated with total infarct volume. Overall, PBM therapy was not associated with a decrease in overall or cortical infarct volume [59].

In the next study [64], they performed a pooled analysis of previously published patients in NEST-1 [56] and NEST-2 [57], which amounted to a total of 778 stroke patients. Baseline characteristics and prognostic factors were balanced between the active and placebo treatment groups. The improvement rate in the active treatment group (n = 410) was significantly higher as compared to the placebo treatment group (n = 368). The distribution of scores on the day 90 mRS was significantly different in the active PBM group compared with placebo group. Also, subgroup analysis revealed that a moderate stroke could be a predictor of better treatment response [64].

The NEST-3 was a double-blind, sham-controlled, randomized (1:1), parallel-group, multicenter trial involving 630 acute stroke patients (314 active and 316 placebo) [58]. All patients were to receive standard medical management based on the American Stroke Association and European Stroke Organization Guidelines. Patients in active and placebo treatment groups were treated with real or sham-PBM between 4.5 and 24 h of stroke onset. An interim analysis of 566 completed patients revealed no significant difference in the primary endpoint (active PBM with 140/282 [49.6%] versus placebo PBM with 140/284 [49.3%]) for good improvement using the modified Rankin Scale, 0–2. The results still remained stable after the inclusion of all 630 patients. Furthermore, the mortality rates did not differ statistically significant (4.8% active versus 6.1% placebo), nor serious adverse events (20.9% active versus 28.0% placebo) [58].

Altogether, although phases 1 and 2 of NEST both showed safety and effectiveness, the phase 3 trial was disappointing and was terminated for futility at the interim analysis stage. Insufficient light penetration, varying thickness of different patient's skulls, only a single irradiation session (rather than repeated daily treatment), inappropriate stroke severity measurement scale, too long an interval between stroke onset and PBM therapy, and irradiation areas which were not optimized for the location of individual patients have all been put forward in the literature as possible explanations for this failure [64].

The failure of the large NEST-3 trial at considerable expense to Photothera, has cast a dark cloud over the future of PBM for acute stroke therapy. This is despite positive results being obtained in NEST-1 and NEST-2, which also suffered from the same drawbacks discussed above. On the other hand, the trials of PBM for rehabilitation of chronic stroke patients have been extremely limited, despite the large number of patients who might benefit from this approach, and the promising results obtained so far. We would recommend that much larger trials in the chronic stroke patient population should be considered in the future (Tables 11.1 and 11.2).

Table 11.1 Summary of in vivo studies on the effects of photobiomodulation therapy in stroke

Study/Year	Animal/model	Light Source	Wavelengths	Irradiation parameters	Irradiation approach/sites	Findings
Leung et al. [40]	Rat (cerebral ischemia model)	Laser, Omega Excel Laser (London, UK)	660 nm	8.8 mW, 2.64, 13.2, or 26.4 J/cm²; with corresponding duration of 1, 5, or 10 min, respectively; one irradiation session, PW at 10-Hz	Transcutaneously; directly through a burr hole 5 mm from the cerebrum	Decreased specific activity of NOS and increased expression of TGF-β 1 at 4 days post-injury (at all regimens); down-regulated expression of three NOS isoforms including eNOS, nNOS, and iNOS at 4 days post-injury (at all regimens)
Lapchak et al. [46]	Rabbit (embolic stroke model)	Laser, Acculaser PhotoThera, Inc (San Diego, CA, USA) coupled with OZ Optics Ltd fiber	808 nm	7 mW/cm² for 2 min (0.84 J/cm²) or 25 mW/cm² for 10 min (15 J/cm²), one irradiation session, CW	Transcranially; holding probe in direct contact with the skin	Improved behavioral performance and decreased effective clot dose for stroke 3 h after clot injection (at cortical fluence of 15 J/cm²)
Oron et al. [42]	Rat (atherothrombotic stroke model)	Laser, GaAs, Photothera, Inc. (San Diego, CA, USA)	808 nm	7.5 mW/cm² at brain tissue level, 0.9 J/cm² per site (total of 2 sites), fiber diameter of 4 mm, 2 min, CW or PW at 70-Hz	Transcranially; at two locations on head (3 mm dorsal to eye and 2 mm anterior to ear) and on contralateral hemisphere to stroke	Improved neurological scores at 14 and 21 days post-stroke (CW mode when applied 24 h post-stroke); increased SVZ cell proliferation and migration (CW mode)

(continued)

Table 11.1 (continued)

Study/Year	Animal/model	Light Source	Wavelengths	Irradiation parameters	Irradiation approach/sites	Findings
DeTaboada et al. [41]	Rat (atherothrombotic stroke model)	Laser, GaAs, Photothera, Inc. (San Diego, CA, USA)	808 nm	7.5 mW/cm² at brain tissue level, 0.9 J/cm² per site (total of 2 sites), fiber diameter of 4 mm, 2 min, one irradiation session	Transcranially; at two locations on head (3 mm dorsal to eye and 2 mm anterior to ear) either ipsilateral, contralateral, or to both sides of stroke	Improved modified neurological score at 14, 21, and 28 days post-stroke (by all irradiated locations in the skull)
Lapchak et al. [47]	Rabbit (embolic stroke model)	Laser, Acculaser coupled with OZ Optics Ltd fiber optic (Berkeley, CA, USA)	808 nm	7.5 mW/cm², 0.9–1.2 J, 2 min, PW at 100-Hz (2 ms pulse duration with duty cycle of 20%) or 1000-Hz (0.3 ms pulse duration with duty cycle of 30%), or CW	Transcranially; holding probe in direct contact with shaved scalp	Improved behavioral performance and decreased effective clot dose for stroke 6 h after clot injection (at both PW mode regimens)
Lapchak et al. [51]	Rabbit (embolic stroke model)	Laser, Acculaser coupled with OZ Optics Ltd fiber optic	808 nm	10 mW/cm², 2 min, one irradiation session, CW	Transcranially; holding probe in direct contact with scalp	No significant effects on hemorrhage incidence, volume or survival rate
Lapchak and DeTaboada, [48]	Rabbit (embolic stroke model)	Laser, Acculaser coupled with OZ Optics Ltd fiber optic	808 nm	7.5, 37.5, or 262.5 mW/cm²; 0.9, 4.5, or 31.5 J/cm² at cortex, spot diameter of 5 mm, 2 min, PW at 100-Hz or CW	Transcranially; holding probe in direct contact with scalp	Increased cortical ATP content (at PW mode, both of 4.5 and 31.5 J/cm²)
Uozumi et al. [74]	Mouse (cerebral ischemia model)	Laser, B&W Tek. Inc. (Newark, DE, USA)	808 nm	1.6 W/cm², 30 min, CW	Transcranially; to the left hemisphere	Improved residual CBF and reduced the numbers of apoptotic cells in the hippocampus

(continued)

Table 11.1 (continued)

Study/Year	Animal/model	Light Source	Wavelengths	Irradiation parameters	Irradiation approach/sites	Findings
Yip et al. [75]	Rat (cerebral ischemia model)	Laser, GaAlAs, Omega Excel Laser (London, UK)	606 nm	8.8 mW, 2.64, 13.20, or 26.40 J/cm^2; with corresponding duration of 1, 5, or 10 min, respectively; PW at 10-Hz	Transcutaneously; directly through a burr hole 5 mm from the cerebrum	Increased expression of Akt, phosphorylated-Akt, Bcl-2 and pBAD (at all regimens); decreased expression of caspase-3 (at all regimens) and caspase-9 (at 2.64 and 13.20 J/cm^2) at 4 days post-stroke
Choi et al. [76]	Rat (focal cerebral ischemia and reperfusion model)	LEDs, Qray Inc. (Seongnam, Korea)	710 nm	0.042 mW/cm^2, 1.796 J/cm^2, irradiation area of 1.13 cm^2, 12 h/day continuous irradiation for 20 days	Whole-body irradiation; randomly over the whole skin	Increased $CD4^+$ (after 2 and 10 days post-stroke) and $CD8^+$ T (after 10 and 20 days post-stroke) cell populations; decreased infarct volume (at day 21); increased expression of IL-10 (at day 20); decreased microglial activation in striatum and cortex (at day 20); increased number of $CD4^+CD25^+$ Treg cells in ischemic core and penumbra; improved neurological severity score and step fault scores

(continued)

Table 11.1 (continued)

Study/Year	Animal/model	Light Source	Wavelengths	Irradiation parameters	Irradiation approach/sites	Findings
Huisa et al. [49]	Rabbit (small clot embolic stroke model)	Laser, Acculaser coupled with OZ Optics Ltd fiber optic	808.5 nm	7.5, 10.8, or 20 mW/cm^2, 2 min, CW	Transcranially; holding probe in direct contact with scalp	Improved behavioral performance and decreased effective clot dose for stroke (at double and triple irradiation regimens)
Jiang et al. [77]	Rat (hypoxic–ischemic brain damage)	LEDs, Biological Engineering Institute of Chongqing University (Chongqing, China)	660 nm	30 mW/cm^2, 30 min daily for 7 days	Transcranially	Improved spatial learning and memory in MWM test; decreased Bax and increased Bcl-2 expression (mRNA and protein) levels in the left hippocampus (damage side)
Fukuzaki et al. [78]	Mouse (transient mild ischemia model)	Laser, Nd:YVO$_4$, SUWTECH, LDC-2500 (China)	532 nm	845 mW/cm^2, 30.4 × 10^2 J/cm^2 at cortex, 60 min, CW	Transcranially; at the surface of temporal skull over left auditory cortex	Promoted the migration of NSPCs into deeper layers of the neocortex; increased phosphorylated-Akt and Akt expression levels at 4 h post-irradiation

(continued)

Table 11.1 (continued)

Study/Year	Animal/model	Light Source	Wavelengths	Irradiation parameters	Irradiation approach/sites	Findings
Na al. [79]	Rat (focal cerebral ischemia model)	Laser, Micro/Nano Fabrication Laboratory of Gwangju Institute of Science and Technology	532 and/or 658 nm	532 nm: 30 mW 658 nm: 60 mW	Laser acupuncture; at various acupoints of HT9, LR1, HT3 and KI10	Acupuncture treatment with 658 nm laser at HT9-LR1 and 532 nm laser at HT3-KI10: resulted in a significant decrease in Bax and cytochrome c levels in the hippocampus, and a significant increase in hemoglobin level, hematocrit, total white blood cell, neutrophil, lymphocyte, monocyte, and erythrocyte counts
Meyer et al. [50]	Rabbit (small clot embolic stroke model)	Laser, coupled with OZ Optics Ltd fiber optic cable (PhotoThera, SanDiego, CA, USA)	808.5 nm	7.5–333 mW/cm^2 at cortex, 2 min, 1 or 3 irradiation sessions, CW or PW at 10- or 100-Hz with duty cycle of 20%	Transcranially; holding probe in direct contact with shaved skull	Improved behavioral performance at 2 h post-stroke (at 1111 mW/cm^2 with 100-Hz PW); no observable tissue necrosis or microscopic neural damage
Lapchak and Boitano, [52]	Rabbit (small clot embolic stroke model)	Laser, GaAlAs, coupled with OZ Optics Ltd fiber optic cable	808 nm	7.5 mW/cm^2, 0.9 J/cm^2 at cortex, 2 min, CW	Transcranially; holding probe in direct contact with scalp	Improved behavioral performance at 2 day post-stroke; increased cortical ATP levels at 6 h post-stroke

(continued)

Table 11.1 (continued)

Study/Year	Animal/model	Light Source	Wavelengths	Irradiation parameters	Irradiation approach/sites	Findings
Lee et al. [80]	Mouse (photothrombotic cerebral focal ischemia model)	LEDs, Color Seven Co. (Seoul, Korea)	610 nm	1.7 mW/cm^2, 2 J/cm^2, spot diameter of 4 mm, 20 min, twice a day for 2 days prior to ischemic event, CW	Transcranially; at the skin via double-sided tape at the right midpoint of the parietal bone and the posterior midline of the seventh cervical vertebra	At 24 h post-stroke: decreased infarct size and edema; improved neurological and motor function; decreased astrocyte and microglia cells; decreased transcription of iNOS, COX-2, TNF-α, IL-1 β, TLR- 2, CCL2 and CXCL10 in the ischemic cortex; decrease in protein levels of COX-2 and TLR2: inhibited p38, JNK and ERK-1/2 MAPK activation; attenuated translocation of the NF-κ B p65 protein subunit from cytosol to nucleus; attenuated MPO protein levels; improved rearrangement of tight junction proteins and attenuated blood–brain barrier disruption

(continued)

Table 11.1 (continued)

Study/Year	Animal/model	Light Source	Wavelengths	Irradiation parameters	Irradiation approach/sites	Findings
Xiong and Li [81]	Rat (focal cerebral ischemia and reperfusion model)	Laser, SUNDOM-300IB/216 semiconductor, Beijing SUNDOM Medical Equipment Co. Ltd (China)	650/810 nm	15 mW, 50 mm^2, 5 min, once a day for 7 days	Laser acupuncture; at various acupoints of Baihui, Mingmen, and left Zusanli	Improved energy metabolism via modulation of enzymatic activity of SDH, Na$^+$-K$^+$-ATPase, and LDH; enhanced the expression of GAP-43 and decreased cerebral infarction volume; increased the levels of expression of serum SOD, and decrease the serum MDA content
Jittiwat, [82]	Rat (focal cerebral ischemia and reperfusion model)	Laser, Weberneedle laser	810 nm	100 mW, 10 min, once daily for 14 days	Laser acupuncture; at the Baihui (GV20) acupoint	Decreased brain infarction volume; decreased brain MDA levels and increased CAT, GSH-Px and SOD activities

(continued)

Table 11.1 (continued)

Study/Year	Animal/model	Light Source	Wavelengths	Irradiation parameters	Irradiation approach/sites	Findings
Lee et al. [53]	Mouse (photothrombotic cerebral focal ischemia model)	LEDs, Color Seven Co. (Seoul, Korea)	610 nm	1.7 mW/cm^2, 2 J/cm^2, spot diameter of 4 mm, 20 min, twice a day for 3 days commencing at 4 h post-ischemia, CW	Transcranially; at the skin via double-sided tape at the right midpoint of the parietal bone and the posterior midline of the seventh cervical vertebra	At 72 h post-stroke: decreased infarct volume; improved neurological function; decreased MPO protein levels as a marker for neutrophil infiltration; decreased microglial activation via decrease of Iba-1(+)/CD68(+) cells; decreased cell death and reduced NLRP3, cleaved caspase-1 and -11, IL-1β and IL-18 levels; decreased TLR-2 protein levels; suppressed phospho-JNK and phospho-ERK; decreased translocation of NF-κ B p65 protein subunit into nucleus

(continued)

Table 11.1 (continued)

Study/Year	Animal/model	Light Source	Wavelengths	Irradiation parameters	Irradiation approach/sites	Findings
Lee et al. [54]	Mouse (photothrombotic cerebral focal ischemia model)	LEDs, Color Seven Co. (Seoul, Korea)	610 nm	1.7 mW/cm^2, 2 J/cm^2, 20 min, once a day for 7 days	Transcranially; at 2 concurrent locations on the head (the right midpoint of the parietal bone and the posterior midline of the seventh cervical vertebra)	Improved functional recovery at 21 and 28 days post-ischemia as assessed by wire-grip test and rotarod tests; no effect on the brain atrophy; regulated the proliferation of astrocytes and microglia; promoted the proliferation and differentiation of neuronal cells; promoted CD31-postive cells in cerebral ischemic cortex; upregulated the BDNF levels in the post–ischemic cerebral cortex
Lee et al. [44]	Mouse (middle cerebral artery occlusion/reperfusion model)	LEDs, Color Seven Co. (Seoul, Korea)	610 nm	1.7 mW/cm^2, 2 J/cm^2, spot diameter of 4 mm, 20 min, twice a day for 2 days prior to ischemic event, CW	Transcranially; at the skin via double-sided tape at the right midpoint of the parietal bone and the posterior midline of the seventh cervical vertebra	At 24 h post-stroke: decreased infarct size and edema; improved neurological function; improved vestibular-motor dysfunction in WGMT; increased CBF during 30 min after reperfusion; increased phosphorylated eNOS and decreased phosphorylated Akt expression levels

(continued)

Table 11.1 (continued)

Study/Year	Animal/model	Light Source	Wavelengths	Irradiation parameters	Irradiation approach/sites	Findings
Yun et al. [83]	Rat (middle cerebral artery occlusion stroke model)	Laser, Ellise-005, Ver. 1.0.1, Wontech (Daejeon, South Korea)	650 nm	30 mW, fiber diameter of 125 m μ, 5 min, once every 2 days for 2 weeks, PW at 100-Hz	Laser acupuncture; at the GV20 (head) and HT7 (right forepaw) acupoints	Improved learning and memory in MWM test; decreased cholinergic neuronal cell loss in the hippocampal CA1 region; upregulated gene expression of CREB, BDNF, and Bcl-2 and downregulated gene expression of Bax
Sanderson et al. [84]	Rat (2-vessel occlusion/ hypotension cerebral ischemia model)	LEDs, (750 nm, LED750-66-60; 810 nm, LED810-66-60; 950 nm, LED950-66-60)	750, 810, or 950 nm	50 mW/cm^2, 120 min 750 nm: 15 mW 810 nm: 22 mW 950 nm: 32 mW	Transcranially; LEDs placed 1.5 cm from the dorsal scalp	Decreased ROS accumulation in the CA1 region of the hippocampus (with 950 nm); Improved neurobehavioral function assessed by radial arm maze test (with 750 + 950 nm); attenuated neuronal loss in the CA1 region (with 750 and/or 950 nm); resulted in a better cell survival when 750 + 950 nm light was applied at reperfusion compared vs. when light applied after 1 h of reperfusion

(continued)

Table 11.1 (continued)

Study/Year	Animal/model	Light Source	Wavelengths	Irradiation parameters	Irradiation approach/sites	Findings
Tucker et al. [85]	Neonatal rat (hypoxia cerebral ischemia model)	Laser, (808M1100, Dragon Lasers)	808 nm	25 mW/cm² (at cortex), 3 J/cm² (at cortex), 2 min, once daily for 7 days	Transcranially; centering the beam 3 mm posterior from the eyes and 2 mm anterior from the ears	Profoundly restored mitochondrial dynamics by suppressing hypoxia ischemia-induced mitochondrial fragmentation; attenuated mitochondrial membrane collapse, accompanied with enhanced ATP synthesis; led to robust inhibition of oxidative damage, manifested by significant reduction in the productions of 4-HNE, P-H2AX (S139), MDA, as well as protein carbonyls; suppressed the activation of mitochondria-dependent neuronal apoptosis, as evidenced by decreased pro-apoptotic cascade 3/9 and TUNEL-positive neurons

(continued)

Table 11.1 (continued)

Study/Year	Animal/model	Light Source	Wavelengths	Irradiation parameters	Irradiation approach/sites	Findings
Yang et al. [55]	Rat (photothrombotic stroke model)	Laser, (808M100, Dragon Lasers)	808 nm	25 mW/cm^2 (at cortex), 3 J/cm^2 (at cortex), 2 min, once daily for 7 days	Transcranially; on the infarct injury area	Remarkably enhanced neurogenesis and synaptogenesis, as evidenced by immunostaining of BrdU, Ki67, DCX, MAP2, spinophilin, and synaptophysin; elevated the release of anti-inflammatory cytokines, CCO activity and ATP production in peri-infarct regions; effectively switched an M1 microglial phenotype to an anti-inflammatory M2 phenotype

(continued)

Table 11.1 (continued)

Study/Year	Animal/model	Light Source	Wavelengths	Irradiation parameters	Irradiation approach/sites	Findings
Yang et al., 2019 [86]	Neonatal rat (hypoxia cerebral ischemia model)	Laser, (808M100, Dragon Lasers)	808 nm	100 mW/cm^2, 12 J/cm^2, 6 h before hypoxia cerebral ischemia	Transcranially; centering the beam 3 mm posterior from the eyes and 2 mm anterior from the ears	Attenuated cognitive impairment as assessed by Barnes Maze Task, NOR, and objection Location test (OLT); decreased volume shrinkage in the brain, neuron loss, dendritic and synaptic injury; restored hypoxia ischemia-induced mitochondrial dynamics and inhibited mitochondrial fragmentation, followed by a robust suppression of cytochrome c release, and prevention of neuronal apoptosis by inhibition of caspases 3 and 9 activation

(continued)

Table 11.1 (continued)

Study/Year	Animal/model	Light Source	Wavelengths	Irradiation parameters	Irradiation approach/sites	Findings
Wang et al. [87]	Rat (global cerebral ischemia model)	Laser, (808M100, Dragon Lasers)	808 nm	8.0 mW/cm^2 (at hippocampus), 1, 2, 3, or 4 daily irradiations	Transcranially; centering the beam 3 mm posterior from the eyes and 2 mm anterior from the ears	Markedly preserved both short-term (a week) and long-term (6 months) spatial learning and memory; preserved healthy mitochondrial dynamics and suppressed substantial mitochondrial fragmentation of CA1 neurons, by reducing the detrimental Drp1 GTPase activity and its interactions with adaptor proteins Mff and Fis1 and by balancing mitochondrial targeting fission and fusion protein levels; reduced mitochondrial oxidative damage and excessive mitophagy and restored mitochondrial overall health status and preserved mitochondrial function; suppressed mitochondria-dependent apoptosome formation/caspase-3/9 apoptosis-processing activities

(continued)

Table 11.1 (continued)

Study/Year	Animal/model	Light Source	Wavelengths	Irradiation parameters	Irradiation approach/sites	Findings
Jittiwat, 2019 [88]	Rat (focal cerebral ischemia and reperfusion model)	Laser, Weberneedle laser	810 nm	100 mW, 10 min, once daily for 14 days	Laser acupuncture; at the Baihui (GV20) acupoint	Enhanced memory performance in MWM test; improved neuron density in CA1 and CA3; improved motor performance using neurological score at 14 days post-stroke; improved GSH-Px and SOD activities, and decreased density ratio of IL-6 to β-actin in the hippocampus
Argibay et al. [89]	Rat (focal cerebral ischemia and reperfusion model)	LEDs	830 nm	10 mW/cm2, 18 J/cm2, 30 min, once daily/week or thrice daily/week during 12 weeks	Transcranially	PBM applied 24 h after the onset of ischemia: no effect on the neuronal recovery determined by fMRI; no effect on the ischemic volume; no effect on the neuronal marker (Fox3), glial marker (GFAP), and neurogenesis

(continued)

Table 11.1 (continued)

Study/Year	Animal/model	Light Source	Wavelengths	Irradiation parameters	Irradiation approach/sites	Findings
Vahabzadeh-Hagh et al. [90]	Mouse (middle cerebral artery occlusion/reperfusion model)	Laser, Marubeni (Tokyo, Japan)	808 nm	37 mW/cm^2, 2 min, 3 times at days 5, 9, and 13 after stroke onset	Transcranially	Activated astrocytes and enhanced GFAP signal in the somatosensory cortex and the corpus callosum; increased HMGB1 expression and endothelial progenitor cell accumulation in peri-infarct cortex
Strubakos et al. [91]	Rat (2-vessel occlusion/hypotension cerebral ischemia model)	LEDs. (750 nm, LED750-66–60; 810 nm, LED810-66–60; 950 nm, LED950-66–60)	750, 810, or 950 nm	200 mW/cm^2, 120 or 240 min 750 nm: 15 mW 810 nm: 22 mW 950 nm: 32 mW	Transcranially; LEDs placed 1.5 cm from the dorsal scalp	120 min irradiation: resulted in a 21% reduction in brain injury at 24 h of reperfusion measured by DWI and a 25% reduction in infarct volume measured by T2WI at 7 and 14 days of reperfusion, respectively; reduced brain injury in the acute phase of brain injury, and 7 and 14 days of reperfusion, demonstrating a > 50% reduction in infarction

(continued)

Table 11.1 (continued)

Study/Year	Animal/model	Light Source	Wavelengths	Irradiation parameters	Irradiation approach/sites	Findings
Salehpour et al. [92]	Mouse (concurrent global ischemia superimposed on a model of aging model)	Laser, Thor Photomedicine (Chesham, UK)	810 nm	200 mW, 6.66 W/cm2, 0.03 cm2, 33.3 J/cm2, PW at 10-Hz with a duty cycle of 88%, once daily for 14 days	Transcranially; over the midline of the dorsal surface of the head in region between eyes and ears	improved spatial learning and memory (assessed by Barnes and Lashley III mazes tasks) and episodic memory (assessed by WWWhich task); decreased ROS and raised ATP and general mitochondrial activity as well as biomarkers of mitochondrial biogenesis, namely, SIRT1, PGC-1a, NRF1, and TFAM; decreased neuroinflammatory responses via decreasing iNOS, TNF-α, and IL-1β levels

(continued)

Table 11.1 (continued)

Study/Year	Animal/model	Light Source	Wavelengths	Irradiation parameters	Irradiation approach/sites	Findings
Lee et al. [93]	Mouse (chronic cerebral hypoperfusion model)	LEDs	810 nm	300–600 mW/cm^2, 30 min, for 3 times per week for one month	Transcranially	Increased CBF assessed by PET
de Jesus Fonseca et al. [94]	Rat (Cerebrovascular accident hemiplegia model)	LEDs, (RL5-09,030, Super Bright LEDs)	904 nm	110 mW; 7 J/cm^2, 63 s	Transcranially; in contact with the frontal region of the brain	Improved muscle resistance; reduced anxiety states via reducing open-field freezing time and the number of fecal bolus pellets on treatment days 7 and 21; induced neurogenesis on treatment days 7 and 21

Abbreviations Akt, protein kinase B; ATP, adenosine triphosphate; Bax, Bcl-2-associated X protein; BDNF, brain-derived neurotrophic factor; BrdU, Bromodeoxyuridine; CBF, cerebral blood flow; CCL2, chemokine (C–C motif) ligand-2; CCO, cytochrome c oxidase; CD31, cluster of differentiation 31; CD4 +, cluster of differentiation 4; COX-2, cyclooxygenase-2; CREB, cAMP response element binding protein; CXCL10, C–X–C motif chemokine-10; CW, continuous wave; Drp1, dynamin-related protein 1; DWI, diffusion-weighted magnetic resonance imaging; eNOS, endothelial nitric oxide synthase; ERK, extracellular signal-regulated kinase; Fis1, fission protein-1; fMRI, functional magnetic resonance imaging; GAP-43, growth-associated protein-43; GFAP, glial fibrillary acidic protein; GSH-Px, glutathione peroxidase; GTPase, guanosine triphosphatase; HMGB1, high mobility group box-1; IL, interleukin; iNOS, inducible nitric oxide synthase; JNK, c-Jun N-terminal kinases; LDH, lactate dehydrogenase; LEDs, light-emitting diodes; MAP2, microtubule-associated protein-2; MAPK, A mitogen-activated protein kinase; MDA, malondialdehyde; Mff, mitochondrial fission factor; MPO, myeloperoxidase; mRNA, massager ribonucleic acid; MWM, Morris water maze; NF-κB, nuclear factor kappa-light-chain-enhancer of activated B cells; NLRP3, Nod-like receptor pyrin domain-containing-3; nNOS, neuronal nitric oxide synthase; NOS, nitric oxide synthase; NRF1, nuclear respiratory factor-1; NSPCs, neural stem/progenitor cells; pBAD, PBAD promoter; PBM, photobiomodulation; PET, positron emission tomography; PGC-1a, peroxisome proliferative activated receptor g coactivator-1a; PW, pulsed wave; ROS, reactive oxygen species; SDH, succinate dehydrogenase; SIRT1, silent information regulator-1; SOD, superoxide dismutase; SVZ, subventricular zone; WWWhich, What-Where-Which Task; T2WI, T2-weighted; TFAM, mitochondrial transcription factor-A; TGF-β1, transforming growth factor-β1; TLR- 2, Toll-like receptor-2; TNF-α, tumor necrosis factor-α; WGMT, wire-grip and motion test.

Table 11.2 Summary of clinical studies on the effects of photobiomodulation therapy in Stroke

Study/Year	Subjects (n)	Light Source	Wavelengths	Irradiation parameters	Irradiation approach/sites	Findings
Naeser et al. [65]	Acute and chronic stroke (7)	Laser, Uni-laser (Denmark)	780 nm	20 mW, 51 and 103 J/cm^2, 1 mm diameter aperture, 20–40 s for each acupoint, 3–5 times per week	Laser acupuncture; several acupuncture points on the arm, leg, and hand and/or face	Resulted in overall improvement in five of the seven patients; improved knee flexion, knee extension and/or shoulder abduction in patients with arm/leg paralysis; improved finger and hand strength in patients with hand paralysis; reduced lesion area as assessed by CT scan
Xiaoa et al. [66]	Cerebral ischemia (35)	Laser, ZJC-480A (China)	632.8 nm	1.5–2.5 mW, 60 min for once a day for 10 days or only 30 min irradiation	Intravascularly; basilic vein or median basilic vein in the paralyzed arm	Improved regional CBF as assessed by SPECT; quantified analysis showed an improvement of brain blood flow function change rate (BFCR%)
Lampl et al. [67]	Acute stroke (120)	Laser, Neurothera PhotoThera Inc. (Carlsbad, CA, USA)	808 nm	10 mW/cm^2, 1.2 J/cm^2 at cortex, 2 min for each site, CW	Transcranially; 20 sites, multiple areas over the entire shaved scalp	Positive effects of irradiation within 24 h of stroke onset evaluated by National Institutes of Health Stroke Scale, modified Rankin Scale, Barthel Index, and Glasgow Outcome Scale

(continued)

Table 11.2 (continued)

Study/Year	Subjects (n)	Light Source	Wavelengths	Irradiation parameters	Irradiation approach/sites	Findings
Zivin et al. [68]	Acute stroke (660)	Laser, Neurothera PhotoThera Inc. (Carlsbad, CA, USA)	808 nm	10 mW/cm^2, 1.2 J/cm^2 at cortex, 2 min for each site, CW	Transcranially; 20 sites, multiple areas over the entire shaved scalp	No significant positive effects of irradiation within 24 h of stroke onset
Boonswang et al. [69]	Chronic stroke (1)	LEDs, XR3T-1 device (THOR, London, UK)	660 + 850 nm	1400 mW, 2.95 J/cm^2 delivered to 32 sites, 1 min for each site, spot size of 0.196 cm^2, 1/week for 8 weeks	Multiple areas; 32 sites, including cerebral cortices, brainstem and cervical spine (8 sites) and core musculature and lymphatics (24 sites)	Positive change in every area of deficits and improved physical clinical signs
Kasner et al. [70]	Acute stroke (640)	Laser, Neurothera PhotoThera Inc. (Carlsbad, CA, USA)	808 nm	10 mW/cm^2, 1.2 J/cm^2 at cortex, 2 min for each site, CW	Transcranially; 20 sites, multiple areas over the entire shaved scalp	No effect on total cortical infarct volume as measured by computed tomography
Hacke et al. [71]	Acute stroke (630)	Laser, Neurothera PhotoThera Inc. (Carlsbad, CA, USA)	808 nm	10 mW/cm^2, 1.2 J/cm^2 at cortex, 2 min for each site, CW	Transcranially; 20 sites, multiple areas over the entire shaved scalp	No measurable neuroprotective effect when irradiation applied within 24 h of stroke onset
Yang et al. [72]	Sub-acute stroke (1)	Laser, YJ-ILIB-5, Bio-ILIB Human Energy Ltd (Taiwan)	632.8 nm	1.79–2.04 W/cm^2, 6428 J/cm2, 60 min for once a day for 10 days	Intravascularly; bilateral median antebrachial vein, cephalic vein, accessory cephalic vein by turns	Regional perfusion SPECT data showed an improvement in crossed cerebellar diaschisis; patient appeared to be more energetic and muscle power was improved to grade 3 in the left extremities; improved patient's functional recovery quantifying by Barthel index

(continued)

Table 11.2 (continued)

Study/Year	Subjects (n)	Light Source	Wavelengths	Irradiation parameters	Irradiation approach/sites	Findings
Naeser et al. [73]	Patients with aphasia due to left hemisphere stroke (6)	LEDs, three cluster heads	633 + 870 nm	500 mW, 22.2 mW/cm², 22.48 cm², 3 ×/week for 6 weeks	Transcranially; **Protocol A:** bilateral LED placements, including midline, at scalp vertex over left and right supplementary motor areas **Protocol B:** LEDs only on ipsilesional, left hemisphere side **Protocol C:** LED placements on ipsilesional left hemisphere side, plus one midline placement over mPFC at front hairline, a cortical node of the DMN **Protocol D:** LED placements on ipsilesional left hemisphere side, plus over two midline nodes of the DMN, mPFC, and precuneus (high parietal) simultaneously	**Protocol A:** improved picture naming and increased activation in left and right hemisphere, including left and right supplementary motor areas **Protocol B:** improved naming ability and activated only ipsilesional left hemisphere side as evidence by fMRI **Protocol C:** resulted in only moderate/poor response, and no increase in functional connectivity on resting-state functional-connectivity MRI **Protocol D:** resulted in a good response with a significant increase in functional connectivity within DMN, salience network, and central executive network

Abbreviations CW, continuous wave; DMN, default mode network; EEG, electroencephalography; LEDs, light-emitting diodes; mPFC, mesial prefrontal cortex; PW, pulsed wave.

References

1. Sacco, R.L., et al. 2013. An updated definition of stroke for the 21st century: A statement for healthcare professionals from the American Heart Association/American Stroke Association. *Stroke* 44 (7): 2064–2089.
2. Mahmoudi, J., et al. 2018. Imidazoline Receptor Agonists for Managing Hypertension May Hold Promise for Treatment of Intracerebral Hemorrhage. *Current molecular medicine* 18 (4): 241–251.
3. Farhoudi, M., et al. 2014. A review on molecular mechanisms of reocclusion following thrombolytic therapy in ischemic stroke patients. *Journal of Experimental and Clinical Neurosciences* 1 (1): 21–25.
4. Aguilar, M.I., and T.G. Brott. 2011. Update in intracerebral hemorrhage. *The Neurohospitalist* 1 (3): 148–159.
5. O'Donnell, M.J., et al. 2010. Risk factors for ischaemic and intracerebral haemorrhagic stroke in 22 countries (the INTERSTROKE study): A case-control study. *The Lancet* 376 (9735): 112–123.
6. Allen, C.L., and U. Bayraktutan. 2009. Oxidative stress and its role in the pathogenesis of ischaemic stroke. *International journal of stroke* 4 (6): 461–470.
7. Li, P., et al. 2018. Oxidative stress and DNA damage after cerebral ischemia: Potential therapeutic targets to repair the genome and improve stroke recovery. *Neuropharmacology* 134: 208–217.
8. Sims, N.R. and H. Muyderman. 2010. Mitochondria, oxidative metabolism and cell death in stroke. *Biochimica et Biophysica Acta (BBA)-Molecular Basis of Disease* 1802(1): 80–91
9. Lakhan, S.E., A. Kirchgessner, and M. Hofer. 2009. Inflammatory mechanisms in ischemic stroke: Therapeutic approaches. *Journal of translational medicine* 7 (1): 97.
10. Walter, S., et al. 2012. Diagnosis and treatment of patients with stroke in a mobile stroke unit versus in hospital: A randomised controlled trial. *The Lancet Neurology* 11 (5): 397–404.
11. Fischer, U., et al. 2005. NIHSS score and arteriographic findings in acute ischemic stroke. *Stroke* 36 (10): 2121–2125.
12. Sulter, G., C. Steen, and J. De Keyser. 1999. Use of the Barthel index and modified Rankin scale in acute stroke trials. *Stroke* 30 (8): 1538–1541.
13. Musuka, T.D., et al. 2015. Diagnosis and management of acute ischemic stroke: Speed is critical. *CMAJ* 187 (12): 887–893.
14. Eliasziw, M., and L. Paddock-Eliasziw. 2005. Comparison of MRI and CT for detection of acute intracerebral hemorrhage. *JAMA* 293 (5): 550–551.
15. Prabhakaran, S., I. Ruff, and R.A. Bernstein. 2015. Acute stroke intervention: A systematic review. *JAMA* 313 (14): 1451–1462.
16. Farhoudi, M., and A. Majdi. 2018. A New and promising perspective on neuroprotective agents in stroke management. *Journal of Experimental and Clinical Neurosciences* 5 (1): 1–2.
17. Fitzpatrick, D., and D. O'Neill. 2018. Stroke and cerebrovascular disease. In *learning geriatric medicine*, 137–143. Springer.
18. Teasell, R., et al. 2012. Time to rethink long-term rehabilitation management of stroke patients. *Topics in stroke rehabilitation* 19 (6): 457–462.
19. Dickey, L., et al. 2010. Incidence and profile of inpatient stroke-induced aphasia in Ontario, Canada. *Archives of physical medicine and rehabilitation* 91 (2): 196–202.
20. Engelter, S.T., et al. 2006. Epidemiology of aphasia attributable to first ischemic stroke: Incidence, severity, fluency, etiology, and thrombolysis. *Stroke* 37 (6): 1379–1384.

21. Yao, J., et al. 2015. Relationship of post-stroke aphasic types with sex, age and stroke types. *World Journal of Neuroscience* 5 (01): 34.
22. Pulvermüller, F., et al. 2001. Constraint-induced therapy of chronic aphasia after stroke. *Stroke* 32(7):1621–1626
23. Pitt, R., et al. 2017. The feasibility of delivering constraint-induced language therapy via the Internet. *Digital health* 3: 2055207617718767.
24. Antal, A., M.A. Nitsche, and W. Paulus. 2001. External modulation of visual perception in humans. *NeuroReport* 12 (16): 3553–3555.
25. Berthier, M.L., et al. 2011. Drug therapy of post-stroke aphasia: A review of current evidence. *Neuropsychology Review* 21 (3): 302.
26. Sadigh-Eteghad, S., et al. 2017. D-galactose-induced brain ageing model: A systematic review and meta-analysis on cognitive outcomes and oxidative stress indices. *PLoS ONE* 12 (8): e0184122.
27. Kumar, A., and V. Gupta. 2016. A review on animal models of stroke: An update. *Brain research bulletin* 122: 35–44.
28. Pulsinelli, W., and A. Buchan. 1988. The four-vessel occlusion rat model: Method for complete occlusion of vertebral arteries and control of collateral circulation. *Stroke* 19 (7): 913–914.
29. Carmichael, S.T. 2005. Rodent models of focal stroke: Size, mechanism, and purpose. *NeuroRx* 2 (3): 396–409.
30. Sugimori, H., et al. 2004. Krypton laser-induced photothrombotic distal middle cerebral artery occlusion without craniectomy in mice. *Brain Research Protocols* 13 (3): 189–196.
31. Durukan, A., and T. Tatlisumak. 2007. Acute ischemic stroke: Overview of major experimental rodent models, pathophysiology, and therapy of focal cerebral ischemia. *Pharmacology Biochemistry and Behavior* 87 (1): 179–197.
32. Sadigh-Eteghad, S., et al. 2018. Intranasal cerebrolysin improves cognitive function and structural synaptic plasticity in photothrombotic mouse model of medial prefrontal cortex ischemia. *Neuropeptides* 71: 61–69.
33. Fluri, F., M.K. Schuhmann, and C. Kleinschnitz. 2015. Animal models of ischemic stroke and their application in clinical research. *Drug design, development and therapy* 9: 3445.
34. Traystman, R.J. 2003. Animal models of focal and global cerebral ischemia. *ILAR journal* 44 (2): 85–95.
35. Oron, U., et al. 2001. Attenuation of infarct size in rats and dogs after myocardial infarction by low-energy laser irradiation. *Lasers in Surgery and Medicine: The Official Journal of the American Society for Laser Medicine and Surgery* 28 (3): 204–211.
36. Oron, U., et al. 2001. Low-energy laser irradiation reduces formation of scar tissue after myocardial infarction in rats and dogs. *Circulation* 103 (2): 296–301.
37. Yaakobi, T., et al. 2001. Long-term effect of low energy laser irradiation on infarction and reperfusion injury in the rat heart. *Journal of Applied Physiology* 90 (6): 2411–2419.
38. Ad, N., and U. Oron. 2001. Impact of low level laser irradiation on infarct size in the rat following myocardial infarction. *International journal of cardiology* 80 (2–3): 109–116.
39. Streeter, J., L. De Taboada, and U. Oron. 2004. Mechanisms of action of light therapy for stroke and acute myocardial infarction. *Mitochondrion* 4 (5–6): 569–576.
40. Leung, M.C., et al. 2002. Treatment of experimentally induced transient cerebral ischemia with low energy laser inhibits nitric oxide synthase activity and up-regulates the expression of transforming growth factor-beta 1. *Lasers in Surgery and Medicine: The Official Journal of the American Society for Laser Medicine and Surgery* 31 (4): 283–288.
41. DeTaboada, L., et al. 2006. Transcranial application of low-energy laser irradiation improves neurological deficits in rats following acute stroke. *Lasers in Surgery and Medicine: The Official Journal of the American Society for Laser Medicine and Surgery* 38 (1): 70–73.

42. Oron, A., et al. 2006. Low-level laser therapy applied transcranially to rats after induction of stroke significantly reduces long-term neurological deficits. *Stroke* 37 (10): 2620–2624.
43. Ilic, S., et al. 2006. Effects of power densities, continuous and pulse frequencies, and number of sessions of low-level laser therapy on intact rat brain. *Photomedicine and Laser Therapy* 24 (4): 458–466.
44. Lee, H.I., et al. 2017. Pretreatment with light-emitting diode therapy reduces ischemic brain injury in mice through endothelial nitric oxide synthase-dependent mechanisms. *Biochemical and biophysical research communications* 486 (4): 945–950.
45. Lapchak, P.A. 2010. Taking a light approach to treating acute ischemic stroke patients: Transcranial near-infrared laser therapy translational science. *Annals of medicine* 42 (8): 576–586.
46. Lapchak, P.A., J. Wei, and J.A. Zivin. 2004. Transcranial infrared laser therapy improves clinical rating scores after embolic strokes in rabbits. *Stroke* 35 (8): 1985–1988.
47. Lapchak, P., et al. 2007. Transcranial near-infrared light therapy improves motor function following embolic strokes in rabbits: An extended therapeutic window study using continuous and pulse frequency delivery modes. *Neuroscience* 148 (4): 907–914.
48. Lapchak, P.A., and L. De Taboada. 2010. Transcranial near infrared laser treatment (NILT) increases cortical adenosine-5'-triphosphate (ATP) content following embolic strokes in rabbits. *Brain research* 1306: 100–105.
49. Huisa, B.N., et al. 2013. Incremental treatments with laser therapy augments good behavioral outcome in the rabbit small clot embolic stroke model. *Lasers in medical science* 28 (4): 1085–1089.
50. Meyer, D.M., Y. Chen, and J.A. Zivin. 2016. Dose-finding study of phototherapy on stroke outcome in a rabbit model of ischemic stroke. *Neuroscience letters* 630: 254–258.
51. Lapchak, P.A., et al. 2008. Safety profile of transcranial near-infrared laser therapy administered in combination with thrombolytic therapy to embolized rabbits. *Stroke* 39 (11): 3073–3078.
52. Lapchak, P.A., and P.D. Boitano. 2016. A novel method to promote behavioral improvement and enhance mitochondrial function following an embolic stroke. *Brain research* 1646: 125–131.
53. Lee, H.I., et al. 2017. Low-level light emitting diode (LED) therapy suppresses inflammasome-mediated brain damage in experimental ischemic stroke. *Journal of biophotonics* 10 (11): 1502–1513.
54. Lee, H.I., et al. 2017. Low-level light emitting diode therapy promotes long–term functional recovery after experimental stroke in mice. *Journal of biophotonics* 10 (12): 1761–1771.
55. Yang, L., et al. 2018. Photobiomodulation therapy promotes neurogenesis by improving post-stroke local microenvironment and stimulating neuroprogenitor cells. *Experimental neurology* 299: 86–96.
56. Lampl, Y., et al. 2007. Infrared laser therapy for ischemic stroke: A new treatment strategy: Results of the NeuroThera Effectiveness and Safety Trial-1 (NEST-1). *Stroke* 38 (6): 1843–1849.
57. Zivin, J.A., et al. 2009. Effectiveness and safety of transcranial laser therapy for acute ischemic stroke. *Stroke* 40 (4): 1359–1364.
58. Hacke, W., et al. 2014. Transcranial laser therapy in acute stroke treatment: Results of neurothera effectiveness and safety trial 3, a phase III clinical end point device trial. *Stroke* 45 (11): 3187–3193.
59. Kasner, S.E., et al. 2013. Transcranial laser therapy and infarct volume. *Stroke* 44 (7): 2025–2027.
60. Xiaoa, X., et al. 2000. A single photon emission computed tomography study of the therapy of intravascular low intensity laser irradiation on blood for brain infarction. *Laser Therapy* 13 (1): 110–113.

61. Yang, W.-H., S.-P. Lin, and S.-T. Chang. 2017. Case report: Rapid improvement of crossed cerebellar diaschisis after intravascular laser irradiation of blood in a case of stroke. *Medicine* 96(2)

62. Naeser, M.A., et al. 1995. Laser acupuncture in the treatment of paralysis in stroke patients: A CT scan lesion site study. *American journal of acupuncture* 23: 13–13.

63. Ab Boonswang, N., et al. 2012. A new treatment protocol using photobiomodulation and muscle/bone/joint recovery techniques having a dramatic effect on a stroke patient's recovery: a new weapon for clinicians. *Case Reports* 2012:bcr0820114689

64. Stemer, A.B., B.N. Huisa, and J.A. Zivin. 2010. The evolution of transcranial laser therapy for acute ischemic stroke, including a pooled analysis of NEST-1 and NEST-2. *Current cardiology reports* 12 (1): 29–33.

65. Naeser, M.A., M.P. Alexander, D. Stiassny-Eder, V. Galler, J. Hobbs, D. Bachman, et al. 1995. Laser acupuncture in the treatment of paralysis in stroke patients: a CT scan lesion site study. *American journal of acupuncture* 23:13

66. Xiaoa, X., J. Donga, X. Chua, J.-L. Jiaob, S. Jiaa, and X. Zhenga, et al. 2000. A single photon emission computed tomography study of the therapy of intravascular low intensity laser irradiation on blood for brain infarction. *Laser Therapy* 13(1): 110–113

67. Lampl, Y., J.A. Zivin, M. Fisher, R. Lew, L. Welin, B. Dahlof, et al. (2007). Infrared laser therapy for ischemic stroke: a new treatment strategy: results of the NeuroThera Effectiveness and Safety Trial-1 (NEST-1). *Stroke* 38(6):1843–1849

68. Zivin, J.A., G.W. Albers, N. Bornstein, T. Chippendale, B. Dahlof, T. Devlin, et al. 2009. Effectiveness and safety of transcranial laser therapy for acute ischemic stroke. *Stroke* 40(4): 1359–1364

69. Ab Boonswang, N., M. Chicchi, A. Lukachek, and D. Curtiss. 2012. A new treatment protocol using photobiomodulation and muscle/bone/joint recovery techniques having a dramatic effect on a stroke patient's recovery: a new weapon for clinicians. *Case Reports* (2012)2012: bcr0820114689

70. Kasner, S.E., D.Z. Rose, A. Skokan, M.G. Walker, J. Shi, and J. Streeter. 2013. Transcranial laser therapy and infarct volume. *Stroke* 44(7): 2025-2027

71. Hacke, W., P.D. Schellinger, G.W. Albers, N.M. Bornstein, B.L. Dahlof and R. Fulton, et al. 2014. Transcranial laser therapy in acute stroke treatment: results of neurothera effectiveness and safety trial 3, a phase III clinical end point device trial. *Stroke* 45(11):3187–3193

72. Yang, W-H, S-P. Lin, and S-T. Chang. 2017. Case report: Rapid improvement of crossed cerebellar diaschisis after intravascular laser irradiation of blood in a case of stroke. *Medicine* 96(2)

73. Naeser, M.A., M.D. Ho, P.I. Martin, M.R. Hamblin, and B-B. Koo. 2019. Increased functional connectivity within intrinsic neural networks in chronic stroke following treatment with red/ Near-infrared transcranial photobiomodulation: case series with improved naming in aphasia. *Photobiomodulation, Photomedicine, and Laser Surgery*

74. Uozumi, Y., et al. 2010. Targeted increase in cerebral blood flow by transcranial near-infrared laser irradiation. *Lasers in surgery and medicine* 42 (6): 566–576.

75. Yip, K., et al. 2011. The effect of low-energy laser irradiation on apoptotic factors following experimentally induced transient cerebral ischemia. *Neuroscience* 190: 301–306.

76. Choi, D.-H., et al. 2012. Effect of 710-nm visible light irradiation on neuroprotection and immune function after stroke. *NeuroImmunoModulation* 19 (5): 267–276.

77. Jiang, W., et al. 2014. Red photon treatment inhibits apoptosis via regulation of bcl-2 proteins and ROS levels, alleviating hypoxic–ischemic brain damage. *Neuroscience* 268: 66–74.

78. Fukuzaki, Y., et al. 2015. 532 nm low-power laser irradiation facilitates the migration of GABAergic neural stem/progenitor cells in mouse neocortex. *PLoS ONE* 10 (4): e0123833.

79. Na, C.-S., et al. 2015. Low-level green and red laser treatment of Shaochong (HT9)·Dadun (LR1) and Shaohai (HT3)·Yingu (KI10) Acupoints in a rat model of focal cerebral ischemia. *Transactions on Electrical and Electronic Materials* 16 (2): 65–69.

80. Lee, H.I., et al. 2016. Pre-conditioning with transcranial low-level light therapy reduces neuroinflammation and protects blood-brain barrier after focal cerebral ischemia in mice. *Restorative neurology and neuroscience* 34 (2): 201–214.

81. Xiong, G., and X. Li. 2017. Effects of laser acupoint irradiation on energy metabolism of brain tissue of rats with cerebral ischemia-reperfusion. *Laser Physics* 27 (12): 125601.

82. Jittiwat, J. 2017. Laser acupuncture at GV20 improves brain damage and oxidative stress in animal model of focal ischemic stroke. *Journal of acupuncture and meridian studies* 10 (5): 324–330.

83. Yun, Y.-C., et al. 2017. Laser acupuncture exerts neuroprotective effects via regulation of Creb, Bdnf, Bcl-2, and Bax gene expressions in the hippocampus. *Evidence-Based Complementary and Alternative Medicine* 2017

84. Sanderson, T.H., et al. 2018. Inhibitory modulation of cytochrome c oxidase activity with specific near-infrared light wavelengths attenuates brain ischemia/reperfusion injury. *Scientific reports* 8 (1): 3481.

85. Tucker, L.D., et al. 2018. Photobiomodulation therapy attenuates hypoxic-ischemic injury in a neonatal rat model. *Journal of Molecular Neuroscience* 65 (4): 514–526.

86. Yang, L., et al. 2019. Photobiomodulation preconditioning prevents cognitive impairment in a neonatal rat model of hypoxia-ischemia. *Journal of biophotonics* 12 (6): e201800359.

87. Wang, R., et al. 2019. Photobiomodulation for global cerebral ischemia: Targeting mitochondrial dynamics and functions. *Molecular neurobiology* 56 (3): 1852–1869.

88. Jittiwat, J. 2019. Baihui point laser acupuncture ameliorates cognitive impairment, motor deficit, and neuronal loss partly via antioxidant and anti-inflammatory effects in an animal model of focal ischemic stroke. *Evidence-Based Complementary and Alternative Medicine* 2019

89. Argibay, B., et al. 2019. Light-emitting diode photobiomodulation after cerebral ischemia. *Frontiers in neurology* 10: 911.

90. Vahabzadeh-Hagh, A., et al. 2019. Near infrared light amplifies endothelial progenitor cell accumulation after stroke. *Conditioning Medicine*

91. Strubakos, C.D., et al. 2019. Non-invasive treatment with near-infrared light: A novel mechanisms-based strategy that evokes sustained reduction in brain injury after stroke. *Journal of Cerebral Blood Flow & Metabolism*:0271678X19845149

92. Salehpour, F., et al. 2019. Photobiomodulation and coenzyme Q10 treatments attenuate cognitive impairment associated with model of transient global brain ischemia in artificially aged mice. *Frontiers in cellular neuroscience* 13

93. Lee, D.-J., et al. 2019. Photobiomodulation therapy in mice with chronic cerebral hypoperfusion using application-specific near-infrared light-emitting diode system. *Transactions on Electrical and Electronic Materials* 20 (5): 420–425.

94. de Jesus Fonseca, E.G., et al. 2019. Study of transcranial therapy 904 nm in experimental model of stroke. *Lasers in Medical Science*, 2019: 1–7

Photobiomodulation Therapy for Tnraumatic Brain Injury

12

12.1 Traumatic Brain Injury

12.1.1 Acute and Chronic Traumatic Brain Injury

Traumatic brain injury (TBI) is a serious public health problem all over the world [1]. According to recent findings, TBI is a leading cause of death, accounting for around one-third of all injury-associated deaths in the United States, and continues to show an increasing trend [2]. TBI affects 1.4 million individuals annually in the US and imposes a significant socioeconomic burden [3–5]. In 2006, the Center for Disease Control and Prevention (CDC) announced that fall-associated TBIs exceeded motor vehicle accident-associated TBI, and were responsible for at least 34% of all TBI-related morbidity and mortality [6].

By definition "TBI is defined as an alteration in brain function, or other evidence of brain pathology, caused by an external force" [7]. However, due to its importance, TBI has different classifications. It is also important to categorize TBI according to its severity as this affects epidemiologic surveys of TBI [8]. Until now, the most commonly applied method to determine the severity of TBI is the Glasgow Coma Scale (GCS) score [9]. According to this scale, which ranges from 3 to 15, three being the most severe and 15 being normal, severe TBI is defined as a GCS score between 3 and 8. Moderate TBI is defined as a GCS score between 8 and 13, and mild TBI, on the other hand, is a GCS score between 13 and 15 [10]. Unfortunately, the rate of severe TBI has risen over the years in the US, while the frequency of other categories of TBI has fallen. Intriguingly, an epidemiological survey revealed that a GCS score of 3 was the most common score, followed by 15, found in patients with TBI [11]. However, this number has decreased over the years from 83% in 1994 to 41% in 2013 [12].

Four major causes of TBI have been described; vehicle-related, violence-related, fall-related, and other causes [13]. Vehicle-associated TBIs occur mainly in the young and active individuals who usually do not have a history of any medical or psychological disorders or drug abuse. The resulting acceleration and deceleration injury leads to diffuse axonal and shear damage to the brain neurons. However, these patients have a tendency to recover their functional abilities one year after the accident, which is more than in other subgroups of TBI [14]. Comparatively, patients with violence-related TBIs tend to be male, non-white, jobless, single, and to be a multidrug abuser [15, 16]. Unfortunately, the severity of TBI in the violence group is not less than in the accidental group, and is usually categorized as moderate or severe TBIs. Also, the violence group tends to have worse outcomes compared to the fall and vehicle groups, one year after injury. In the fall group, patients tend to be older than in other groups. They are more likely to be married and retired. Fall patients with moderate TBI, however, tend to show some degree of improvement one year after the damage. The severity of the injury ranges between moderate and severe at the time of damage [13]. The trend of TBIs has shifted from vehicle accidents toward fall injuries over recent years. It has been found that a majority of fall-related TBI cases are due to alcohol intake (29%). These patients tend to be male and younger than others, and are more likely to suffer from a subdural hematoma (SDH) [12].

Pathophysiologically, TBI is characterized by overproduction of reactive oxygen species (ROS) and impairment of anti-oxidant scavenging system. This leads to an increase in the permeability of the blood–brain barrier and indirect activation of matrix metalloproteinases (MMPs) and aquaporins. The final result of these harmful signaling pathways is vascular and cellular edema along with neuroinflammation. Neuroinflammation is also caused by the production of inflammatory cytokines and growth factors such as IL-1β, tumor necrosis factor-α (TNF-α), and transforming growth factor-beta (TGF-β). One result of neuroinflammation is apoptotic cell damage via activation of caspase-1/3 [17, 18].

Studies have linked TBI to abnormal frontal lobe activation or damage to the temporal lobes [4]. Focal and diffuse brain damage are two different types of TBI based on the presence or absence of focal lesions. Focal damage dominates in ventral and polar frontal, and lateral anterior temporal lobe regions, which cause cognitive impairment. However, diffuse injury is more global leading to white matter atrophy and destruction of the afferent connections from and to the frontal and temporal brain regions [4]. The PFC and the anterior cingulate gyrus are the two important regions within the frontal lobes, which are susceptible to damage during TBI. Damage to the olfactory bulb and tract can also occur along with ventral frontal damage in TBI [19, 20].

Unfortunately, patients with mild TBI often do not present any specific signs or symptoms. On the other hand, symptoms reported by TBI patients can often be non-specific. Another issue which complicates the semiology of TBI is the presence of secondary gain. This casts doubt over the authenticity of the patient self-reported symptoms patients. In a study by Dikmen et al. [21], the most common symptoms of TBI reported by the patients were as follows; memory impairment, trouble concentrating, fatigue, anxiety, dizziness, headache, blurred vision, irritation by noise, and trouble sleeping. These were present at one month and one year after the injury to a different extent. The comparison between the two periods revealed that cognitive symptoms and fatigue tend to persist for at least one year post-injury, while other symptoms are more likely to disappear [21]. Generally speaking, moderate to severe TBI patients undergo loss of consciousness, the duration of which can range from minutes to hours. They suffer from a constant headache or a headache which worsens over time, repeated nausea or vomiting, mydriasis both unilateral or bi-lateral, rhinorrhea, otorrhea, seizures or convulsions, and difficulty waking from sleep. These symptoms and signs are less severe in patients with mild TBI, and may range from a bad taste in the mouth to loss of consciousness, ranging from a few seconds to minutes [22]. TBI may be also complicated by neuropsychiatric disorders. These include (but are not limited to) post-traumatic amnesia, post-traumatic stress disorder, epilepsy, anxiety, psychosis, fatigue/apathy, suicide, dementia, aphasia, personality disorders, depression, and mania [23]. These psychiatric disorders may further complicate the treatment and management of TBI patients.

An appropriate diagnostic procedure is vital in the acute management of TBI patients. Accordingly, accurate diagnosis helps the attending physician to classify the severity of the damage, and perform subsequent necessary procedures to help the injured patients recover [24]. Taking a history and physical examination are of high value in the diagnosis of TBI. As mentioned earlier, GCS is the clinical cornerstone to categorize the severity of TBI. This test relies on the clinical manifestations of TBI rather than its pathophysiology. However, some studies have argued that the clinical GCS test may be unreasonably complex for this purpose [25]. Neuroimaging modalities such as computed tomography (CT), and magnetic resonance imaging (MRI) may be needed in case of moderate to severe TBI. Nevertheless, their value in mild TBI cases is questionable [26]. Several complications of TBI are not detectable via CT or MRI. Diffuse axonal injury (DAI), a major type of brain damage in 40% to 50% of the cases, is one important complication of TBI which is cannot easily be detected by CT or MRI, but can be readily diagnosed using newer modalities such as diffusion tensor imaging (DTI) [14, 27]. Newer imaging modalities, such as DTI, positron emission tomography (PET), and high definition fiber tracking (HDFT) can show increased sensitivity and specificity for TBI diagnosis [28].

The therapeutic treatments in the emergency department for patients presenting with acute TBI will not be discussed here. In line with the progress and development in the diagnosis of TBI, its management has undergone a dramatic change over recent years. Stem cell-based and nanotechnology-based approaches are emerging novel therapeutic alternatives. Pharmaceutical interventions such as tauroursodeoxycholic acid, erythropoietin, cyclosporine, or steroids are also promising for the inhibition or reversal of TBI damage [28, 29]. All of these therapeutic strategies are aimed at the prevention of secondary damage, and preservation of cerebral perfusion, and brain oxygenation, one way or another. Also, early nutritional intervention has been linked to a better outcome, and enteral or parenteral administration of nutrients is desirable [30]. Besides, these patients are at risk for hyperglycemia, seizures, and thromboembolic events, which further complicate their prognosis [31]. These complications should be prevented using deep vein thrombosis prophylaxis, seizure prophylaxis, and management of blood glucose [32]. One emerging therapeutic option for the treatment of TBI is transcranial magnetic stimulation (TMS). TMS has both prognostic and therapeutic value in TBI management. Due to its effects on synaptic plasticity and cortical reorganization, repetitive TMS (rTMS) has been shown to provide positive effects in TBI patients [33]. A recent clinical trial showed that ten sessions of rTMS were effective in cognitive rehabilitation in TBI patients [34].

12.1.2 Animal Models of Traumatic Brain Injury

Due to its high incidence and prevalence, and serious long-term effects on patients, TBI models have been constructed in several animals. Controlled cortical impact (CCI) is a mechanical TBI, commonly induced by the acceleration of a metal rod, which impacts the exposed cortical surface through a craniotomy. The rod can be accelerated by pneumatic or electromagnetic force. CCI has been successfully applied in rats (parietal cortex; midline), mice (parietal cortex), pigs (frontal lobe), and primates (frontal lobe) [35, 36]. The CCI lesion has a high degree of precision and reproducibility. Moreover, the severity of the injury can be controlled by adjusting the depth of tissue damage. However, this model has several limitations, such as not being able to produce a diffuse impact, and unavoidable mechanical variation [36, 37]. Another model that has been used to mimic TBI in animals, is cryogenic trauma/injury which has been employed used in rats, mice, and rabbits. It also has a high reproducibility rate, but induces milder TBI as compared to the CCI model. Other models such as fluid percussion, closed-head injury (CHI) model, rigid indentation, and rotational injury also exist, and have been used to study different aspects of TBI [38].

12.2 Photobiomodulation Therapy for Traumatic Brain Injury

12.2.1 Photobiomodulation Therapy for Traumatic Brain Injury in Animal Models

Historically, the Oron team from Israel [39] was the first to show that single irradiation of the mouse head (a few hours after creation of a TBI lesion) with a NIR laser could improve neurological performance and decrease the size of the brain lesion. A weight-drop device was used to induce a closed-head TBI in mice. An 808 nm diode laser with a fluence value calculated at the surface of the brain (1.22.4 J/cm^2 provided by 2 min of irradiation with 200 mW laser power to the scalp) was transcranially delivered to the head at 4 h post-TBI. The neurological performance of the mice was evaluated using the neurological severity score (NSS). There was no significant difference in NSS between the two irradiances (10 versus 20 mW/cm^2) or significant difference between the control and PBM-treated group at early time points (24 and 48 h) post-TBI. But, there was a significant improvement (27% lower NSS score) in the PBM-treated group between 5 days and 4 weeks. PBM-treated animals also showed a smaller loss of cortical tissue compared to sham-treated animals [39]. In another study, Oron and colleagues [40] varied the pulse parameters [continuous wave (CW), 100 Hz, or 600 Hz pulsed wave (PW)] and tested whether transcranial PBM was equally effective when delivered at 4, 6, or 8 h post-TBI. They first showed that a calculated fluence to the cortical surface of 1.2 J/cm^2 808 nm laser at 200 mW applied to the head, was more effective when delivered at 6 h post-TBI than at 8 h. They then selected an even shorter time post-TBI (4 h) and compared CW with PW mode at 100 and 600 Hz frequency. At 56 days, more mice in the 100 Hz PW group (compared to the CW and 600 Hz PW groups) had fully recovered. The 600 Hz PW group had a better NSS mean score than the CW and 100 Hz PW groups up to 20 days. MRI analysis also showed significantly smaller lesion volume in treated mice (all PBM regimens) compared to control animals.

Khuman et al. [41] delivered 800 nm laser light to mice, either directly to the injured brain tissue (through a craniotomy) or else transcranially, beginning 60–80 min after CCI TBI. At a fluence of 60 J/cm^2, the mice showed increased performance in the Morris water maze task compared to non-treated animals. When PBM therapy was delivered via open craniotomy there was decreased microglial activation at 48 h (IbA-1 1 cells). Nevertheless, non-significant effects of PBM on post-injury cognitive function were observed using lower or higher fluences, a 4-h administration time point, or 60 J/cm2 at 7-day post-TBI.

Quirk et al. [42] used Sprague–Dawley rats with a severe CCI TBI allocated into three groups: real-TBI, sham surgery, and anesthesia only. Each group received either real or sham PBM therapy consisting of 670 nm LED treatment (50 mW/cm^2 for 5 min, providing 15 J/cm^2) given two times per day for 3 days (chemical analysis) or 10 days (neurobehavioral analysis using a TruScan nose-poke device). Significant differences in the number of task entries, repeat entries, and task errors were observed in the TBI rats treated with PBM versus untreated TBI mice, and in sham surgery mice treated with PBM versus untreated sham mice. A significant reduction was observed in the pro-apoptotic marker Bax, and an increase in the anti-apoptotic marker Bcl-2, along with reduced glutathione (GSH) levels in PBM-treated TBI mice.

Moreira et al. [43] used a different model of TBI. Wistar rats received a craniotomy, and a copper probe cooled in liquid nitrogen was applied to the surface of the brain to create a standardized cryogenic injury. Animals were treated with either a 780 or 660 nm laser at one of two different fluences (3 or 5 J/cm^2) twice (once immediately after the injury and again 3 h later). Animals were sacrificed 6 and 24 h after the injury. The 780 nm laser was better at reducing levels of pro-inflammatory cytokines (e.g., TNF-α, IL-6, and IL-1β), particularly at early time points [43]. In a follow-up study using 3 J/cm^2 [44], these researchers investigated the healing of the injury in these rats at time points 6 h, 1, 7, and 14 days after the last PBM session. The cryogenic injury created focal lesions in the cortex, characterized by necrosis, edema, hemorrhage, and inflammatory infiltrate. The most prominent finding was that PBM-treated lesions showed less tissue loss than control lesions at 6 h. During the first 24 h, the number of viable neurons was significantly higher in the PBM-treated rats. PBM decreased the amount of glial fibrillary acidic protein (GFAP, a marker of astrogliosis), and the numbers of leukocytes and lymphocytes, thus confirming an anti-inflammatory effect.

In addition to the above-mentioned research from various laboratories, a series of studies from Michael Hamblin's laboratory has also confirmed the beneficial neuroprotective effects of transcranial PBM therapy in both closed-head and CCI TBI mouse models [45, 46]. In this respect, Wu et al. [47] first investigated the effect of varying the laser wavelength on the efficiency of PBM in CHI TBI in mice. Closed-head injury was induced via a weight-drop apparatus. Mice were randomly allocated to a PBM-treated group with a particular wavelength, or to a sham-treated TBI group as a control. To analyze the severity of the TBI, the NSS was assessed. The injured mice were then treated with one of four different wavelengths (665, 730, 810, or 980 nm) at a scalp fluence of 36 J/cm^2. A single irradiation to the head was delivered at 4 h post-TBI. Both 665 and 810 nm laser groups demonstrated significant improvement in NSS when compared to the control group between days 5 and 28. By contrast, the 730 and 980 nm laser groups did not show any significant improvement in NSS. Ando et al. [48] next used the 810 nm diode laser delivered at similar parameters to those used in the Wu study [47], and varied the pulsing parameters of the laser. These were either PW at 10-Hz or 100 Hz (50% duty cycle), or CW laser. They used a different mouse model of TBI that was induced with a CCI device.

The mice were allowed to recover after the surgery, before being tested for NSS at 1-h post-CCI. A single 810 nm PBM treatment with an average irradiance of 50 mW/m^2 and a fluence of 36 J/cm^2 was given to the closed-head in mice at 4 h post-CCI. At 48 h to 28 days post-TBI, all PBM-treated mice had significant improvements in the measured NSS when compared to the non-treated TBI mice. Although all PBM-treated mice had similar NSS improvement rates up to day 7, the 10 Hz PW group began to show even greater improvement beyond this point. At day 28, the forced swim test for depression was performed and the results showed a significant decrease in the immobility time for the 10 Hz PW group. Similarly, in the tail suspension test, which also measures depression, there was also a significant decrease in the immobility time at day 28, and also at day 1, in the 10 Hz PW group. The next series of studies from Hamblin's laboratory employed the same mouse CCI model as explained above, and used the same 810 nm laser. However, in these studies [49–52], they tested the hypothesis that multiple daily applications of transcranial PBM therapy would be more effective than a single application delivered at 4 h post-CCI. But, because single irradiation of 36 J/cm^2 delivered at 50 mW/cm^2 had been found so effective [48], they decreased the total fluence to test the effect of multiple PBM administration. Hence, they used an irradiance of 25 mW/cm^2 for the same 12 min to deliver scalp fluence of 18 J/cm^2 given either in a single application (1× PBM regimen, 4 h post-TBI), or given three times, once a day for 3 days beginning 4 h post-TBI (3× PBM regimen), or given 14 times once a day for 14 days beginning 4 h post-TBI (14× PBM regimen) [49]. The 3× PBM regimen resulted in significantly better NSS scores than a single application (1 3 PBM), over 4 weeks post-TBI. Nonetheless, it was surprising that the same benefits were not found in the 14 × PBM regimen mice. PBM could decrease the number of neurons undergoing degeneration (shown by Fluoro-Jade staining) [49]. Also, memory and learning as assessed by the Morris water maze task was enhanced by PBM therapy [50]. In the final study, Xuan et al. [52] followed the mice that were treated with the three various regimens (1×, 3×, or 14 × PBM regimens) for as long as 8 weeks. They observed that the 3× PBM regimen significantly improved neuromuscular performance and cognitive functioning compared to non-treated animals at 4 weeks, and this improvement continued up to 8 weeks. Nevertheless, when the PBM was repeated for 14 days (14× PBM regimen), there was a decline in cognitive performances at 2 weeks and it was not until week 4 that the NSS caught up with the non-treated TBI animals. But, the 14× PBM regimen group improved at a relatively faster rate from week 4 up to week 8. Although the 14× PBM regimen group did not catch up to the 3× PBM regimen group, it does appear that the detrimental effect of the excessive 14× PBM applications was only temporary rather than permanent. This was attributed to the excessive glial activation produced by the 14 × PBM fading over time, while the stimulation to neurogenesis still remained.

To further improve the therapeutic effectiveness of transcranial NIR PBM therapy, Dong et al. [53] combined PBM with three different mitochondria-improving metabolic agents, glucose, lactate, and pyruvate, which are all components of the mitochondrial tricarboxylic acid (TCA) cycle. None of the three substrates could increase ATP generation in normoxic cultures, but they exhibited varying effects in hypoxic cultures, with the most predominant effect of lactate on ATP production. On the other hand, when combined with 810 nm laser PBM therapy (150 mW/cm^2 for 4 min, providing a scalp fluence of 36 J/cm^2), glucose or pyruvate significantly increased oxidative phosphorylation in hypoxic cells compared to lactate. Besides, a combination of PBM therapy with glucose, pyruvate, or lactate, could protect neurons from hypoxia-induced death significantly better than either single modality. Remarkably, the effect of PBM therapy on neuronal survival was more prominent in the presence of pyruvate than in the presence of glucose or lactate. These findings suggest that PBM and energy metabolic modulators in combination can increase ATP production and cell survival in hypoxic conditions.

Recently, a novel study has shown that after fluid percussion injury, PBM of rats with non-invasive optoacoustic laser therapy improved cognitive function and prevents TBI-induced impaired maturation and aberrant migration of neural progenitor cells in the hippocampus dentate gyrus. Their data have also shown that PBM exerted its neuroprotective effects at least in part by mitigating TBI-driven dysregulation of specific microRNAs known to modulate neurogenesis in the hippocampus [54].

12.2.2 Photobiomodulation Therapy for Traumatic Brain Injury in Human Studies

Patients who recover from moderate or severe acute head injuries are at risk of developing a range of long-lasting symptoms, including cognitive deficits (e.g., poor memory, impaired executive function, and difficulties focusing), headaches, as well as sleep and mood disturbances [55]. Up to now, although the majority of animal studies in vivo have been conducted on acute TBI models, most of the clinical trials have been performed on chronic TBI patients who have suffered head injuries at different times in the past (sometimes several years ago).

In 2011, the first open study in two chronic TBI patients was conducted by Naeser et al. [56] who applied 500 mW CW LEDs (combined 660 nm and 870 nm) with an irradiance of 22.2 mW/cm^2 for 10 min over the whole head of the patients. Following eight weekly PBM treatments, the first case reported that she could concentrate on tasks for a longer period of time. She had also a better ability to remember what she read, and better mathematical skills. The second case had significant improvement in neuropsychological tasks (Stroop test for executive function and Wechsler Memory scale test for the logical memory test) after 9 months of weekly PBM therapy [56]. Naeser and her colleagues [57] then went on to report a case series containing a further 11 TBI patients. This was also an open protocol study that investigated whether transcranial application of red/NIR LEDs could enhance cognitive function in patients with chronic and mild TBI, who suffered from persistent cognitive impairment following various accidents. The trauma had been caused by sports-related events, motor vehicle accidents, and for one patient, an improvised explosive device blast. The PBM treatment plan consisted of 18 sessions and was commenced anywhere from 10 months to 8 years post-trauma. A total of 11 LED cluster arrays (500 mW, 22.2 mW/cm^2) were applied to the head for 10 min per set (5 or 6 LEDs placements per set, Set A and then Set B, in each session) providing a scalp fluence of 13 J/cm^2 per set. Neuropsychological tasks were conducted pre-PBM therapy and 1 week, 1 month, and 2 months after the last treatment session. According to their data, there was a significant positive linear trend for the Stroop Test for executive function, in trial 3 inhibition; Stroop, trial 4 inhibition switching; California Verbal Learning Test (CVLT)-II, total trials 1–5; CVLT-II, long delay free recall. Along with these cognitive improvements, better sleep quality and fewer post-traumatic stress disorder (PTSD) symptoms were reported after PBM therapy. In addition, patients and caregivers or family members reported better social functioning and improved ability to carry out interpersonal and occupational activity [57]. Further studies from Naeser et al. [58] tested an intranasal LED PBM applicator in two mild TBI patients. Two small intranasal devices (one red [633 nm] and the other NIR [810 nm]) were clipped into each nostril. The first mild TBI patient who had sustained four separate sports-related accidents, received intranasal PBM three times per week for 6 consecutive weeks. Significant improvements in executive function, verbal memory, attention and verbal fluency, as well as total time asleep and sleep efficiency were reported. The second case sustained a mild TBI in a motor vehicle accident 30 years prior to receiving the intranasal PBM therapy series. The patient showed significant improvement on the Controlled Oral Word Association-FAS Test after PBM therapy. His sleep data also showed he was already a good sleeper at entry [58].

In 2012, Nawashiro et al. [59] used transcranial PBM to treat a severe TBI patient who had been in a persistent vegetative state for 8 months after an accident. The patient had no spontaneous limb movement, and showed a focal low-density area in the right frontal lobe of the brain (according to a CT scan at 8 months post-accident). The PBM device had 23 individual 850 nm LEDs with a total power of 299 mW. Transcranial PBM treatment consisted of 30 min irradiation per session delivering 20.5 J/cm^2 fluence over the right and left forehead regions, repeated twice daily for 73 days. Five days after starting the treatment, the patient began to spontaneously move his left arm and hand, which had not occurred during the previous 8 months. The brain scan showed a focal increase (20% higher) in cerebral blood flow (CBF) in the uninjured left anterior frontal lobe 30 min after the final PBM session, compared to pre-PBM treatment [59].

In a pilot case study on two patients (1 female) with moderate TBI and persistent cognitive dysfunction, Bogdanova et al. [60] administered 18 sessions of transcranial PBM therapy (3x/week for 6 weeks) using a mixed red/NIR LED cluster, and conducted neuropsychological tests for executive function, memory, depression, PTSD, and sleep quality in patients pre-(T1), mid-(T2), and one week (T3) post-PBM treatment. Both patients showed noticeable improvement in sleep (actigraphy total sleep time) 1 week post-therapy (T3), as compared to pre-therapy (T1). One of the patients also showed an improvement in executive function, verbal memory, and sleep efficiency; whereas the other patient showed an improvement in measures of PTSD and depression [60].

In another study by Henderson and Morries [61], they used a high-power laser device (10–15 W, 810 plus 980 nm) delivered transcranially to the head, to treat one patient with moderate TBI. The participant received 20 PBM sessions over a 2-month period. They performed anatomical magnetic resonance imaging (MRI) and perfusion single-photon emission computed tomography (SPECT). The patient showed improvement in symptoms, such as depression, anxiety, headache, and insomnia, while cognitive function and quality of life also improved, along with positive changes in SPECT imaging [61].

In the same year, Henderson and Morries reported another study [62] conducted on 10 chronic TBI patients (with an average time since the injury of 9.3 years). These patients received ten PBM treatments over 2 months with one of two high-power laser devices (13.2 W at 810 nm; or 9 W at 980 nm). Overall results showed improvements in headache, sleep efficiency, cognitive performance, mood dysregulation, anxiety, and irritability. They suggested that high-power NIR lasers are better for transcranial brain PBM therapy, because the light energy can better reach the cerebral cortex [62].

Besides the aforementioned studies, 785 nm laser irradiation (a somewhat uncommon wavelength for PBM therapy) has been delivered transcranially to TBI patients with severe disorders of consciousness, to improve the alertness and awareness of the participants [63].

In the most recent case report, Chao et al. [64] also showed the beneficial effects of combined PBM therapy, with 810 nm light pulsed at 10-Hz or 40-Hz delivered by an intranasal LED applicator, combined with four transcranial LED arrays, in a 23-year old professional hockey player with a history of multiple concussions. After 8 weeks of therapy, changes in neuroimaging measures (including increased brain volume, improved functional connectivity, and cerebral perfusion), as well as improvements in behavior and neuropsychological test scores were reported [64] (Tables 12.1 and 12.2)

Table 12.1 Summary of in vivo studies on the effects of photobiomodulation therapy in traumatic brain injury

Study/Year	Animal/model	Light source	Wavelengths	Irradiation parameters	Irradiation approach/sites	Findings
Oron et al. [39]	Mouse (closed-head injury model)	Laser, PhotoThera, Inc. (Carlsbad, CA, USA)	808 nm	10 or 20 mW/cm², 1.2 or 2.4 J/cm², fiber diameter of 3 mm, 2 min, CW	Transcranially; at the sagittal suture located 4 mm caudal to coronal suture	Improved neurological severity score at 5–28 days post-TBI (at both regimens); decreased lesion volume at 28 days post-TBI (at both regimens)
Moreira et al. [43]	Rat (cryogenic brain injury model)	Laser, MM Optics Ltda (São Carlos, SP, Brazil)	660 or 780 nm	40 mW, 3 or 5 J/cm² per site (total of 2 sites); irradiation area of 0.04 cm², 3 or 5 s, CW	Direct irradiation; at the injury site	Brain: decreased IL-1β at 24 h post-TBI compared to 6 h (at 5 J/cm² 660 nm, or 3 J/cm² 780 nm); Blood: increased TNF-α at 24 h vs. 6 h (at 3 J/cm² 660 nm, 5 J/cm² 660 nm, or 5 J/cm² 780 nm); increased IL-6 at 24 h vs. 6 h (at 3 J/cm² 660 nm, 5 J/cm² 660 nm, or 5 J/cm² 780 nm); decreased IL-6 at 6 or 24 h vs. control (at 3 J/cm² 660 nm)
Ando et al. [48]	Mouse (controlled cortical-impact model)	Laser, DioDent Micro 810, HOYA ConBio (Fremont, CA, USA)	810 nm	50 mW/cm², 36 J/cm², 12 min, one irradiation session, CW or PW at 10- or 100-Hz with duty cycle of 50%	Transcutaneously; on the left front-parietal cortex (3 mm anterior to lambda and 2.5 mm left of midline)	Improved neurological severity score and increased body weight (at all regimens); improved depressive-like behaviors at 28 days post-TBI in FST (at 10-Hz) and TST (at 10- and 100-Hz); decreased lesion size at 15 and 28 days post-TBI (at 10-Hz)

(continued)

Table 12.1 (continued)

Study/Year	Animal/model	Light source	Wavelengths	Irradiation parameters	Irradiation approach/sites	Findings
Moreira et al. [44]	Rat (cryogenic brain injury)	Laser, MM Optics Ltda (São Carlos, SP, Brazil)	780 nm	40 mW, 3 J/cm² per site (total of 2 sites); irradiation area of 0.04 cm², 3 s, CW	Direct irradiation; at the injury site	Decreased lesion size at 6 h post-irradiation; increased neuronal survival at 6 and 24 h post-irradiation; increased the amount of GFAP at 14 days decreased leukocytes and lymphocytes in lesion area at the 24 h post-irradiation
Khuman et al. [41]	Mouse (controlled cortical-impact model)	Laser, Thor Photomedicine Ltd. (Chesham, Buckinghamshire, UK)	800 nm	250–1000 mW/cm², 60–210 J/cm², spot area of 1.32 cm², 2 or 7 min, CW	A) via an open craniotomy; holding probe at 1 cm above head B) transcranially; at the right and left parieto-temporal region	Improved cognitive performance in MWM test (at 60 J/cm² by transcranially or via an open craniotomy); decreased microglial activation at 48 h post-TBI
Wu et al. [47]	Mouse (closed-head injury model)	Laser, BWF-665-1, B&W-Tek; 730/6, Diomed Inc.; DioDent Micro 810 and V-Raser, ConBio (USA)	665, 730, 810, or 980 nm	150 mW/cm², 36 J/cm², spot diameter of 1 cm, 4 min, one irradiation session, CW	Transcranially; over the sutured incision	Improved neurological severity score and decreased small deficits in brain at 4 weeks post-irradiation (at 665 or 810 nm)
Oron et al. [40]	Mouse (closed-head injury model)	Laser, PhotoThera, Inc. (Carlsbad, CA, USA)	808 nm	21 mW, 10 mW/cm² (at cortex), 1.2 J/cm² (at cortex), fiber diameter of 3 mm, 2 min, CW or PW at 100-Hz or 600-Hz, one single session at 4, 6, or 8 h post-TBI	Transcranially; at the sagittal suture located 4 mm caudal to coronal suture	Improved behavioral performance assessed by NSS; increased the percentage of surviving mice that demonstrated full recovery at 56 days post-stroke (with 100-Hz PW); decreased infarct lesion volumes as assessed by MRI

(continued)

Table 12.1 (continued)

Study/Year	Animal/model	Light source	Wavelengths	Irradiation parameters	Irradiation approach/sites	Findings
Quirk et al. [42]	Rat (closed-head injury model)	LEDs	670 nm	50 mW/cm², 15 J/cm², 5 min, 2 sessions/day for 72 h or 10 days, CW	Transcranially; holding probe at 0.5 cm above the head	Improved locomotor activity in OFT (at 10 days post-irradiation); increased Bcl-2 and GSH, and decreased Bax expression levels
Xuan et al. [49]	Mouse (controlled cortical impact focal TBI model)	Laser, DioDent Micro 810, HOYA ConBio (Fremont, CA, USA)	810 nm	25 mW/cm², 18 J/cm², spot diameter of 1 cm, 12 min, for 1, 3, or 14 days, CW	Transcranially; centrally on top of the head	By 1 or 3 × days: improved neurological severity scores and WGT scores; decreased brain lesions sizes at 14 and 28 days post-TBI. By 3 × days: decreased degeneration at 14 and increased neurogenesis at 28 days post-TBI
Giacci et al. [74]	Rat (lateral fluid percussion TBI model)	LEDs, VET75, Quantum Devices Inc (Barneveld, WI, USA)	670 or 830 nm	28.4 J/cm² (670 nm) or 22.6 J/cm² (830 nm), 30 min, once daily	Transcranially	No effect on the motor and sensory function as well as lesion size at 7 days post-TBI
Zhang et al. [75]	Mouse (controlled cortical-impact model)	Laser, Acculaser, PhotoThera (Carlsbad, CA, USA)	810 nm	150 mW/cm², 36 J/cm², spot diameter of 1 cm, 4 min, PW at 10-Hz (50 ms pulse duration)	Transcranially; holding probe in direct contact with scalp at contusion site	Improved neurological severity score and increased body weight gain (at 3–28 days post-TBI); decreased IL-1 β (at 6 h) and IL-6, CCL2, and CXCL10 (at 6 h and 28 days); upregulated TNF-α; ; decreased morphological deficits such as necrotic and apoptotic cells in neocortex and hippocampus (at 28 days); increased cortical ATP (at 6 h and 28 days)

(continued)

Table 12.1 (continued)

Study/Year	Animal/model	Light source	Wavelengths	Irradiation parameters	Irradiation approach/sites	Findings
Xuan et al. [50]	Mouse (controlled cortical-impact model)	Laser, DioDent Micro 810, HOYA ConBio (Fremont, CA, USA)	810 nm	25 mW/cm^2, 18 J/cm^2, spot diameter of 1 cm, 12 min, for 1 or 3 consecutive days, CW	Transcranially; entire head irradiation	At both 1 and 3 × days: improved motor functions in WGT (at days 21, and 28 post-TBI); improved learning and memory in MWM (at day 28); decreased caspase-3 expression in lesion region (at day 4); increased neurogenesis in dentate gyrus and SVZ (at days 7 and 28); up-regulated migrating neuroprogenitor cells and neuronal differentiation in dentate gyrus and SVZ (at days 7 and 28)
Xuan et al. [51]	Mouse (controlled cortical impact focal TBI model)	Laser, PhotoThera, Inc. (Carlsbad, CA, USA)	810 nm	50 mW/cm^2, 36 J/cm^2 at scalp, spot diameter of 1 cm, 12 min, for 1 or 3 days, CW	Transcranially	At 1 × day: improved neurological severity score (at 21 and 28 days post-TBI); increased expression of BDNF in dentate gyrus and SVZ By 3 × days: improved neurological severity score (at 14, 21 and 28 days); increased expression of BDNF in dentate gyrus and SVZ area, and synapsin-1 in SVZ and lesion area (at 28 days)

(continued)

Table 12.1 (continued)

Study/Year	Animal/model	Light source	Wavelengths	Irradiation parameters	Irradiation approach/sites	Findings
Dong et al. [53]	Mouse (controlled cortical-impact model)	Laser, Acculaser, PhotoThera (Carlsbad, CA, USA)	810 nm	150 mW/cm², 36 J/cm², 4 min, PW at 10-Hz (50 ms pulse duration)	Transcranially; holding probe in direct contact with scalp at contusion site	Prevented the loss of hippocampal tissues; protected the hippocampus from secondary damage; increased ATP and deceased ROS production in injured cortex (at 5 h post-TBI); decreased lesion size (at 3 and 7 days)
Xuan et al. [52]	Mouse (controlled cortical-impact model)	Laser, DioDent Micro 810, HOYA ConBio (Fremont, CA, USA)	810 nm	25 mW/cm², 18 J/cm², spot diameter of 1 cm, 12 min, for 3 or 14 days, CW	Transcranially; covered the entire skull	By 3 × days: improved neurological severity score; improved cognitive performance in MWM test; decreased lesion size (at 2–8 weeks post-TBI); increased expression of GFAP in perilesional cortex, Dentate Gyrus and SVZ (at 8 weeks) By 14 × days: decreased lesion size (at 8 weeks); increased expression of GFAP in perilesional cortex, Dentate Gyrus and SVZ (at 8 weeks)

(continued)

Table 12.1 (continued)

Study/Year	Animal/model	Light source	Wavelengths	Irradiation parameters	Irradiation approach/sites	Findings
Esenaliev et al.. [76]	Rat (blast-induced neurotrauma model)	Laser	808 nm	300 J/cm^2, 5 min, 10 ns pulses and pulse repetition rate of 20-Hz	Transcranially	Improved motor performance assessed by the vestibulomotor test; improved learning and memory assessed by MWM: up-regulated mRNA encoding BDNF and downregulated the pro-apoptotic protein caspase-3 in cortical neurons; inhibited microglia activation and reduced the number of cortical neurons expressing activated caspase-3; increased expression of BDNF in the hippocampus and the number of proliferating progenitor cells in the dentate gyrus

Abbreviations: ATP–adenosine triphosphate; Bax–Bcl-2-associated X protein; BDNF–brain-derived neurotrophic factor; CCL2, chemokine (C–C motif) ligand-2; CW, continuous wave; CXCL10, Spinal IFN-γ-induced protein-10; FST, forced swimming test; GFAP, glial fibrillary acidic protein; GSH, glutathione; IL, interleukin; LEDs, light-emitting diodes; MRI, magnetic resonance imaging; mRNA, messenger ribonucleic acid; MWM, Morris water maze; NSS, neurologic severity score; OFT, open field test; PW, pulsed wave; ROS, reactive oxygen species; SVZ, subventricular zone; TBI, traumatic brain injury; TNF-α, tumor necrosis factor-α; TST, tail suspension test; WGMT, wire grip motion test

Table 12.2 Summary of clinical studies on the effects of photobiomodulation therapy in traumatic brain injury

Study/Year	Subjects (n)	Light Source	Wavelengths	Irradiation parameters	Irradiation approach/sites	Findings
Naeser et al. [65]	Chronic TBI (1 with depression) (2)	LEDs, three cluster heads	633 + 870 nm	19.39 mW/cm² and 22.48 mW/cm², 13.3 J/cm², 10 min per site, 1 ×/week for 6 years or 1 day × for 4 months, CW	Transcranially; bilateral left and right forehead (and multiple other areas)	Improved executive function and memory; decreased post-traumatic stress disorder symptoms
Stephan et al. [66]	TBI (1)	Laser, Theralase TLC 900 (Toronto, Canada)	905 + 660 nm	25 and 100 mW, 2.5 min, 5 irradiations over a 2 week	Transcranially; 4 areas on the scalp (midline occipital region just below the lamboidal suture, superior aspect of the nape to target the Circle of Willis and over the mastoid processes bilaterally)	Reduced pain, improved mood and sleep
Naeser et al. [67]	Chronic TBI (2 with depression) (11)	LEDs, MedX Health Model 1100 (Toronto, Canada)	633 + 870 nm	500 mW, 22.48 mW/cm², 13 J/cm², 10 min per site, 3 ×/week for 6 weeks, CW	Transcranially; 11 sites, midline and bilateral forehead	Improved sleep quality; decreased post-traumatic stress disorder symptoms; improved performance in social, interpersonal, and occupational functions
Bogdanova et al.						
Henderson and Morries [68]	Moderate TBI (1)	Laser, class 4, Diowave 810, (West Palm Beach, FL, USA)	810 nm	55 to 81 J/cm², 20 irradiation over 2 months	Transcranially	Improved gait, speech, sleep, mood, and resolution of recurrent headaches; resulted in a better judgment and less erratic; increased the perfusion in the left and right temporal cortices as indicated by SPECT neuroimaging

(continued)

Table 12.2 (continued)

Study/Year	Subjects (n)	Light Source	Wavelengths	Irradiation parameters	Irradiation approach/sites	Findings
Morries et al. [69]	Chronic TBI (6 with MDD) (10)	Laser, LiteCure LT1000, (Newark, DE, USA) Diowave 810 (Diowave, Riviera Beach, FL, USA)	810 and 980 nm	10 and 15 W, 14.8–28.3 J/cm^2, 8–12 min per site, 2–3 ×/week for 8 weeks, PW at 10-Hz	Transcranially; 2 sites, bilateral (forehead) 3 bilateral (prefrontal and temporal)	Improved symptoms of headache, sleep disturbance, cognition, mood dysregulation, anxiety, and irritability
Hesse et al. [70]	TBI with disorders of consciousness (5)	Laser, Power Twin 21 by MKW Laser system	785 nm	10 mW/cm^2, 10 min, 5 ×/week for 6 weeks, CW	Transcranially; 5 sites, on the level of the superior crest of the fossa sphenoidale on the forehead	Improved alertness and awareness; occurred epileptic fits as a side effect
Chao [71]	Gulf War Veterans with a history of TBI (2)	LEDs, Neuro Alpha device, Vielight, Inc. (Toronto, Canada)	810 nm	Transcranial: 75 and 100 mW, 75 and 100 mW/cm^2, 20 min, PW at 10-Hz with duty cycle of 50%, every other day for 12 weeks Intranasal: 25 mW, 25 mW/cm^2, 20 min, PW at 10-Hz with duty cycle of 50%, every other day for 12 weeks	Transcranially; multiple areas, bilateral mesial prefrontal cortex, precuneus/posterior cingulate cortex, angular gyrus (correspond to Fpz, Cz, T3 and T4 EEG points) Intranasally; left nose	Decreased chronic multisymptom illness (CMI) and the Kansas GWI case rates from severe to mild-moderate; improved mood, cognitive function, pain, sleep, and fatigue symptoms
Hipskind et al. [72]	Symptomatic military Veterans with chronic TBI (12)	LEDs (total 180 red and 222 infrared diodes), InLight Wellness Systems, Inc. (Addison, TX, USA)	629 + 850 nm	3.3 W, 6.4 mW/cm^2, 7.7 J/cm^2, 20 min, PW at 73-, 587-, and 1175-Hz with duty cycle of 35%, 3 ×/week for 6 consecutive weeks	Transcranially; whole skull irradiation with LED helmet	Improved neuropsychological scores; increased regional CBF as measured by qualitative and quantitative SPECT analysis

(continued)

Table 12.2 (continued)

Study/Year	Subjects (n)	Light Source	Wavelengths	Irradiation parameters	Irradiation approach/sites	Findings
Carneiro et al. [72]	Cranioencephalic trauma (14)	LEDs (13 sets of 4 LEDs), Medical Systems, Sao Jose dos Campos (Brasil)	630 nm	25.73 mW/m^2, 3.74 J/ cm^2, 30 min, 3 ×/week for 6 weeks, CW	Transcranially; whole head irradiation with LED cap	Improved cerebral blood flow and cerebral oxygenation; improved performance in some neuropsychological tests such as complex Rey figure; RAVLT, verbal fluency test F-A-S, semantic, symbol digit, Stroop test, TMT A, and TMT B

Abbreviations CBF, cerebral blood flow; CW, continuous wave; EEG, electroencephalography; GWI, Gulf War Illness; LEDs, light-emitting diodes; MDD, major depressive disorder; PW, pulsed wave; RAVLT, Rey Auditory Verbal Learning Test; SPECT, single-photon emission computed tomography; TBI, traumatic brain injury; TMT, Trail making test

References

1. Roozenbeek, B., A.I. Maas, and D.K. Menon. 2013. Changing patterns in the epidemiology of traumatic brain injury. *Nature Reviews Neurology* 9 (4): 231.
2. Faul, M., and V. Coronado. 2015. Epidemiology of traumatic brain injury. In *Handbook of clinical neurology*, 3–13. Elsevier.
3. Sosin, D.M., J.E. Sniezek, and D.J. Thurman. 1996. Incidence of mild and moderate brain injury in the United States, 1991. *Brain injury* 10 (1): 47–54.
4. Ahmed, S., et al. 2017. Traumatic brain injury and neuropsychiatric complications. *Indian journal of psychological medicine* 39 (2): 114.
5. Faul, M., et al. 2010. Traumatic brain injury in the United States: National estimates of prevalence and incidence, 2002–2006. *Injury Prevention* 16 (Suppl 1): A268–A268.
6. Faul, M., et al. 2010. *Traumatic brain injury in the United States; emergency department visits, hospitalizations, and deaths, 2002–2006*
7. Menon, D.K., et al. 2010. Position statement: Definition of traumatic brain injury. *Archives of physical medicine and rehabilitation* 91 (11): 1637–1640.
8. Savitsky, B., et al. 2016. Traumatic brain injury: It is all about definition. *Brain injury* 30 (10): 1194–1200.
9. Teasdale, G., et al. 2014. The Glasgow Coma Scale at 40 years: Standing the test of time. *The Lancet Neurology* 13 (8): 844–854.
10. Summers, C.R., B. Ivins, and K.A. Schwab. 2009. Traumatic brain injury in the United States: An epidemiologic overview. *Mount Sinai Journal of Medicine: A Journal of Translational and Personalized Medicine: A Journal of Translational and Personalized Medicine* 76 (2): 105–110.
11. Gennarelli, T.A., et al. 1994. Comparison of mortality, morbidity, and severity of 59,713 head injured patients with 114,447 patients with extracranial injuries. *The Journal of trauma* 37 (6): 962–968.
12. Salottolo, K., et al. 2017. The epidemiology, prognosis, and trends of severe traumatic brain injury with presenting Glasgow Coma Scale of 3. *Journal of critical care* 38: 197–201.
13. Bushnik, T., et al. 2003. Etiology of traumatic brain injury: Characterization of differential outcomes up to 1 year postinjury. *Archives of physical medicine and rehabilitation* 84 (2): 255–262.
14. Meythaler, J.M., et al. 2001. Current concepts: Diffuse axonal injury–associated traumatic brain injury. *Archives of physical medicine and rehabilitation* 82 (10): 1461–1471.
15. Bogner, J.A., et al. 2001. A comparison of substance abuse and violence in the prediction of long-term rehabilitation outcomes after traumatic brain injury. *Archives of physical medicine and rehabilitation* 82 (5): 571–577.
16. Sorenson, S.B. and J.F. Kraus. 1991. Occurrence, severity, and outcomes of brain injury. *The Journal of Head Trauma Rehabilitation*
17. Abdul-Muneer, P., N. Chandra, and J. Haorah. 2015. Interactions of oxidative stress and neurovascular inflammation in the pathogenesis of traumatic brain injury. *Molecular neurobiology* 51 (3): 966–979.
18. Simon, D.W., et al. 2017. The far-reaching scope of neuroinflammation after traumatic brain injury. *Nature Reviews Neurology* 13 (3): 171.
19. Cicerone, K.D., and L.N. Tanenbaum. 1997. Disturbance of social cognition after traumatic orbitofrontal brain injury. *Archives of clinical neuropsychology* 12 (2): 173–188.
20. Levin, H., W. High, and H. Eisenberg. 1985. Impairment of olfactory recognition after closed head injury. *Brain* 108 (3): 579–591.

21. Dikmen, S., et al. 2010. Rates of symptom reporting following traumatic brain injury. *Journal of the International Neuropsychological Society* 16 (3): 401–411.
22. Hachinski, V., et al. 2006. National Institute of Neurological Disorders and Stroke-Canadian stroke network vascular cognitive impairment harmonization standards. *Stroke* 37 (9): 2220–2241.
23. Reeves, R.R., and R.L. Panguluri. 2011. Neuropsychiatric complications of traumatic brain injury. *Journal of psychosocial nursing and mental health services* 49 (3): 42–50.
24. Borg, J., et al. 2004. Diagnostic procedures in mild traumatic brain injury: Results of the WHO Collaborating Centre Task Force on Mild Traumatic Brain Injury. *Journal of rehabilitation medicine* 36: 61–75.
25. Gill, M., et al. 2005. A comparison of the Glasgow Coma Scale score to simplified alternative scores for the prediction of traumatic brain injury outcomes. *Annals of emergency medicine* 45 (1): 37–42.
26. Levin, H.S., and R.R. Diaz-Arrastia. 2015. Diagnosis, prognosis, and clinical management of mild traumatic brain injury. *The Lancet Neurology* 14 (5): 506–517.
27. Inglese, M., et al. 2005. Diffuse axonal injury in mild traumatic brain injury: A diffusion tensor imaging study. *Journal of neurosurgery* 103 (2): 298–303.
28. Reis, C., et al. 2015. What's new in traumatic brain injury: Update on tracking, monitoring and treatment. *International journal of molecular sciences* 16 (6): 11903–11965.
29. Gronbeck, K.R., et al. 2016. Application of tauroursodeoxycholic acid for treatment of neurological and non-neurological diseases: Is there a potential for treating traumatic brain injury? *Neurocritical care* 25 (1): 153–166.
30. Dinsmore, J. 2013. Traumatic brain injury: An evidence-based review of management. *Continuing Education in Anaesthesia, Critical Care & Pain* 13 (6): 189–195.
31. Brain, T.F. 2007. Guidelines for the management of severe traumatic brain injury. Introduction. *Journal of Neurotrauma* 24:S1
32. Carney, N., et al. 2017. Guidelines for the management of severe traumatic brain injury. *Neurosurgery* 80 (1): 6–15.
33. Stockbridge, M.D., and R.S. Newman. 2017. Translating neurodevelopmental findings into predicted outcomes and treatment recommendations for language skills in children and young adults with brain injury. *Translational Issues in Psychological Science* 3 (1): 104.
34. Neville, I.S., et al. 2015. Repetitive Transcranial Magnetic Stimulation (rTMS) for the cognitive rehabilitation of traumatic brain injury (TBI) victims: Study protocol for a randomized controlled trial. *Trials* 16 (1): 440.
35. Osier, N.D., and C.E. Dixon. 2016. The controlled cortical impact model: Applications, considerations for researchers, and future directions. *Frontiers in neurology* 7: 134.
36. Osier, N., and C.E. Dixon. 2016. The controlled cortical impact model of experimental brain trauma: Overview, research applications, and protocol. In *Injury Models of the Central Nervous System*, 177–192. Springer.
37. Frankowski, J.C. and R.F. Hunt. 2018. Modeling traumatic brain injury using controlled cortical impact injury. *KOPF Carrier* 93
38. Leker, R., E. Shohami, and S. Constantini. 2002. Experimental models of head trauma. In *Research and Publishing in Neurosurgery*, 49–54. Springer.
39. Oron, A., et al. 2007. Low-level laser therapy applied transcranially to mice following traumatic brain injury significantly reduces long-term neurological deficits. *Journal of neurotrauma* 24 (4): 651–656.
40. Oron, A., et al. 2012. Near infrared transcranial laser therapy applied at various modes to mice following traumatic brain injury significantly reduces long-term neurological deficits. *Journal of neurotrauma* 29 (2): 401–407.

41. Khuman, J., et al. 2012. Low-level laser light therapy improves cognitive deficits and inhibits microglial activation after controlled cortical impact in mice. *Journal of neurotrauma* 29 (2): 408–417.
42. Quirk, B.J., et al. 2012. Near-infrared photobiomodulation in an animal model of traumatic brain injury: Improvements at the behavioral and biochemical levels. *Photomedicine and laser surgery* 30 (9): 523–529.
43. Moreira, M.S., et al. 2009. Effect of phototherapy with low intensity laser on local and systemic immunomodulation following focal brain damage in rat. *Journal of photochemistry and photobiology B: Biology* 97 (3): 145–151.
44. Moreira, M.S., et al. 2011. Effect of laser phototherapy on wound healing following cerebral ischemia by cryogenic injury. *Journal of Photochemistry and Photobiology B: Biology* 105 (3): 207–215.
45. Wu, M.X., and M.R. Hamblin. 2019. Photobiomodulation and mitochondria for traumatic brain injury in mouse models. In *Photobiomodulation in the Brain*, 169–187. Elsevier.
46. Hamblin, M.R. 2019. Photobiomodulation for traumatic brain injury in mouse models. In *Photobiomodulation in the Brain*, 155–168. Elsevier.
47. Wu, Q., et al. 2012. Low-level laser therapy for closed-head traumatic brain injury in mice: Effect of different wavelengths. *Lasers in surgery and medicine* 44 (3): 218–226.
48. Ando, T., et al. 2011. Comparison of therapeutic effects between pulsed and continuous wave 810-nm wavelength laser irradiation for traumatic brain injury in mice. *PLoS ONE* 6 (10): e26212.
49. Xuan, W., et al. 2013. Transcranial low-level laser therapy improves neurological performance in traumatic brain injury in mice: Effect of treatment repetition regimen. *PLoS ONE* 8 (1): e53454.
50. Xuan, W., et al. 2014. Transcranial low-level laser therapy enhances learning, memory, and neuroprogenitor cells after traumatic brain injury in mice. *Journal of biomedical optics* 19 (10): 108003.
51. Xuan, W., et al. 2015. Low-level laser therapy for traumatic brain injury in mice increases brain derived neurotrophic factor (BDNF) and synaptogenesis. *Journal of biophotonics* 8 (6): 502–511.
52. Xuan, W., L. Huang, and M.R. Hamblin. 2016. Repeated transcranial low-level laser therapy for traumatic brain injury in mice: Biphasic dose response and long-term treatment outcome. *Journal of biophotonics* 9 (11–12): 1263–1272.
53. Dong, T., et al. 2015. Low-level light in combination with metabolic modulators for effective therapy of injured brain. *Journal of Cerebral Blood Flow & Metabolism* 35 (9): 1435–1444.
54. Mocciaro, E., et al. 2020. Non-invasive transcranial nano-pulsed laser therapy ameliorates cognitive function and prevents aberrant migration of neural progenitor cells in the hippocampus of rats subjected to traumatic brain injury. *Journal of neurotrauma* 37 (8): 1108–1123.
55. Scholten, A. 2016. *Outcome after traumatic brain injury*
56. Naeser, M.A., et al. 2011. Improved cognitive function after transcranial, light-emitting diode treatments in chronic, traumatic brain injury: Two case reports. *Photomedicine and laser surgery* 29 (5): 351–358.
57. Naeser, M.A., et al. 2014. Significant improvements in cognitive performance post-transcranial, red/near-infrared light-emitting diode treatments in chronic, mild traumatic brain injury: Open-protocol study. *Journal of neurotrauma* 31 (11): 1008–1017.
58. Naeser, M.A., et al. 2016. Transcranial, red/near-infrared light-emitting diode therapy to improve cognition in chronic traumatic brain injury. *Photomedicine and laser surgery* 34 (12): 610–626.

59. Nawashiro, H., et al. 2012. Focal increase in cerebral blood flow after treatment with near-infrared light to the forehead in a patient in a persistent vegetative state. *Photomedicine and laser surgery* 30 (4): 231–233.

60. Bogdanova, Y., et al. 2014. LED therapy improves sleep and cognition in chronic moderate TBI: Pilot case studies. *Archives of Physical Medicine and Rehabilitation* 95 (10): e77.

61. Henderson, T.A., and L.D. Morries. 2015. SPECT perfusion imaging demonstrates improvement of traumatic brain injury with transcranial near-infrared laser phototherapy. *Advances in mind-body medicine* 29 (4): 27–33.

62. Morries, L.D., P. Cassano, and T.A. Henderson. 2015. Treatments for traumatic brain injury with emphasis on transcranial near-infrared laser phototherapy. *Neuropsychiatric disease and treatment* 11: 2159.

63. Hesse, S., C. Werner, and M. Byhahn. 2015. Transcranial low-level laser therapy may improve alertness and awareness in traumatic brain injured subjects with severe disorders of consciousness: A case series. *International Archives of Medicine* 8

64. Chao, L.L., et al. 2020. Changes in brain function and structure after self-administered home photobiomodulation treatment in a concussion case. *Frontiers in neurology* 11: 952.

65. Naeser, M.A. 2011. A Saltmarche, MH Krengel, MR Hamblin, and JA Knight, "Improved cognitive function after transcranial, light-emitting diode treatments in chronic, traumatic brain injury: Two case reports." *Photomedicine and Laser Surgery* 29 (5): 351–358.

66. Stephan, W., L.J. Banas., M. Bennett., and H. Tunceroglu. 2012. Efficacy of super-pulsed 905 nm Low Level Laser Therapy (LLLT) in the management of Traumatic Brain Injury (TBI): A case study. *World Journal of Neuroscience* 2(04): 231

67. Naeser, M.A., R. Zafonte., M.H. Krengel., P.I. Martin., J. Frazier., M.R. Hamblin, et al. 2014. Significant improvements in cognitive performance post-transcranial, red/near-infrared light-emitting diode treatments in chronic, mild traumatic brain injury: open-protocol study. *Journal of Neurotrauma* 31(11): 1008–1017

68. Henderson, T., L. Morries. 2015. SPECT perfusion imaging demonstrates improvement of traumatic brain injury with transcranial near-infrared laser phototherapy. *Advances in mind-body medicine* 29(4): 27–33

69. Morries, L.D., P. Cassano, T.A. Henderson. 2015. Treatments for traumatic brain injury with emphasis on transcranial near-infrared laser phototherapy. *Neuropsychiatric disease and treatment* 2

70. Hesse, S., C. Werner, M. Byhahn. 2015. Transcranial low-level laser therapy may improve alertness and awareness in traumatic brain injured subjects with severe disorders of consciousness: a case series. *International Archives of Medicine* 8

71. Chao, L.L. (2019). Improvements in Gulf war illness symptoms after near-infrared transcranial and intranasal photobiomodulation: two case reports. *Military Medicine*

72. Hipskind, S.G., F.L. Grover Jr., T.R. Fort., D. Helffenstein., T.J. Burke., S.A. Quint, et al. 2018. Pulsed transcranial red/near-infrared light therapy using light-emitting diodes improves cerebral blood flow and cognitive function in veterans with chronic traumatic brain injury: a case series. *Photomedicine and laser surgery*

73. Carneiro, A.M.C., G.C. Poiani, AL Zaninnoto, R Lazo Osorio, MdL Oliveira, WS Paiva, et al. 2019. Transcranial Photobiomodulation Therapy in the Cognitive Rehabilitation of Patients with Cranioencephalic Trauma. *Photobiomodulation, photomedicine, and laser surgery* 37(10): 657–666

74. Giacci, M.K., et al. 2014. Differential effects of 670 and 830 nm red near infrared irradiation therapy: A comparative study of optic nerve injury, retinal degeneration, traumatic brain and spinal cord injury. *PLoS ONE* 9 (8): e104565.

75. Zhang, Q., et al. 2014. Low-level laser therapy effectively prevents secondary brain injury induced by immediate early responsive gene X-1 deficiency. *Journal of Cerebral Blood Flow & Metabolism* 34 (8): 1391–1401.

76. Esenaliev, R.O., et al. 2018. Nano-pulsed laser therapy is neuroprotective in a rat model of blast-induced neurotrauma. *Journal of neurotrauma* 35 (13): 1510–2152.

Photobiomodulation Therapy for Psychiatric Disorders

13

13.1 Depression

13.1.1 The Problem of Depression

Depression is a common psychiatric disorder that unfortunately often goes undiagnosed and untreated. Depression is a broad diagnostic entity, including major depressive disorder (MDD), dysthymic disorder, cyclothymic disorder, seasonal affective disorder, and episodic depression. Depression may also be part of an associated mood, bipolar, or psychotic disorder. Depression affects more than 300 million people worldwide. The prevalence of MDD ranges between 5 and 9% among adults; however, its lifetime prevalence is as high as 17%. Several risk factors have been reported for depression, including female gender, history of anxiety or eating disorders, positive family history of mood disorders, drug or alcohol abuse, and domestic or sexual abuse [1, 2].

Depression is associated with severe and persistent symptoms leading to changes in mood, thoughts, behavior, and physical health [3]. It is characterized by loss of interest in pleasurable activities (anhedonia), depressed mood, low energy or fatigue, sleep and psychomotor disturbances, low self-esteem, decreased capacity to carry out even the simplest of daily tasks, and suicidal tendencies [4]. The DSM-V Criteria for Major Depression have been extensively discussed elsewhere [5]. Moreover, thorough history taking and physical examination should be conducted to detect any other comorbid conditions which may cause or exacerbate depression, such as dementia, chronic liver or renal disease, hypothyroidism, adrenal insufficiency, and congestive heart failure, especially in older adults [6]. Until now, no objective method has yet been developed to evaluate the severity, endophenotype, or response to treatment in MDD. Recently, it has been shown that peripheral/serum BDNF, IGF-1, and cytokines could serve as biomarkers of MDD, and its response to treatment [2]. Due to the high morbidity and mortality rate, depression should be treated as soon as it is diagnosed, because any delay could result in increased death, adverse outcomes, and prolonged impairment of normal functioning.

It is widely accepted that structural alterations in the brain and neurochemical imbalances underlie the pathophysiology of mood disorders, particularly MDD. The neurochemical imbalances which have been implicated in the pathophysiology of depression, include dysfunction of the hypothalamus-pituitary axis (HPA), oxidative stress-mediated neurotoxicity, monoamine deficiency (serotonin, norepinephrine, and dopamine) as well as over-activation of inflammatory responses [7–10]. Several studies have also shown that depression is linked to regional brain abnormalities, especially those involving circuits or connectivity. Brain imaging studies have revealed several structural and functional abnormalities in depressed patients. These include atrophy in the frontal lobes, reduced metabolism/blood flow in particular brain regions (frontal lobe, dorsolateral prefrontal cortex (dlPFC), and the basal ganglia), volume reductions in the frontal lobe, HIP, vmPFC, OFC, lateral PFC, insula, amygdala, and striatum, as well as abnormalities in corticolimbic connectivity [7, 8, 11–13].

Treatment mainly focuses on altering serotonin and norepinephrine levels (monoamine deficiency hypothesis). These psychiatric medications include selective serotonin reuptake inhibitors (SSRIs), serotonin norepinephrine reuptake inhibitors (SNRIs), tricyclic antidepressants (TCAs), maprotiline, mirtazapine, and nefazodone. Apart from pharmacologic intervention, psychotherapy and electroconvulsive therapy (ECT) also play an important role in the treatment of depression [14]. Treatment of depression is necessary over a long period to reduce the morbidity and improve the prognosis.

13.1.2 Animal Models of Depression

Psychiatric symptoms of depression cannot be fully recapitulated in animal models, nevertheless, some features of this syndrome have been mimicked in rodents [15]. Animal models used in psychiatric studies should be validated in terms of sensitivity and selectivity for the specific animal species selected, and they should also be simple and easy to replicate [16]. It is noteworthy that animal models of depression are valid since they completely rely on innate social behavior [15].

There are several pharmacological drugs (e.g., reserpine, apomorphine, clonidine, and yohimbine) which have been used for the induction of depression in animal models [16]. Of note, the reserpine model is the most frequently used pharmacological model in rodents. Reserpine is an irreversible inhibitor of the vesicular monoamine transporter 2, which could diminish brain monoamines [17]. Studies have shown that the acute [18] or chronic [17] administration of reserpine could recapitulate behavioral characteristics of depression in animals. Intraperitoneal administration of high doses of reserpine (6 or 8 mg/kg) could mimic the disease symptoms over periods of 24–48 h [18, 19], whereas, lower doses of the drug (0.1 or 0.2 mg/kg) displayed the same phenomena after 14 days of administration in rodents [20].

Exposure to chronic stress (or even to a single brief episode of stress) can potentially cause depression-like behavioral changes in rodents [21]. The unpredictable chronic mild stress (UCMS) protocol involves a variety of mild stressors delivered in a random order over several weeks [22]. The unpredictable stressors typically include, food and water deprivation, exposure to an empty water bottle, alternative isolation and paired housing, overnight stroboscopic illumination, exposure to white noise, cold stress, cage tilting, soiled bedding, restraint, and forced swimming [23].

Restraint stress is another important paradigm of psychological depression models. In this model, the animal is restrained in a narrow tube, restraint apparatus, or a small cage, for a period of 15 min–6 h once daily [21]. This model depends on the daily restraint time and may take from 1–2 weeks to 2–3 months for full induction [24, 25].

Among all the animal models of depression, genetic models allow the study of the underlying biological mechanisms. Evidence indicates that a mutation of the Abelson helper integration site-1 (AHI1) gene is linked with neurodevelopmental and psychiatric symptoms [26]. It has been shown that a conditional Ahi1 gene knockout may cause depression-like behavior in mice, likely through dysregulation of the TrkB signaling pathway [27].

There is evidence that post-traumatic stress disorder (PTSD) is common following traumatic brain injury (TBI) [28]. Moreover, depression is the most frequent psychiatric manifestation after TBI [29]. Additional rodent models of depression including, "Social Stress", "Early Life Stress", "Learned Helplessness" and "Fear Conditioning" are also used to replicate different aspects of stress and depression.

13.1.3 Photobiomodulation Therapy for Depression

13.1.3.1 Photobiomodulation Therapy for Depression in Animal Models

In 2011, Tanaka et al. [30] used an infrared light device (600–1600 nm) to irradiate the naïve rat head and assessed its antidepressant and anti-anxiety effects. After 10 days of treatment, depression-like behavior was meaningfully diminished as evaluated by an increased mobility time in the forced swimming test. Both a single PBM session and chronic (10 sessions) application of PBM increased the number of BrdU-positive cells, a marker for neurogenesis, in the hippocampus (CA1 region). Furthermore, anxiety-like behavior was noticeably lower in the chronic PBM group as measured by improved performance in the elevated plus maze and light/dark tasks.

In 2011, Ando et al. [31] assessed the possible antidepressant effects of transcranial 810 nm NIR laser therapy using various frequencies (CW and PW at 10- or 100 Hz modes) on depression signs after TBI in a mouse model. After a single PBM session with a scalp fluence of 36 J/cm^2, behavioral performance was improved more after 10 Hz frequency compared to 100 Hz and CW frequencies, measured by a decrease in the immobility time in the forced swimming test and tail suspension test at 4 weeks post-TBI.

In 2012, Wu et al. [32] evaluated the therapeutic effects of transcranial 810 nm laser therapy pulsed at 100 Hz in a chronic mild stress rat model of depression, and compared the results of PBM therapy with fluoxetine drug therapy. They used a much higher scalp fluence (120 J/cm^2) and PBM sessions (nine sessions) as compared to the study of Ando et al. [31]. The authors reported that the antidepressant effect of PBM therapy was as effective as fluoxetine as shown by significant improvement in depression-like behavior in the forced swimming test (with decreased immobility and increased swimming time) at 3 weeks post-treatment.

In 2016, Salehpour et al. [33] compared the antidepressant effects of transcranial PBM therapy using different lasers (630 and 810 nm) at 10 Hz pulse frequency with drug therapy (citalopram), in a chronic mild stress rat model of depression. The results showed that after 12 sessions of PBM therapy (1.2 J/cm^2 on the cortex), only 810 nm laser exhibited antidepressant-like effects similar to citalopram as shown by increased swimming time and decreased immobility time in the forced swimming test. In addition to the neurobehavioral improvement, 810 nm laser was able to promote body weight gain, and 630 nm laser could decrease serum cortisol levels at the end of the experiment.

In 2016, Mohammed [34] applied an 804 nm NIR laser at CW mode transcranially in a rat model of pharmacological depression. Fourteen daily PBM sessions to six points on the scalp (38.4 J/cm^2 per point) significantly decreased immobility time in forced swimming test and also modulated electrocorticogram spectra. Moreover, the power of 80 mW showed the best therapeutic effects compared to higher powers (e.g., 200 and 400 mW), for the relief of depression-like symptoms. In this respect, the biphasic dose-response of low, intermediate, and high powers on the rat swimming activity showed that the optimal (38.4 J/cm^2) and the highest (190.8 J/cm^2) fluences of PBM had stimulatory and inhibitory effects, respectively.

In 2017, Xu et al. [35] reported that transcranial 808 nm laser therapy using CW mode had antidepressant effects in two mouse models of depression caused by restraint stress or genetic mutation. Even though mitochondrial complex I–III activity levels did not show any statistically significant changes in the hippocampus, PFC, and hypothalamus regions, a significant increase in ATP production and improved mitochondrial complex IV activity levels within the PFC were observed after 28 daily PBM sessions using a scalp fluence of 41.4 J/cm^2. In addition to molecular markers, neurobehavioral data also showed a reduction in depression-like behavior in the forced swimming test and tail suspension test after 28 days of PBM for the restraint stress model, and after 14 days of PBM for the genetic mutation model.

In 2019, Salehpour et al. [36] used a mouse restraint stress model of depression, and showed that 810 nm laser PBM and/or CoQ$_{10}$ treatment could suppress stress-induced activation of microglia and the expression of pro-inflammatory markers (such as NF-κB, p38, and JNK) in the PFC and hippocampus regions, along with a decrease in serum levels of IL-6 and TNF-α. Moreover, both 810 nm laser PBM and CoQ_{10} treatments significantly reduced levels of superoxide dismutase, glutathione peroxidase, and malondialdehyde, and increased total antioxidant capacity in the hippocampus and PFC areas, as well as decreased serum glutathione levels.

In 2019, Banqueri et al. [37] showed that the application of PBM could restore cognitive flexibility and oxidative metabolism levels, which had been altered in new born rats subjected to stress. They conducted maternal separation (10 days, 4 h per day) in new born rats, and when they reached adulthood they evaluated the spatial navigation ability and cognitive flexibility. PBM treatment was carried out with 1064 nm light (30 mW, 60 cycles) on the rat head. Results showed that PBM could rescue cognitive flexibility and restore the oxidative energy metabolism balance as indicated by cytochrome c oxidase levels in different brain regions.

In 2019, Eshaghi et al. [38] studied the therapeutic effects of three different doses of transcranial 810 nm laser PBM therapy (4, 8, and 16 J/cm^2) on behavioral performance in a mouse stress model of depression. Both depression- and anxiety-like behavior exhibited a biphasic dose-response pattern to PBM therapy (with a peak response at 8 J/cm^2). Furthermore, PBM therapy at both fluences of 8 and 16 J/cm^2 increased the levels of serotonin and nitric oxide in both the PFC and hippocampus regions (a peak response at 8 J/cm^2). PBM therapy at all fluences significantly reduced serum cortisol levels.

In 2020, Meynaghizadeh-Zargar et al. [39] used PBM therapy coupled with the administration of methylene blue in a mouse model of unpredictable mild chronic stress. Transcranial laser irradiation was conducted three times a week for 4 weeks. The methylene blue group received a daily dose of 0.5 mg/kg intraperitoneally for 4 consecutive weeks. Neurobehavioral tests showed that PBM, methylene blue, and the combination, reversed the impairment of learning and memory. Also, PBM and methylene blue significantly increased the levels of NO, ROS, and superoxide dismutase activity in the brain, as well as serum cortisol levels. They concluded that both PBM and methylene blue, either individually or in combination had significant therapeutic benefits on learning and memory impairment.

13.1.3.2 Photobiomodulation Therapy for Depression in Human Clinical Studies

In the first open study on transcranial LED PBM therapy for MDD, Schiffer et al. [40] showed that CW mode 810 nm LED irradiation (250 mW/cm^2 per site over 4 min, 60 J/cm^2 on the scalp) on to the forehead of patients (EEG sites F3 and F4) meaningfully improved depression symptoms after 2 weeks. Their study also suggested that a single PBM session with a LED device might have been insufficient to provide a significant long term benefit [40].

In another open study, Cassano et al. [41] used a laser device to clarify the efficacy and safety of laser PBM therapy in MDD patients. Multiple transcranial PBM sessions (two sessions-a-week for 3 weeks) with a CW mode 808 nm diode laser (700 mW/cm^2

per site over 2 min, 84 J/cm^2 on the scalp) to four bilateral sites on the forehead, led to 50% remission in depression symptoms as assessed by the Hamilton Depression Rating Scale—17 items (HAM-D17) at 6 and 7 weeks post-PBM.

In a randomized controlled trial, researchers from the University of Texas at Austin assessed the effects of transcranial PBM in combination with attention bias modification (ABM) in patients with depressive symptoms [42]. The subjects ($n = 51$) were randomized to receive two sessions on the left forehead, right forehead, or sham transcranial PBM therapy with a 1064 nm laser for 4 min per site. The interval between the sessions was 48 h, and all subjects received one session of ABM before PBM, and one after each session of transcranial PBM. In the subjects who responded to the ABM, an improvement in depression severity (CES-D total score) was produced by the right transcranial PBM, whereas no significant effect was produced by the left or the sham transcranial PBM. On the other hand, there was no significant difference in the overall change of the CES-D total scores across all the groups, which could be attributed to transcranial PBM (personal communication with Dr. Christopher Beevers).

In another randomized controlled trial from Massachusetts General Hospital, Cassano et al. [43] evaluated the effectiveness of transcranial PBM therapy in 21 MDD patients. Transcranial LED irradiation was performed with 823 nm light to the bilateral forehead (EEG sites F3 and F4) or sham therapy twice-a-week for 8 weeks. The irradiation time of the initial session was 20 min, and subsequent sessions could be extended up to 30 min based on clinical judgment. The reduction in depression severity assessed by the HAM-D17 (from baseline to endpoint) was better in the transcranial PBM group compared to the sham group. Response (decrease in HAM-D17 scores $\geq 50\%$) was observed in 50% of patients who received the real-PBM therapy and in 27% of patients in the sham group. In addition to the measures of depression, a secondary analysis showed an improvement in sexual dysfunction in the patients receiving the real-PBM therapy regardless of the antidepressant effects [44].

In a subsequent randomized controlled trial study, Cassano and colleagues investigated the antidepressant effects of transcranial PBM therapy in 54 MDD patients [45]. Patients were enrolled in a double-blind, sham-controlled, sequenced parallel comparison trial. The real-PBM treatment was delivered using 830 nm LEDs (CW mode; 51 mW/cm^2 per site over 20 min, 61.2 J/cm^2 on the scalp) over two spots on the forehead. Although PBM therapy with 51 mW/cm^2 irradiance per spot was well tolerated, it was not better than sham-PBM treatment to improve depression symptoms, suggesting that there may be minimum fluence or dose thresholds, which should be exceeded for the efficacy of transcranial PBM therapy for MDD [45].

Nevertheless, little is known about the antidepressant effects of PBM therapy over the long term. In this respect, one case report on anxiety and depression complicated by Takotsubo cardiomyopathy described the result in a patient receiving PBM therapy for 31 months, as an add-on to antidepressant medications to treat MDD with anxiety [46]. During the first 22 months, intranasal PBM applicators (810 nm LEDs, 10-Hz PW) were applied in both nostrils providing a fluence of 10.65 J/cm^2 per nostril. Intranasal PBM alone did not relieve the depression symptoms, until an additional transcranial 830 nm laser device was applied to the forehead over the following 9 months [46].

In addition to transcranial irradiation, which is the most common approach for PBM of the brain, some studies have suggested there might be a potential antidepressant effect of light delivered to peripheral tissues. It should be mentioned that the hypothesis of systemic PBM being able to treat depression symptoms is intriguing, but the available evidence must be interpreted with caution. In this respect, the only two randomized controlled trials published on systemic PBM for MDD were carried out using the laser acupuncture method, which could be mechanistically different from systemic PBM delivered over large skin areas, and through supravenous and intravenous access. These two randomized controlled trials ($n = 30$ [47] and $n = 47$ [48]) were conducted by Quah-Smith et al. and used an 808 nm laser delivered to acupuncture-points on the limbs and trunk (focusing on five primary depression acupoints) providing the total energy of 3–5 J per session. Although this light energy dose was seemingly low, the total energy delivered over the skin per session was sufficient to produce significant antidepressant effects in the two studies, after up to 12 PBM sessions over 8 consecutive weeks [47, 48].

The remaining reports on systemic PBM therapy for MDD come exclusively from two open studies from Russian researchers who applied a combination of red (630 nm) and NIR lasers (890 nm) intravenously and transcutaneously (supravenous sites) [49, 50]. In their first report, 93 patients with treatment–resistant depression episodes received at least 16–20 sessions of systemic PBM treatment divided into two cycles, an addition to ongoing pharmacological treatment. A significant reduction in the mean HAM-D (23.4 ± 1.9 to 2.9 ± 0.1) and Hamilton Anxiety Rating Scale (HAM-A) (23.1 ± 0.6 to 3.9 ± 0.3) total scores at the end of the therapy (8 weeks or more) were observed. The same participants experienced better clinical outcomes when compared to non-randomized and non-blinded controls ($n = 87$) [49]. In the second report, the same team used a similar treatment protocol to evaluate the prophylactic effect of systemic PBM after remission from a depressive disorder [MDD ($n = 34$), bipolar depression ($n = 17$) or "mixed anxiety and depressive disorders" ($n = 28$)]. The first cycle of PBM was applied once remission was achieved using drug therapy, and a second cycle after a 3–4 month interval. One year after the initial remission, relapse or recurrence occurred in 23% of the participants. In non-randomized and non-blinded controls with similar characteristics, the rate of relapse or recurrence at one year was 50%. No side-effects or complications attributable to systemic PBM were observed over the one year follow-up [50].

Furthermore, in a study by Gabel et al. [51], five patients with lower-back pain and concurrent self-reported depression were treated for five weeks with physical therapy (5 sessions) and concurrent PBM therapy (3 sessions), and were matched to five control patients treated with physical therapy alone (5 sessions). The PBM therapy consisted used red (660 nm) and NIR LEDs (850 nm) delivered to 12 symmetrical posterior sites (thoracic, lumbar, and thighs), for 30 s of irradiation per spot. The average irradiance was 100 mW/cm^2 and fluence was 3 J/cm^2, and the total energy per session was 0.72 kJ. Patients who received systemic PBM showed a bigger reduction in their depression score [Orebro Musculoskeletal Screening Questionnaire SF 12 (OMSQ-12), item No. 6]. Both treatment groups had similar improvements in the functional status score (back pain), suggesting that group differences in the antidepressant effect were independent of any improvement in their back pain.

Additionally, although up to now, many clinical studies have shown the efficacy of transcranial, intranasal, or systemic PBM therapy, as well as laser acupuncture for treatment of MDD patients as discussed above, some studies have also shown the beneficial effects of this technique on depression states or post-traumatic stress disorder (PTSD) in TBI patients. For example, in this regard, Naeser et al. [52] reported that transcranial LED therapy might be a useful technique for decreasing PTSD depression levels in TBI patients. They used red/NIR LED arrays (combined 633/870 nm, at 500 mW per site over 10 min, 13.3 J/cm^2 on the scalp) in CW mode to deliver light bilaterally and to the midline sagittal region. After a few months of PBM therapy, there were improvements in self-awareness, self-regulation in social functioning, as well as sleep quality. In a follow-up case series, using the same LED cluster and treatment parameters, Naeser et al. [53] showed antidepressant effects of PBM in eleven chronic TBI patients with PTSD. Although a significant positive trend for antidepressant response was noticed after a week of PBM therapy (0.4 J/cm^2 fluence on the cortex), the results after two months post-PBM did not show an overall linear trend response. It is worthwhile to note that because the studies from both Schiffer et al. [40] and Naeser et al. [53] showed that the post-PBM depression levels did not return to pre-PBM levels, it could be suggested that the unsatisfactory treatment outcomes might be due to the following reasons: (1) an inadequately delivered fluence/dose to the cortical surface due to using only one PBM session; (2) the shorter penetration depth of LED light compared to coherent laser light. Moreover, in another open study on ten TBI patients with secondary and/or comorbid PTSD, depression, and anxiety symptoms, Morries et al. [54] used high-power lasers (a combination of 980 nm at 10 W and 810 nm at 15 W) with scalp fluences of 14.8–28.3 J/cm^2 per site at 10 Hz PW mode, and showed substantial improvement in depression score (down to the non-depressed range), cognitive function, anxiety, irritability, and sleep quality of the participants. Although they used high-power laser devices onto the temporal-bilateral and frontal regions, they also used a continuous handheld motion procedure, so that the scalp temperature changes never exceeded 3 °C.

13.2 Anxiety

13.2.1 The Problem of Anxiety

Anxiety disorders are the most common type of psychiatric disorder, characterized by excessive and uncontrollable fear and worry about the future, difficulty in concentration, irritability, as well as somatic symptoms, such as palpation, sweating, and gastrointestinal disturbance [55]. Billions of dollars are spent every year throughout the world in an attempt to decrease the burden of anxiety in individuals. Accordingly, the annual cost of anxiety disorders is estimated to be $42.3 billion in the United States in 1999 [56]. Anxiety disorders have a high prevalence in all population subgroups over the whole world. A recent study showed that the general prevalence of all anxiety disorders was high (3.8–25%), mainly in women (5.2–8.7%); youth population (2.5–9.1%); patients with chronic illnesses (1.4–70%); and subjects with Euro/Anglo origins (3.8–10.4%). Some population groups are more vulnerable to this anxiety, including older people and their caregivers, pregnant women, lesbians, gay or bisexual individuals, and self-harming patients. Therefore special attention should be paid to these groups [57].

Symptoms of anxiety disorders have been attributed to the disruption of brain circuits and regions involved in the control of emotional responses. The amygdala is the main brain structure involved in the response to anxiety-inducing stimuli. Anatomical and neuroimaging studies have shown that patients with anxiety disorders often show functional and structural brain changes, such as hyperactivity of the amygdala, vmPFC, and HIP, abnormalities in prefrontal-limbic activation, decreased connectivity between the amygdala and PFC, smaller temporal lobe volume, reduced cortical thickness in the OFC, inferior frontal gyrus, and pregenual anterior cingulate cortex, while still having normal hippocampal volumes [58, 59]. Imaging studies have also revealed increased blood flow in the right parahippocampal gyrus, and smaller temporal lobes compared to healthy subjects. There was decreased serotonin type 1A receptor binding in the anterior and posterior cingulate and raphe of subjects suffering from panic disorders (an entity classified as an anxiety disorder) [60]. The principal neurotransmitters playing a role in the development and progression of anxiety disorders are thought to be serotonin, dopamine, norepinephrine, and gamma-aminobutyric acid (GABA). Hormones such as corticotropin-releasing factors have also been found to mediate at least some of the symptoms of anxiety disorders [61].

Due to their high burden on society, the accurate diagnosis and treatment of anxiety disorders are crucial. Several types of pharmaceutical drugs, such as SSRIs, SNRIs, TCAs, benzodiazepines, pregabalin, gabapentin, and antipsychotic agents have been successfully used to treat anxiety disorders. Psychotherapy is another alternative approach that can be used alone or in combination with drugs [62]. For instance, cognitive-behavioral therapy (CBT) is a well-known non-pharmacologic strategy used for the treatment of anxiety disorders, such as social anxiety disorder or generalized anxiety disorder. Studies have shown the efficacy of CBT to treat anxiety and improve the quality of life [62, 63]. Other techniques such as relaxation, behavioral self-control, anxiety-management, self-control desensitization, biofeedback, group anxiety-management training, computer therapy programs, and psychodynamic therapy have also been used to treat patients suffering from anxiety disorder [64].

13.2.2 Photobiomodulation Therapy for Anxiety in Human Studies

Schiffer et al. [40] studied the effect of a single session of transcranial 810 nm LED PBM on ten MDD patients of which nine had a comorbid anxiety disorder. PBM was administered to two forehead regions (EEG sites F3 and F4) with an irradiance of 250 mW/cm^2 and a fluence of 60 J/cm^2 for 4 min exposure per site. The effect of treatment on anxiety symptoms was assessed by the Hamilton Anxiety Scale (HAM-A). The lowest symptom scores occurred 2 weeks post-PBM when the mean HAM-A decreased 14.9 ± 9.6 points from baseline. At week 4, the mean HAM-A was still significantly lower than baseline (a decrease of 9.0 ± 7.5 points from baseline) but significantly higher than at week 2, suggesting that the effects of a single session of transcranial PBM on anxiety were transient [40].

In a recent case report study, Caldieraro et al. [46] treated a patient with anxious depression complicated by Takotsubo cardiomyopathy with combined intranasal and transcranial PBM therapy for 31 months. During the first 22 months, intranasal PBM therapy (810 nm LED, 10 Hz PW mode) was administered to both nostrils providing a fluence of 10.65 J/cm^2 per nostril. With a progressive increase in the frequency of intranasal PBM therapy sessions, from twice a week to daily and then to twice daily, the patient's anxiety symptoms gradually improved with roughly a three-fold decrease in Anxiety Symptom Questionnaire scores [46].

Recently, researchers from Massachusetts General Hospital carried out a pilot study to test the anxiolytic effect of transcranial NIR PBM therapy in patients suffering from generalized anxiety disorder. In the study by Maiello et al. [65], fifteen participants were recruited for an 8-week study using daily self-application of transcranial 830 nm PBM therapy for 20 min. At each treatment session, PBM (30 mW/cm^2, 36 J/cm^2, for a total energy of 2.9 kJ per session) was bilaterally administered to the forehead with a LED cluster (Cerebral Sciences). The anxiety severity, evaluated by the Structured Interview Guide for the Hamilton Anxiety Scale (SIGH-A), reduced from 16.75 ± 5.14 to 6.83 ± 3.79 at the endpoint. Despite the small sample size, the observed effect size was large (Cohen's d effect size = 1.70) [65, 66].

Most recently, Kerppers et al. [67] assessed the effects of transcranial PBM therapy using 945 nm LEDs in 22 university students with anxiety and depression. PBM was applied for 1 min and 25 s (9.35 J/cm^2) to the frontal sinus region. Patients were evaluated at baseline and after one month with the Hospital Anxiety and Depression Scale (HADS), the faces anxiety scale, a drawing test, as well as the grip strength test. PBM therapy significantly improved results in the HADS and memory test. However, although the means of faces and drawing scale scores were smaller than before the tests, this reduction was not statistically significant. According to the authors, the small sample size and non-randomized nature of the study may limit the reliability of their trial [67].

Recently, Brochado et al. [68] conducted a randomized clinical trial on 51 patients with temporomandibular disorder to assess the possible therapeutic effects of either systemic PBM therapy, manual therapy (a type of physiotherapy), or their combination. As a secondary treatment outcome, the effect of these treatments on anxiety levels was evaluated by the Beck depression inventory. PBM therapy was applied using a NIR laser device (with an irradiance of 3.33 W/cm^2, fluence of 133 J/cm^2, for 40-s irradiation per point, and 4 J of total energy per point) to five points on the temporomandibular joint (TMJ) area and also to seven points on muscles related to the TMJ, three times a week for 4 weeks. A significant improvement in anxiety levels was observed in all of the treatment groups [68].

In another randomized clinical trial, Da Silva et al. [69] assessed the effectiveness of systemic PBM therapy, exercise training, or both combined, as an add-on to pharmacological treatment in 160 patients with fibromyalgia. The PBM device applied for this trial combined one NIR laser (905 nm), four red LEDs (640 nm), and four NIR LEDs (875 nm). The treatment consisted of 10 weeks of twice-a-week sessions (real- PBM, sham-PBM + exercise, real-PBM + exercise, or control/sham-PBM and no exercise). The PBM therapy was administered to ten tender points, including one in the head (the temporomandibular joint). Each point was irradiated for 300 s, with a total energy of 39.3 J per session delivered to each point. Anxiety levels were evaluated by the anxiety item of the Fibromyalgia Impact Questionnaire. After 10 weeks, the three treatment groups showed a significant reduction in anxiety as compared to the control group. The improvement was similar in the PBM and exercise groups, and the combination therapy was superior to each of them individually [69].

13.3 Post-traumatic Stress Disorder

13.3.1 The Problem of Post-traumatic Stress Disorder

PTSD is defined by continuous extreme reactions to any type of cues related to traumatic events, involving mood disturbance, a feeling of forthcoming threat, altered sleep, and hypervigilance [70]. Previously, PTSD was categorized as an anxiety disorder; however, DSM-5 omitted PTSD from the anxiety disorders category and reclassified it under trauma-related disorder [70]. PTSD occurs in about 5–10% of the population and its prevalence is double in women compared to men [71]. Symptoms of PTSD may include hopelessness, nightmares, severe anxiety, feelings of extreme fear, irritability, hyper-arousal, sleep disturbance, and lack of concentration, as well as uncontrollable and recurrent thoughts about the precipitating event, which could develop within one month or sometimes years after a traumatic event. The symptoms of PTSD are believed to result from an inadequate adaptation response of the brain to exposure to severe stressors.

Although the traumatic event is the main precipitating factor of PTSD, several bio-logical and psychosocial risk factors have been identified that may predispose certain individuals to PTSD, and can affect the severity, chronicity, and onset [72]. For exam-ple, variations in the HPA axis-associated genes, such as nuclear receptor subfamily 3 group C member 1 (NR3C1; encoding the glucocorticoid receptor) and FKBP5 (encod-ing FK506-binding protein 5) could be involved in PTSD pathophysiology. Cognitive factors and neurocircuitry also play an important role in this regard [73, 74]. PTSD diag-nosis is straightforward, and can be achieved by asking some questions in concert with DSM-V criteria. Neuroimaging studies are not necessary for diagnosis. However, they have shown changes in the size and function of the medial prefrontal cortex, hippocam-pus, and amygdala [75]. This could be due to an increase in glucocorticoid levels during stress, raised levels of excitatory amino acids, inhibition of neurogenesis, and a decrease in brain-derived neurotrophic factor [76, 77].

Psychotherapy approaches such as trauma-focused cognitive-behavioral therapy/exposure therapy (TFCBT) or stress management combined with anxiolytic or antidepres-sant drugs can be effectively used to reduce symptoms [78]. Evidence supports the use of SSRIs as a first line therapy in the pharmacotherapy of PTSD [79]. Non-pharmacological approaches could also be used as adjunctive or second-line treatments. These approaches include, but are not limited to, exposure-based therapy, stress inoculation training, cog-nitive therapy (see above), and eye movement desensitization and reprocessing therapy [80]. Patients with complex PTSD should be treated by specialist multidisciplinary care teams.

13.3.2 Photobiomodulation Therapy for Post-traumatic Stress Disorder in Humans

In an open study, 11 patients with mild chronic TBI were treated with 18 sessions of transcranial PBM therapy [52]. Four of the 11 patients met diagnostic criteria for PTSD. During each 20 min session (three times a week for 6 consecutive weeks), a LED device (combined 870 nm and 633 nm) was applied to 11 sites on the scalp, on the midline and bilaterally on the frontal, parietal, and temporal regions. The PBM treatment protocol included an irradiance of 22.2 mW/cm^2, and a fluence of 13.2 J/cm^2, with a total energy of 3.26 kJ per session. The four patients with comorbid PTSD showed a significant average decrease in the PCL-C (PTSD Checklist-Civilian) scores of 21 (range 9–30) [52].

Stephan and colleagues treated 50 patients with PTSD using transcranial 905 nm laser PBM therapy. After the treatments, all patients showed a remarkable clinical improvement in emotional stability and quality of life, as evaluated by standardized testing scales [81].

13.4   Autism

13.4.1 The Problem of Autism

Pervasive developmental disorders (PDD) are a subtype of neurodevelopmental disorders characterized by deficiencies in mutual social interaction, communication, limited interests, and monotonous or repetitive behavior. PDD includes autistic disorders, Rett syndrome, childhood disintegrative disorder, Asperger's syndrome, and pervasive developmental disorder-not otherwise specified (PDD-NOS) [82]. Autism spectrum disorders (ASD) is the term applied to cover their variable manifestations [83]. Although the exact cause is unknown, genetic factors are thought to play an important role in their pathophysiology [84]. Various risk factors predispose patients to ASD such as familial autoimmune disease, maternal infection during pregnancy, autoantibodies to fetal brain proteins, maternal exposure to drugs such as valproic acid, thalidomide, and SSRIs, advanced parental age, air pollution, gastrointestinal dysfunction with an altered microbiome [85].

Different anatomical regions are involved in ASD. Social impairment results from the involvement of the orbitofrontal cortex, anterior cingulate cortex, fusiform gyrus, superior temporal sulcus, inferior frontal gyrus, mirror neuron regions of the amygdala, and the posterior parietal cortex. Also, communication deficits are caused by the functional and structural impairment of the inferior frontal gyrus, superior temporal sulcus, supplementary motor area, basal ganglia, substantia nigra, pontine nuclei cerebellum, and thalamus. Further, repetitive behavior results from changes in the orbitofrontal cortex, thalamus, basal ganglia, and anterior cingulate cortex [86].

Several therapeutic strategies have been shown to be effective with some ASD patients; however, no treatment completely alleviates the symptoms in ASD children, or is efficacious for all patients with the disorder [87]. Early and intensive application of treatment strategies such as the University of California, Los Angeles (UCLA)/Lovaas model and Early Start Denver Model (ESDM) for enhancing language ability, cognitive function, and adaptive behavior is strongly recommended in autistic children. CBT may also be useful in some patients [88]. Atypical antipsychotics, especially aripiprazole, and risperidone, are helpful in decreasing stereotypy, irritability, and hyperactivity. Methylphenidate, atomoxetine, and alpha-2 agonists seem to be effective in decreasing attention-deficit hyperactivity disorder (ADHD) symptoms [89]. Newer therapies such as rTMS and transcranial direct current stimulation (tDCS) have been successfully tried to treat ASD. A recent study found that deep rTMS to the bilateral dorsomedial prefrontal cortex resulted in a decrease in communication deficit and social anxiety disorder in autistic patients [90]. Another study showed that short-term application of tDCS over the F3 region could be a useful therapeutic intervention in autistic patients. This was confirmed by changes in Autism Treatment Evaluation Checklist (ATEC), Childhood Autism Rating Scale (CARS) and Children's Global Assessment Scale (CGAS) scores [91].

13.4.2 Photobiomodulation Therapy for Autism

13.4.2.1 Photobiomodulation Therapy for Autism in Animal Studies

In two experimental studies conducted by Khongrum et al. [92, 93] in a valproic acid-induced rat model of autism, laser acupuncture was carried out once daily for 10 min at the HT7 acupoint on both the right and the left sides of the animals (405 nm, 100 mW, with a diameter of 500 mm). According to the first study [92], laser acupuncture at HT7 significantly enhanced behavioral outcomes and markers of oxidative stress in the cortex, striatum, as well as the hippocampus. In their follow-up study [93], laser acupuncture significantly decreased oxidative stress and the levels of the pro-inflammatory cytokine IL-6 in the cerebellum. Moreover, an improvement in Purkinje cell survival as well as increased GABAergic function were observed.

13.4.2.2 Photobiomodulation Therapy for Autism in Human Studies

It has been suggested that PBM could be an effective therapeutic strategy for decreasing irritability and other symptoms and behavior accompanying ASD in human patients. In this respect, Leisman et al. [94] reported for the first time that PBM therapy could ameliorate ASD symptoms in children and adolescents between 5 and 17 years of age. Twenty-one of the 40 subjects received eight sessions of 5 min PBM therapy using a pulsed 635 nm laser (15 mW) applied to the base of the skull and temporal areas over 4 consecutive weeks. Subjects were assessed by a variety of tests, including the Aberrant Behavior Checklist (ABC) and the Clinical Global Impressions (CGI) Scale, across four different time points including, at baseline, week 2 (interim), week 4 (endpoint), and week 8 (post-PBM). According to the data, transcranial PBM therapy significantly decreased irritability and other symptoms and behavior associated with ASD (e.g., agitation, lethargy, social withdrawal, stereotypic behavior, hyperactivity, noncompliance, and inappropriate speech) in participants, with positive changes maintained and increased over time. In their follow-up study [95], the same group of researchers reported follow-up assessment up to 12 months after completion of PBM therapy, showing that the improvement in symptoms continued in those participants initially randomized to the real-PBM group, with no change observed for placebo participants (sham-PBM group). They suggested that transcranial PBM therapy might gradually rearrange the anatomical and functional brain connectivity, and modulate those neural networks associated with complex symptoms in ASD subjects.

13.5 Insomnia

13.5.1 The Problem of Insomnia

Insomnia or disturbed sleep is a common complaint often encountered by physicians in clinical settings, that can manifest itself with or without other medical or psychiatric disorders [96]. It could be considered both as a symptom and as a sign. Around 30% of the adult population in most countries report one or more of the symptoms of insomnia [97]. Insomnia may impose considerable health-care and job-related costs, and strongly increases the risk of cardiovascular disease and mental disorders [97, 98]. Based on DSM IV criteria insomnia is the difficulty initiating or maintaining sleep, waking up too early, or poor quality sleep, which continues for at least 1 month, is not associated with another mental disorder, and is not the direct physiological consequence of substance abuse or medical/physical illness [99].

Several psychological and physiological factors can cause insomnia such as stressful events, menopause, aging, some drugs, emotional or physical discomfort, physical inactivity, and irregular bedtimes [100]. The pathophysiology of insomnia involves hyperarousal in both the central and peripheral nervous systems, possibly due to emotional responses, hyperactivity of the HPA axis, and sympathetic nervous system [101]. Important brain structures involved in the pathophysiology of insomnia have been identified using neuroimaging studies, including HIP, thalamus, amygdala, frontal cortex, caudate nucleus, and anterior cingulate cortex [98]. Indeed, decreased brain activity in the brainstem, thalamus, and PFC is associated with the normal transition from wakefulness to sleeping [102]. Patients with insomnia have lesser reductions in brain activity, as well as cerebral glucose metabolism during sleep, relative to resting wakefulness [100]. Neuroimaging studies have revealed structural abnormalities in several brain structures, including the frontal cortex, HIP, anterior cingulate cortex, reduced volume of HIP and frontal lobe, gray matter deficits in the left OFC, pericentral and lateral temporal areas, and precuneus [103, 104].

Benzodiazepines and CBT are the mainstays of treatment. However, evidence for the long-term efficacy of benzodiazepines is limited, and their prolonged use is associated with adverse effects. Thus, combination therapy with both modalities is desirable [96].

13.5.2 Photobiomodulation Therapy for Insomnia

13.5.2.1 Photobiomodulation Therapy for Insomnia in Animal Models

Salehpour et al. [105] demonstrated that sleep deprivation in mice could substantially trigger oxidative stress in the hippocampus region, and subsequent memory impairment. They measured the antioxidant effects of NIR PBM therapy in this animal model. PBM therapy with 810 nm was delivered (once a day for 3 days) transcranially to the entire head. Mice in the real-PBM group showed better spatial memory performance, as indicated by the Barnes maze and the what-where-which tasks. In addition, acute PBM significantly improved hippocampal total antioxidant capacity and antioxidant enzyme activity (e.g., SOD and GPx), as well as decreased levels of ROS and lipid peroxidation (MDA).

13.5.2.2 Photobiomodulation Therapy for Insomnia in Humans

There is emerging evidence for the therapeutic benefit of PBM therapy in some sleep disorders. According to Zhao et al., PBM therapy using 658 nm wavelength (30 J/cm^2) using a whole-body LED machine enhanced sleep quality and serum levels of melatonin in healthy adult subjects [106]. Sleep disturbance is the most common side-effect observed in patients with chronic TBI. An enhancement in sleep quality following transcranial PBM therapy has been reported by many of the TBI patients who were treated with PBM [52–54]. PBM therapy with combined transcranial and intranasal devices also improved sleep performance in AD patients [107] and TBI [108] patients. According to some subjective reports, being sleepy was a common side-effect after the use of a 10 Hz intranasal PBM applicator, possibly due to the pulse rate of the light interfering with the brain alpha wave band (8–12 Hz) [109]. Besides, in studies from Chinese researchers, the beneficial effects of intranasal PBM therapy using a He–Ne laser (632.8 nm) on the sleep quality of patients with insomnia were reported [110–112]. Even though the underlying action mechanisms involved in PBM therapy for sleep improvement remain unclear, modulation of circadian rhythms through an increase in serum levels of melatonin [106, 109, 113], and stimulation of a systemic homeostatic response through the blood circulatory system [114] have been proposed as possible mechanistic pathways (Tables 13.1 and 13.2).

Table 13.1 Summary of in vivo studies on the effects of photobiomodulation therapy in psychiatric disorders

Study/year	Animal/model	Light source	Wavelengths	Irradiation parameters	Irradiation approach/sites	Findings
Rossetti et al. [120]	Rat (space restriction depression model)	Laser	632.8 nm	5 mW, 0.12 J, 1.08 J/cm^2,	Transcranially; on the sinciput, none symmetrical spots (total of 1 cm^2)	Increased total SOD and decreased cytosolic AST; no significant effect on the mitochondrial AST and GIDH
Cassone et al. [121]	Rat (space restriction depression model)	Laser	632.8 nm	5 mW, 0.12 J, 1.08 J/cm^2,	Transcranially; on the sinciput, none symmetrical spots (total of 1 cm^2)	Resulted in increase of serotonin in striatum and hippocampus, a small but significant decrease of noradrenaline in cortex; no effect on the dopamine levels
Tsai et al. [122]	Intact mouse	Infrared emitter, Lucybelle Biological Technology Inc. (Taoyuan, Taiwan)	3000–15,000 nm	33 J/cm^2, 60 min, once daily for 28 days	Whole-body irradiation	Decreased immobility time by the end of the 3rd and 4th weeks
Tanaka et al. [30]	Intact rat	Infrared emitter	600–1600 nm	1800–2200 mW, 3 min, one single session or once daily for 10 days	Transcranially	Decreased indicators of depression- and anxiety-like behavior (in chronic group); no significant effect on general locomotor activity; increased the number of BrdU-positive cells in CA1 of the hippocampus was significantly increased (in both acute and chronic groups)

(continued)

Table 13.1 (continued)

Study/year	Animal/model	Light source	Wavelengths	Irradiation parameters	Irradiation approach/sites	Findings
Wu et al. [32]	Rat (chronic stress depression model)	Laser, PhotoThera, Inc. (Carlsbad, CA, USA)	810 nm	350 mW, 120 J/cm², probe diameter of 3 mm, 2 min, 3x/week for 3 weeks, PW at 100 Hz with duty cycle of 20%	Transcranially; at the midline of the dorsal surface of the shaved head in region between eyes and ears	Decreased depressive-like behaviors in FST
Salehpour et al. [33]	Rat (chronic stress depression model)	Laser, Mustang 2000 + (Moscow, Russia)	630 or 810 nm	89 or 562 mW/cm², 6 or 10.7 J/cm²; probe diameter of 3 mm, 4 × / week for 3 weeks, PW at 10 Hz with duty cycle of 50%	Transcranially; at the midline of the dorsal surface of the shaved scalp in prefrontal region	810 nm: decreased depressive-like behaviors in FST; increased body weight; decreased blood glucose levels 630 nm: decreased serum cortisol and blood glucose levels
Mohammed [34]	Rat (pharmacological depression model)	Laser, Lasotronic Inc. (Zug, Switzerland)	804 nm	640 mW/cm², 38 J/cm² per point, spot diameter of 4 mm, 1 min per point, for 7 days. CW	Transcranially; at six points arranged symmetrically, three on each side of the skull	Decreased depressive-like behaviors in FST; regulated EEG in all wave frequency bands such as delta, alpha, beta-1, and beta-2 except for theta wave
Xu et al. [8]	Mouse (space restriction or Ahi1 KO depression model)	Laser, quartz-silica fiber, Shenzhen Fuzhe Technology Co., Ltd. (Shenzhen, China)	808 nm	23 mW/cm², 41.4 J/cm², spot diameter of 1 cm, 30 min/day for 28 consecutive days, CW	Transcranially	Restraint-induced stress mouse: decreased depressive-like behaviors in FST and TST (at days 14, 21, and 28); increased ATP synthesis in PFC, increased CCO levels and activity in PFC Ahi1 KO mouse: decreased depressive-like behaviors in FST and TST (at day 28)

(continued)

Table 13.1 (continued)

Study/year	Animal/model	Light source	Wavelengths	Irradiation parameters	Irradiation approach/sites	Findings
Meynaghizadeh-Zargar et al. [123]	Rat (chronic stress model)	Laser, Thor Photomedicine (Chesham, UK)	810 nm	200 mW, 4.75 W/cm², 8 J/cm² (at cortex), PW at 10 Hz with duty cycle of 88%	Transcranially; covered the entire brain	Improved learning and memory assessed by NOR and Barnes maze tasks; decreased anxiety states as assessed by EPM test, decreased NO, ROS, and SOD activity and increased TAC and GPx activity in brain; decreased serum cortisol levels
Salehpour et al. [36]	Mouse (space restriction depression model)	Laser, Thor Photomedicine (Chesham, UK)	810 nm	200 mW, 4.75 W/cm², 8 J/cm² (at cortex), PW at 10 Hz with duty cycle of 88%, once daily for 5 days	Transcranially; covered the entire brain	Ameliorated depressive-like behaviors as indicated by decreased immobility time in both the FST and TST; decreased MDA and enhanced TAC, GSH levels, GPx and SOD activities in both PFC and hippocampus; suppressed neuroinflammatory responses as indicated by decreased NF-kB, p38, JNK levels in both PFC and hippocampus; down-regulated intrinsic apoptosis biomarkers, Bax, Bcl-2, cytochrome c release, and caspase-3 and -9; decreased serum levels of cortisol, corticosterone, TNF-α, and IL-6

(continued)

Table 13.1 (continued)

Study/year	Animal/model	Light source	Wavelengths	Irradiation parameters	Irradiation approach/sites	Findings
Banqueri et al. [37]	Rat (maternal separation stress model)	Laser	1064 nm	30 mW, 20 J/cm², 1 h, once daily for 5 days	Transcranially; on the first third of the head just behind the eyes, trying to target mostly frontal areas	Improved spatial learning and memory as assessed by MWM; modulated cytochrome oxidase units in various brain regions including septum, thalamus, retrosplenial cortex, amygdala, hippocampus, entorhinal cortex, ventral tegmental area, and supramammillary nucleus
Eshaghi et al. [38]	Mouse (space restriction depression model)	Laser, Thor Photomedicine (Chesham, UK)	810 nm	200 mW, 4.75 W/cm², 4, 8, or 16 J/cm² (at cortex), PW at 10 Hz with duty cycle of 88%, 3 ×/week for 3 weeks	Transcranially; covered the entire brain	With an optimum dose of 8 J/cm²: improved anxiety- and depression-like behaviors; decreased serum cortisol levels; increased serotonin levels and decreased NO levels in the PFC and hippocampus areas

Abbreviations: AST—Aminotransferase; ATP—Adenosine triphosphate; Bax—Bcl-2-associated X protein; BrdU—Bromodeoxyuridine; CCO—Cytochrome c oxidase; CW—Continuous wave; EEG—Electroencephalography; EPM—Elevated plus maze; FST—Forced swimming test; GIDH—Glutamate dehydrogenase; GPx—Glutathione peroxidase; GSH—Glutathione; IL—Interleukin; JNK—C-Jun N-terminal kinases; LEDs—Light-emiting diodes; MDA—Malondialdehyde; MWM—Morris water maze; NF-kB—Nuclear factor kappa-light-chain-enhancer of activated B cells; NO—Nitric oxide; NOR—Novel recognition test; PFC—Prefrontal cortex; PW—Pulsed wave; ROS—Reactive oxygen species; SOD—Superoxide dismutase; TAC—Total antioxidant capacity; TNF-α—Tumor necrosis factor-α; TST—Tail suspension test

Table 13.2 Summary of clinical studies on the effects of photobiomodulation therapy in psychiatric disorders

Study/year	Subjects (n)	Light source	Wavelengths	Irradiation parameters	Irradiation approach/sites	Findings
Kartelishev et al. [49]	Treatment resistant major depressive episode (180)	Laser, (ALOU-2) and Uzor-2 K and Uley-2 km	630 or 890 nm	630 nm laser: 1–2 mW for 10–15 min, 7–10 session per cycle, 2 or more cycles (2–4 weeks apart), CW 890 nm laser: 40–60 min, 8–10 session per cycle 2 or more cycles (2–4 weeks apart), PW at 80- and 1500 Hz	Intravenously with 630 nm laser Transcutaneously with 890 nm laser	Decreased depression and anxiety rates at the end of the therapy cycles (8 weeks or more of trial) assessed by Hamilton Depression Rating Scale and Hamilton Anxiety Rating Scale
Quah-Smith et al. [47]	Elevated depressive symptoms (30)	Laser, Meyer (Melbourne, Australia)	804 nm	100 mW, 5 J per point, 5 s, twice weekly for 4 weeks then weekly for a further 4 weeks	Laser acupuncture; 6 primary acupuncture points [LR14 on the right, CV15 and CV14 in the anterior midline (all being alarm points), HT7 at the wrists (for calming) and LR8 on the left (Liver Yin)]	Decreased depression at the end of the treatment period assessed by Beck Depression Inventory (BDI); no significant effect on the Hospital Anxiety and Depression Scale (HADS); resulted in some adverse effects such as fatigue, insomnia, dry mouth, and headache
Kolupaev et al. [50]	Depressive disorders (106)	Laser, (ALOU-2) and Uzor-2 K and Uley-2 km	630 or 890 nm	4–5 W, 10–12 sessions per cycle, 2 cycles (3–4 months apart), once a day	Transcutaneously	Decreased the rate of relapses or recurrences one year after the initial remission
Schiffer et al. [40]	Major depressive disorder (10)	LEDs, Marubeni America Corp. (Santa Clara, CA, USA)	810 nm	250 mW/cm^2, 60 J/cm^2, 4 min, one irradiation session, CW	Transcranially; 2 sites, bilateral (right and left forehead at EEG map sites: F3, F4)	Decreased depression and anxiety rates at 2-week post-irradiation assessed by Hamilton Depression Rating Scale (HDRS) and Hamilton Anxiety Rating Scale (HARS); no significant effects on cerebral blood flow

(continued)

Table 13.2 (continued)

Study/year	Subjects (n)	Light source	Wavelengths	Irradiation parameters	Irradiation approach/sites	Findings
Quah-Smith et al. [115]	Major depressive disorder (10)	Laser, Moxla prototype fiber-optic infrared laser, Euryphaessa AB (Stockholm, Sweden)	808 nm	25 mW, 4 J per acupoint	Laser acupuncture; 4 primary acupuncture points (LR8, LR14, CV14, HT7, KI3)	Each acupoint laser irradiation resulted in a different activation size and pattern of neural activity; activated and deactivated fronto–limbic–striatal brain regions; there was no significant activation or deactivation with KI3
Quah-Smith et al. [116]	Major depressive disorder (20)	Laser, Moxla prototype fiber-optic infrared laser, Euryphaessa AB (Stockholm, Sweden)	808 nm	25 mW, 4 J per acupoint	Laser acupuncture; 4 primary acupuncture points (LR14, LR8, CV14, and HT7)	Laser acupuncture on LR8, LR14, and CV14 stimulated both the anterior and posterior DMN in both the nondepressed and depressed participants; in the nondepressed participants, there was consistently outstanding modulation of the anterior DMN at the medial frontal gyrus across LR8, LR14, and CV14 acupoints; in the depressed participants, there was wider posterior DMN modulation at the parieto–temporal–limbic cortices

(continued)

Table 13.2 (continued)

Study/year	Subjects (n)	Light source	Wavelengths	Irradiation parameters	Irradiation approach/sites	Findings
Cassano et al. [41]	Major depressive disorder (4)	Laser, Neurothera PhotoThera Inc. (Carlsbad, CA, USA)	808 nm	5 W, 700 mW/cm^2, 84 J/cm^2, 2 min per site, 2 ×/week for 3 weeks, CW	Transcranially; 4 sites, bilateral (right and left forehead center at 20 and 40 mm from sagittal line)	Decreased depression rate at 6 to 7 weeks post-irradiation assessed by Hamilton Depression Rating Scale-17 items
Disner et al. [42]	Patients with elevated depression symptoms (51)	Laser, CG-5000, Cell Gen Therapeutics (Dallas, TX, USA)	1064 nm	250 mW/cm^2, 60 J/cm^2, 4 min per site, for 2 sessions, CW	Transcranially; 2 sites, medial and lateral parts of the left or right side of the forehead	Right prefrontal irradiation improved attention bias modification intervention effects and depression symptoms
Henderson and Morries [117]	Depression comorbid to TBI (39)	Laser, LT1000 (LiteCure, Newark, DE, USA); Diowave 810 (Diowave, Riviera Beach, FL, USA); Aspen Laser (Denver, CO, USA)	810/980 nm	8–15 W, 55 to 81 J/cm^2, 30 min, 8–34 irradiation sessions over 4–8 weeks, depending upon individual patient improvement	Transcranially; forehead and temporal regions bilaterally	Decreased depression rate assessed by Quick Inventory of Depression Symptomatology-Self Report (QIDS) and Hamilton Depression Rating Scale; resolved suicidal ideation
Gabel et al. [51]	Patients with low back pain and depressive states (10)	LEDs, Thor-UK:DDII, LED-104 instrument (UK)	660 + 850 nm	100 mW/cm^2, 3 J/cm^2, 6 min per session (30 s/site), 3 × over 12 days	Remote tissue irradiation; 12 sites, symmetrical bilateral: 8 thoracic and 4 posterior-thigh sites	Decreased depression rate assessed by one-question depression-scale; improved functional status assessed by Advise-Rehab Global Scale

(continued)

Table 13.2 (continued)

Study/year	Subjects (n)	Light source	Wavelengths	Irradiation parameters	Irradiation approach/sites	Findings
Caldieraro et al. [46]	Anxious depression complicated by Takotsubo cardiomyopathy (1)	LEDs, Vielight, Inc. (Toronto, Canada); Omnilux New U, Photomedex, Inc. (Montgomeryville, PA, USA)	Transcranial: 830 nm Intranasal: 810 nm	Transcranial: 1 W, 33.2 mW/cm^2, 49.8 J/cm^2, 28.7 cm^2, 25 min, CW Intranasal: 14.2 mW, 14.2 mW/cm^2, 10.65 J/cm^2; 1 cm^2, 25 min per nostril, 10 Hz with a duty cycle of 50%	Transcranially; 2 sites, bilateral (right and left forehead at EEG map sites: F3, F4) Intranasally; both nostrils	Intranasal PBM alone did not prevent the recurrence of moderate depressive symptoms; transcranial PBM twice a week on F3/F4 was also not associated with MDD symptom reduction; however, with daily transcranial PBM on Fpz, MDD symptoms were becoming mild; improved anxiety symptoms, despite short-term fluctuation; no effect on the cognitive functioning
Cassano et al. [43]	Major depressive disorder (21)	LEDs, Omnilux New U, Photomedex, Inc. (Montgomeryville, PA, USA)	823 nm	36.2 mW/cm^2, up to 65.2 J/cm^2, 28.7 × 2 cm^2; 20–30 min, twice a week for 8 weeks, CW	Transcranially; 2 sites, bilateral (right and left forehead at EEG map sites: F3, F4)	Resulted in antidepressant action with a medium to large effect size assessed by Hamilton Depression Rating Scale and self-rated Quick Inventory of Depressive Symptomatology (QIDS)

(continued)

Table 13.2 (continued)

Study/year	Subjects (n)	Light source	Wavelengths	Irradiation parameters	Irradiation approach/sites	Findings
Cassano et al. [44]	Major depressive disorder complicated by sexual dysfunction (20)	LEDs, Omnilux New U, Photomedex, Inc. (Montgomeryville, PA, USA)	823 nm	36.2 mW/cm^2, up to 65.2 J/cm^2, 28.7 × 2 cm^2; 20–30 min, twice a week for 8 weeks, CW	Transcranially: 2 sites, bilateral (right and left forehead at EEG map sites: F3, F4)	Improved overall sexual functioning assessed by Systematic Assessment for Treatment-Emergent Effects (SAFTEE) sex total score: improved sexual interest, sexual arousal, and delayed or absent orgasm
Cassano et al. [118]	Major depressive disorder (18)	LEDs, Omnilux New U, Photomedex, Inc. (Montgomeryville, PA, USA)	823 nm	36.2 mW/cm^2, up to 65.2 J/cm^2, 28.7 × 2 cm^2; 20–30 min, twice a week for 8 weeks, CW	Transcranially: 2 sites, bilateral (right and left forehead at EEG map sites: F3, F4)	Resulted in short lasting but non-significant side effects, resolving in 1 or 2 weeks and not requiring interruption of the treatment; resulted in non-significant weight gain; resulted a significant mild increase in diastolic blood pressure
Mannu et al. [119]	Bipolar-I disorder (4)	LEDs, Omnilux New U, Photomedex, Inc. (Montgomeryville, PA, USA)	830 nm	33.2 mW/cm^2, 40 J/cm^2, 28.7 × 2 cm^2; 20 min, twice a week for 4 weeks, CW	Transcranially: 2 sites, bilateral (right and left forehead at EEG map sites: F3, F4)	Increased serum lithium levels; decreased anhedonia/apathy and increased libido; improved anxiety, sleep, irritability, and impulsivity

(continued)

Table 13.2 (continued)

Study/year	Subjects (n)	Light source	Wavelengths	Irradiation parameters	Irradiation approach/sites	Findings
Maiello et al. [65]	Generalized anxiety disorder (15)	LEDs, *Cerebral Science* headband device	830 nm	2.4 W, 30 mW/cm², 36 J/cm², 80 cm², 20 min	Transcranially; forehead (EEG sites: Fp1, Fp2, F7, F8, and Fpz)	Decreased anxiety symptoms as reduction in the total scores of Structured Interview Guide for the Hamilton Anxiety Scale and the Clinical Global Impressions-Severity (CGI-S) subscale; improved sleep assessed by Pittsburgh Sleep Quality Index (PSQI)
Stephan et al. [81]	Post-traumatic stress disorder (6)	Laser, Theralase TLC 900 (Toronto, Canada)	905 + 660 nm	60 mW, 2.5 min, PW with pulse duration of 200 nanosec, an average of 4 times irradiation over an 8-day period	Transcranially; various areas including prefrontal cortex, temporal lobe, hippocampus, and circle of Willis	Patients scored no emotional problems after 3–5 treatments and all experienced overall sense of well-being; one patient experienced return of ability to smell he had not had for 5 years

Abbreviations: CW—Continuous wave; DMN—Default mode network; EEG—Electroencephalography; LEDs—Light-emitting diodes; MDD—Major depressive disorder; PBM—Photobiomodulation; PW—Pulsed wave; TBI—Traumatic brain injury

References

1. McCarter, T. 2008. Depression overview. *American Health & Drug Benefits* 1 (3): 44.
2. Schmidt, H.D., R.C. Shelton, and R.S. Duman. 2011. Functional biomarkers of depression: Diagnosis, treatment, and pathophysiology. *Neuropsychopharmacology* 36 (12): 2375.
3. Lim, G.Y., et al. 2018. Prevalence of depression in the community from 30 countries between 1994 and 2014. *Scientific Reports* 8 (1): 2861.
4. Brigitta, B. 2002. Pathophysiology of depression and mechanisms of treatment. *Dialogues in Clinical Neuroscience* 4 (1): 7–20.
5. Friedman, R.A. 2012. Grief, depression, and the DSM-5. *New England Journal of Medicine* 366 (20): 1855–1857.
6. Krishnan, K.R.R., et al. 2002. Comorbidity of depression with other medical diseases in the elderly. *Biological psychiatry* 52 (6): 559–588.
7. Hasler, G. 2010. Pathophysiology of depression: Do we have any solid evidence of interest to clinicians? *World Psychiatry : Official Journal of the World Psychiatric Association (WPA)* 9 (3): 155–161.
8. aan het Rot, M., S.J. Mathew, and D.S. Charney. 2009. Charney, Neurobiological mechanisms in major depressive disorder. *CMAJ: Canadian Medical Association Journal = Journal de l'Association Medicale Canadienne* 180 (3): 305–313.
9. Miller, A.H., and C.L. Raison. 2016. The role of inflammation in depression: From evolutionary imperative to modern treatment target. *Nature Reviews. Immunology* 16 (1): 22–34.
10. Rawdin, B.J., et al. 2013. Dysregulated relationship of inflammation and oxidative stress in major depression. *Brain, Behavior, and Immunity* 31: 143–152.
11. Bench, C.J., et al. 1992. The anatomy of melancholia–focal abnormalities of cerebral blood flow in major depression. *Psychological Medicine* 22 (3): 607–615.
12. Pandya, M., et al. 2012. Where in the brain is depression? *Current Psychiatry Reports* 14 (6): 634–642.
13. Sheline, Y.I., B.L. Mittler, and M.A. Mintun. 2002. The hippocampus and depression1To be presented at ECNP Barcelona, 5–9 October 2002, during the symposium "A new pharmacology of depression: The concept of synaptic plasticity.". *European Psychiatry* 17: 300–305.
14. McClintock, S.M., et al. 2011. A systematic review of the combined use of electroconvulsive therapy and psychotherapy for depression. *The Journal of ECT* 27 (3): 236.
15. Krishnan, V., and E.J. Nestler. 2011. Animal models of depression: Molecular perspectives. In *Molecular and functional models in neuropsychiatry*, 121–147. Springer.
16. Porsolt, R., A. Lenegre, and R. McArthur. 1991. Pharmacological models of depression. In *Animal models in psychopharmacology*, 137–159. Springer.
17. Antkiewicz-Michaluk, L., et al. 2014. Antidepressant-like effect of tetrahydroisoquinoline amines in the animal model of depressive disorder induced by repeated administration of a low dose of reserpine: Behavioral and neurochemical studies in the rat. *Neurotoxicity Research* 26 (1): 85–98.
18. Huang, Q.-J., et al. 2004. Brain IL-1beta was involved in reserpine-induced behavioral depression in rats. *Acta Pharmacologica Sinica* 25 (3): 293–296.
19. Minor, T.R., and T.C. Hanff. 2015. Adenosine signaling in reserpine-induced depression in rats. *Behavioural Brain Research* 286: 184–191.
20. Ikram, H., and D.J. Haleem. 2017. Repeated treatment with reserpine as a progressive animal model of depression. *Pakistan Journal of Pharmaceutical Sciences* 30 (3).
21. Stepanichev, M., et al. 2014. Rodent models of depression: neurotrophic and neuroinflammatory biomarkers. *BioMed Research International* 2014.

22. Willner, P. 2017. The chronic mild stress (CMS) model of depression: History, evaluation and usage. *Neurobiology of Stress* 6: 78–93.
23. Overstreet, D.H. 2012. Modeling depression in animal models. In *Psychiatric disorders*, 125–144. Springer.
24. Lee, B., et al. 2013. Chronic administration of baicalein decreases depression-like behavior induced by repeated restraint stress in rats. *The Korean Journal of Physiology & Pharmacology* 17 (5): 393–403.
25. Nagata, K., et al. 2009. Consumption of molecular hydrogen prevents the stress-induced impairments in hippocampus-dependent learning tasks during chronic physical restraint in mice. *Neuropsychopharmacology* 34 (2): 501.
26. Ren, L., et al. 2014. Loss of Ahi1 impairs neurotransmitter release and causes depressive behaviors in mice. *Plos One* 9 (4): e93640.
27. Xu, X., et al. 2010. Neuronal Abelson helper integration site-1 (Ahi1) deficiency in mice alters TrkB signaling with a depressive phenotype. *Proceedings of the National Academy of Sciences* 107 (44): 19126–19131.
28. Perez-Garcia, G., et al. 2018. PTSD-related behavioral traits in a rat model of blast-induced mTBI are reversed by the mGluR2/3 receptor antagonist BCI-838. *ENeuro* 5 (1).
29. Fleminger, S., et al. 2003. The neuropsychiatry of depression after brain injury. *Neuropsychological Rehabilitation* 13 (1–2): 65–87.
30. Tanaka, Y., et al. 2011. Infrared radiation has potential antidepressant and anxiolytic effects in animal model of depression and anxiety. *Brain Stimulation* 4 (2): 71–76.
31. Ando, T., et al. 2011. Comparison of therapeutic effects between pulsed and continuous wave 810-nm wavelength laser irradiation for traumatic brain injury in mice. *Plos One* 6 (10): e26212.
32. Wu, X., et al. 2012. Pulsed light irradiation improves behavioral outcome in a rat model of chronic mild stress. *Lasers in Surgery and Medicine* 44 (3): 227–232.
33. Salehpour, F., et al. 2016. Therapeutic effects of 10 HzPulsed wave lasers in rat depression model: a comparison between near-infrared and red wavelengths. *Lasers in Surgery and Medicine* 48 (7): 695–705.
34. Mohammed, H.S. 2016. Transcranial low-level infrared laser irradiation ameliorates depression induced by reserpine in rats. *Lasers in Medical Science* 31 (8): 1651–1656.
35. Xu, Z., et al. 2017. Low-level laser irradiation improves depression-like behaviors in mice. *Molecular Neurobiology* 54 (6): 4551–4559.
36. Salehpour, F., et al. 2019. Near-infrared photobiomodulation combined with coenzyme Q10 for depression in a mouse model of restraint stress: Reduction in oxidative stress, neuroinflammation, and apoptosis. *Brain Research Bulletin* 144: 213–222.
37. Banqueri, M., et al. 2019. Photobiomodulation rescues cognitive flexibility in early stressed subjects. *Brain Research* 146300.
38. Eshaghi, E., et al. 2019. Transcranial photobiomodulation prevents anxiety and depression via changing serotonin and nitric oxide levels in brain of depression model mice: A study of three different doses of 810 nm laser. *Lasers in Surgery and Medicine* 51 (7): 634–642.
39. Meynaghizadeh-Zargar, R., et al. 2020. Effects of transcranial photobiomodulation and methylene blue on biochemical and behavioral profiles in mice stress model. *Lasers in Medical Science* 35 (3): 573–584.
40. Schiffer, F., et al. 2009. Psychological benefits 2 and 4 weeks after a single treatment with near infrared light to the forehead: A pilot study of 10 patients with major depression and anxiety. *Behavioral and Brain Functions* 5 (1): 1–13.
41. Cassano, P., et al. 2015. Near-infrared transcranial radiation for major depressive disorder: proof of concept study. *Psychiatry Journal* 2015.

42. Disner, S.G., C.G. Beevers, and F. Gonzalez-Lima. 2016. Transcranial laser stimulation as neuroenhancement for attention bias modification in adults with elevated depression symptoms. *Brain Stimulation* 9 (5): 780–787.

43. Cassano, P., et al. 2018. Transcranial photobiomodulation for the treatment of major depressive disorder. The ELATED-2 pilot trial. *Photomedicine and Laser Surgery* 36(12): 634–646.

44. Cassano, P., et al. 2019. Effects of transcranial photobiomodulation with near-infrared light on sexual dysfunction. *Lasers in Surgery and Medicine* 51 (2): 127–135.

45. Caldieraro, M.A., F. Salehpour, and P. Cassano, Transcranial and systemic photobiomodulation for the enhancement of mitochondrial metabolism in depression. In *Clinical bioenergetics*, 635–651. Elsevier.

46. Caldieraro, M.A., et al. 2018. Long-term near-infrared photobiomodulation for anxious depression complicated by Takotsubo cardiomyopathy. *Journal of Clinical Psychopharmacology* 38 (3): 268–270.

47. Quah-Smith, J.I., W.M. Tang, and J. Russell. 2005. Laser acupuncture for mild to moderate depression in a primary care setting—A randomised controlled trial. *Acupuncture in Medicine* 23 (3): 103–111.

48. Quah-Smith, I., et al. 2013. Laser acupuncture for depression: A randomised double blind controlled trial using low intensity laser intervention. *Journal of Affective Disorders* 148 (2–3): 179–187.

49. Kartelishev, A., et al. 2004. Laser technologies used in the complex treatment of psychopharmacotherapy resistant endogenic depression. *Voenno-Meditsinskii Zhurnal* 325 (11): 37–42.

50. Kolupaev, G., et al. 2007. Technologies of laser prophylaxis of depressive disorder relapses. *Voenno-Meditsinskii Zhurnal* 328 (2): 31–34.

51. Gabel, C.P., et al. 2018. A case control series for the effect of photobiomodulation in patients with low back pain and concurrent depression PBM for low back pain and depression. *Laser Therapy* 27 (3): 167–173.

52. Naeser, M.A., et al. 2011. Improved cognitive function after transcranial, light-emitting diode treatments in chronic, traumatic brain injury: Two case reports. *Photomedicine and Laser Surgery* 29 (5): 351–358.

53. Naeser, M.A., et al. 2014. Significant improvements in cognitive performance posttranscranial, red/near-infrared light-emitting diode treatments in chronic, mild traumatic brain injury: Open-protocol study. *Journal of Neurotrauma* 31 (11): 1008–1017.

54. Morries, L.D., P. Cassano, and T.A. Henderson. 2015. Treatments for traumatic brain injury with emphasis on transcranial near-infrared laser phototherapy. *Neuropsychiatric Disease and Treatment* 11: 2159.

55. Craske, M.G., et al. 2011. What is an anxiety disorder? *Focus* 9 (3): 369–388.

56. Greenberg, P., T. Sisitsky, and R. Kessler. 2001. The economic burden of anxiety disorders in the 1990s. *Year Book of Psychiatry and Applied Mental Health* 2001 (1): 186–187.

57. Remes, O., et al. 2016. A systematic review of reviews on the prevalence of anxiety disorders in adult populations. *Brain and Behavior* 6 (7): e00497.

58. Engel, K., et al. 2009. Neuroimaging in anxiety disorders. *Journal of Neural Transmission* (Vienna, Austria: 1996) 116 (6): 703–716.

59. Andreescu, C., et al. 2017. Brain structural changes in late-life generalized anxiety disorder. *Psychiatry Research. Neuroimaging* 268: 15–21.

60. Vythilingam, M., et al. 2000. Temporal lobe volume in panic disorder—A quantitative magnetic resonance imaging study. *Psychiatry Research: Neuroimaging* 99 (2): 75–82.

61. Freitas-Ferrari, M.C., et al. 2010. Neuroimaging in social anxiety disorder: A systematic review of the literature. *Progress in Neuro-Psychopharmacology and Biological Psychiatry* 34 (4): 565–580.

62. Koen, N., and D.J. Stein. 2011. Pharmacotherapy of anxiety disorders: A critical review. *Dialogues in Clinical Neuroscience* 13 (4): 423.
63. Heimberg, R.G. 2002. Cognitive-behavioral therapy for social anxiety disorder: Current status and future directions. *Biological Psychiatry* 51 (1): 101–108.
64. Falsetti, S.A., and J. Davis. 2001. The nonpharmacologic treatment of generalized anxiety disorder. *Psychiatric Clinics of North America* 24 (1): 99–117.
65. Maiello, M., et al. 2019. Transcranial photobiomodulation with near-infrared light for generalized anxiety disorder: A pilot study. *Photobiomodulation, Photomedicine, and Laser Surgery* 37 (10): 644–650.
66. Caldieraro, M.A., et al. 2020. Transcranial photobiomodulation for anxiety disorders and post-traumatic stress disorder. In *Clinical handbook of anxiety disorders*, 283–295. Springer.
67. Kerppers, F.K., et al. (2020) Study of transcranial photobiomodulation at 945-nm wavelength: Anxiety and depression. *Lasers in Medical Science* 1–10.
68. Brochado, F.T., et al. 2018. Comparative effectiveness of photobiomodulation and manual therapy alone or combined in TMD patients: A randomized clinical trial. *Brazilian Oral Research* 32.
69. da Silva, M.M., et al. 2018. Randomized, blinded, controlled trial on effectiveness of photobiomodulation therapy and exercise training in the fibromyalgia treatment. *Lasers in Medical Science* 33 (2): 343–351.
70. Shalev, A., I. Liberzon, and C. Marmar. 2017. Post-traumatic stress disorder. *New England Journal of Medicine* 376 (25): 2459–2469.
71. Yehuda, R., et al. 2015. Post-traumatic stress disorder. *Nature Reviews Disease Primers* 1: 15057.
72. Association, A.P. 2013. *Diagnostic and statistical manual of mental disorders (DSM-5®)*. American Psychiatric Pub.
73. Zoladz, P.R., and D.M. Diamond. 2013. Current status on behavioral and biological markers of PTSD: A search for clarity in a conflicting literature. *Neuroscience & Biobehavioral Reviews* 37 (5): 860–895.
74. Yehuda, R. 2002. Post-traumatic stress disorder. *New England Journal of Medicine* 346 (2): 108–114.
75. Bremner, J.D. 2002. Neuroimaging studies in post-traumatic stress disorder. *Current Psychiatry Reports* 4 (4): 254–263.
76. Gould, E., et al. 1998. Proliferation of granule cell precursors in the dentate gyrus of adult monkeys is diminished by stress. *Proceedings of the National Academy of Sciences* 95 (6): 3168–3171.
77. Nibuya, M., S. Morinobu, and R.S. Duman. 1995. Regulation of BDNF and trkB mRNA in rat brain by chronic electroconvulsive seizure and antidepressant drug treatments. *Journal of Neuroscience* 15 (11): 7539–7547.
78. Bisson, J., and M. Andrew. 2007. Psychological treatment of post-traumatic stress disorder (PTSD). *Cochrane Database of Systematic Reviews* (3).
79. Stein, D.J., et al. 2006. Pharmacotherapy for post traumatic stress disorder (PTSD). *Cochrane Database of Systematic Reviews* (1).
80. Hendriksen, H., B. Olivier, and R.S. Oosting. 2014. From non-pharmacological treatments for post-traumatic stress disorder to novel therapeutic targets. *European Journal of Pharmacology* 732: 139–158.
81. Stephan, W., et al. 2017. Management of post-traumatic stress (PTSD) dementia and other neuro-degenerative disease with photo-medicine: Clinical experience and case studies. *Open Journal of Psychiatry* 7 (4): 386–394.

82. Edition, F. 2013. *Diagnostic and statistical manual of mental disorders*. Arlington: American Psychiatric Publishing.
83. Faras, H., N. Al Ateeqi, and L. 2010. Tidmarsh, Autism spectrum disorders. *Annals of Saudi medicine* 30 (4): 295.
84. Muhle, R., S.V. Trentacoste, and I. Rapin. 2004. The genetics of autism. *Pediatrics* 113 (5): e472–e486.
85. Matelski, L., and J. Van de Water. 2016. Risk factors in autism: Thinking outside the brain. *Journal of Autoimmunity* 67: 1–7.
86. Amaral, D.G., C.M. Schumann, and C.W. Nordahl. 2008. Neuroanatomy of autism. *Trends in Neurosciences* 31 (3): 137–145.
87. Stahmer, A.C., L. Schreibman, and A.B. Cunningham. 2011. Toward a technology of treatment individualization for young children with autism spectrum disorders. *Brain Research* 1380: 229–239.
88. Warren, Z., et al. 2011. *Therapies for children with autism spectrum disorders*.
89. Ji, N.Y., and R.L. Findling. 2015. An update on pharmacotherapy for autism spectrum disorder in children and adolescents. *Current Opinion in Psychiatry* 28 (2): 91–101.
90. Enticott, P.G., et al. 2014. A double-blind, randomized trial of deep repetitive transcranial magnetic stimulation (rTMS) for autism spectrum disorder. *Brain Stimulation* 7 (2): 206–211.
91. Amatachaya, A., et al. 2014. Effect of anodal transcranial direct current stimulation on autism: A randomized double-blind crossover trial. *Behavioural Neurology* 2014.
92. Khongrum, J., and J. Wattanathorn. 2015. Laser acupuncture improves behavioral disorders and brain oxidative stress status in the valproic acid rat model of autism. *Journal of Acupuncture and Meridian Studies* 8 (4): 183–191.
93. Khongrum, J., and J. Wattanathorn. 2017. Laser acupuncture at HT7 improves the cerebellar disorders in valproic acid-rat model of autism. *Journal of Acupuncture and Meridian Studies* 10 (4): 231–239.
94. Leisman, G., et al. 2018. Effects of low-level laser therapy in autism spectrum disorder. In *Clinical medicine research*, 111–130. Springer.
95. Machado, C., Y. Machado, and M. Chinchilla. 2019. *Follow-up assessment of autistic children 6 months after finishing low lever laser therapy*.
96. Morin, C.M., and R. Benca. 2012. Chronic insomnia. *The Lancet* 379 (9821): 1129–1141.
97. Roth, T. 2007. Insomnia: Definition, prevalence, etiology, and consequences. *Journal of clinical sleep medicine: JCSM: Official publication of the American Academy of Sleep Medicine* 3 (5 Suppl): S7.
98. Riemann, D., et al. 2015. The neurobiology, investigation, and treatment of chronic insomnia. *The Lancet Neurology* 14 (5): 547–558.
99. Ohayon, M.M. 1997. Prevalence of DSM-IV diagnostic criteria of insomnia: Distinguishing insomnia related to mental disorders from sleep disorders. *Journal of psychiatric research* 31 (3): 333–346.
100. Levenson, J.C., D.B. Kay, and D.J. Buysse. 2015. The pathophysiology of insomnia. *Chest* 147 (4): 1179–1192.
101. Roth, T. 2007. Insomnia: Definition, prevalence, etiology, and consequences. *Journal of Clinical Sleep Medicine: JCSM: Official Publication of the American Academy of Sleep Medicine* 3 (5 Suppl): S7–S10.
102. Basta, M., et al. 2007. Chronic insomnia and the stress system. *Sleep Medicine Clinics* 2 (2): 279–291.
103. Joo, E.Y. 2015. Structural brain neuroimaging in primary insomnia. *Sleep Medicine Research* 6 (2): 50–53.

104. Altena, E., et al. 2010. Reduced orbitofrontal and parietal gray matter in chronic insomnia: A voxel-based morphometric study. *Biological Psychiatry* 67 (2): 182–185.
105. Salehpour, F., et al. 2018. Transcranial near-infrared photobiomodulation attenuates memory impairment and hippocampal oxidative stress in sleep-deprived mice. *Brain Research* 1682: 36–43.
106. Zhao, J., et al. 2012. Red light and the sleep quality and endurance performance of Chinese female basketball players. *Journal of Athletic Training* 47 (6): 673–678.
107. Saltmarche, A.E., et al. 2017. Significant improvement in cognition in mild to moderately severe dementia cases treated with transcranial plus intranasal photobiomodulation: Case series report. *Photomedicine and Laser Surgery* 35 (8): 432–441.
108. Bogdanova, Y., et al. 2017. Transcranial LED treatment for cognitive dysfunction and sleep in chronic TBI: Randomized controlled pilot trial. *Archives of Physical Medicine and Rehabilitation* 98 (10): e122–e123.
109. Salehpour, F., et al. 2020. Therapeutic potential of intranasal photobiomodulation therapy for neurological and neuropsychiatric disorders: A narrative review. *Reviews in the Neurosciences* 31 (3): 269–286.
110. Xu, C., et al. 2001. Endonasal low energy He–Ne laser treatment of insomnia. *Qianwei Journal of Medicine & Pharmacy* 18 (5): 337–338.
111. Xu, C., et al. 2002. The effects of endonasal low energy He–Ne laser treatment of insomnia on sleep EEG. *Practice and Journal of Medicine Pharmacy* 19: 407–408.
112. Chen, Y., and H. Cheng. 2004. Clinical observation of the integrated therapy of intranasal low intensity He-Ne laser therapy and herb therapy on insomnia. *Journal of Traditional Chinese Medicine* 24: 38.
113. Liu, T.C.-Y., et al. 2010. Applications of intranasal low intensity laser therapy in sports medicine. *Journal of Innovative Optical Health Sciences* 3 (01): 1–16.
114. Moshkovska, T., and J. Mayberry. 2005. It is time to test low level laser therapy in Great Britain. *Postgraduate Medical Journal* 81 (957): 436–441.
115. Quah-Smith, I., W. Wen, X. Chen, M.A. Williams, and P.S. Sachdev. 2012. The brain effects of laser acupuncture in depressed individuals: an fMRI investigation. *Medical Acupuncture* 24 (3): 161–71.
116. Quah-Smith, I, C. Suo, M.A. Williams, and P.S. Sachdev. 2013. The antidepressant effect of laser acupuncture: a comparison of the resting brain's default mode network in healthy and depressed subjects during functional magnetic resonance imaging. *Medical Acupuncture* 25 (2): 124–133.
117. Henderson, T.A., and L.D. Morries. 2017.Multi-watt near-infrared phototherapy for the treatment of comorbid depression: an open-label single-arm study. *Frontiers in Psychiatry* 8: 187.
118. Cassano, P., M.A. Caldieraro, R. Norton, D. Mischoulon, N.-H. Trinh, M. Nyer, et al. 2019.Reported side effects, weight and blood pressure, after repeated sessions of transcranial photobiomodulation. *Photobiomodulation, Photomedicine, and Laser Surgery* 37 (10): 651–656.
119. Mannu, P., L.F. Saccaro, V. Spera, and P. Cassano. 2019. Transcranial photobiomodulation to augment Lithium in Bipolar-I disorder. *Photobiomodulation, Photomedicine, and Laser Surgery* 37 (10): 577–578.
120. Rossetti, V., et al. 1991. Rat brain metabolism enzyme activity variations following He–Ne laser irradiation. *Molecular and Chemical Neuropathology* 15 (2): 185–191.
121. Cassone, M., et al. 1993. Effect of in vivo He–Ne laser irradiation on biogenic amine levels in rat brain. *Journal of Photochemistry and Photobiology B: Biology* 18 (2–3): 291–294.

122. Tsai, J.-F., S. Hsiao, and S.-Y. Wang. 2007. Infrared irradiation has potential antidepressant effect. *Progress in Neuro-Psychopharmacology and Biological Psychiatry* 31 (7): 1397–1400.
123. Meynaghizadeh-Zargar, R., et al. 2019. Effects of transcranial photobiomodulation and methylene blue on biochemical and behavioral profiles in mice stress model. *Lasers in Medical Science* 1–12.

Photobiomodulation Therapy for Other Brain-Related Disorders

14

14.1 Disorders of Consciousness and Vegetative State

14.1.1 Introduction

A large amount of research has been devoted to explore consciousness. Nevertheless, while these studies have led to a crucial change in our concept of consciousness, they have not yielded sufficient understanding of even the most basic facts, for example, why does the cerebral cortex play such an important role in consciousness, whereas the cerebellum does not. This is despite the fact that the cerebellum has even more neurons and is complex enough to be involved in such a process [1]. The definition of consciousness varies across the literature. In one view, consciousness is defined as "the selection of information for global broadcasting, thus making it flexibly available for computation and report (C1, consciousness in the first sense), and the self-monitoring of those computations, leading to a subjective sense of certainty or error (C2, consciousness in the second sense)" [2]. The process of consciousness allows the animal to plan and prepare itself for an upcoming event. Others have defined consciousness as the state of wakefulness (level of consciousness) and awareness (content of consciousness) [3]. The most important parts of the brain, which contribute to consciousness are the thalamus and cerebral cortex. However, not all parts of the thalamus and cortex contribute equally to consciousness, and an injury to large portions of these structures does not necessarily render the affected animal unconscious [4]. However, other injuries to the mentioned parts, which disrupt particular processes, could finally lead to disorders of consciousness. Disorders of consciousness, in general, can be categorized into coma, vegetative state (now called unresponsive wakefulness syndrome, UWS), and minimally conscious state (MCS). By far, the most common cause of disorders of consciousness is a traumatic brain injury.

Vegetative state, or apallic syndrome is a condition during which vegetative nervous functions such as sleep–wake cycles, thermoregulation, respiration, and digestion are maintained to some extent, however, the patient is not aware of their surroundings [5, 6]. The Multi-Society Task Force on persistent vegetative state defines it as a condition during which the patient is not aware, maintains sleep–wake cycles, and displays no purposeful movement, or experiences suffering. EEG monitoring shows diffuse generalized polymorphic delta or theta activity which is not influenced by sensory stimulation [5]. Vegetative state is distinguished from coma by the presence of intermittent wakefulness and sleep–wake cycles. On the other hand, as opposed to MCS in which there remains a fluctuating level of awareness, no detectable trace of awareness can be found in persistent vegetative state. Locked-in syndrome is differentiated from persistent vegetative state by the presence of awareness and wakefulness. Pathophysiologically persistent vegetative state is thought to be caused by damage to the cerebral cortex or thalamus, rather than to the brain stem [7]. From an etiological viewpoint, persistent vegetative state results from insults which can cause coma, but are less severe, or from which the patient has partially recovered [8].

Accurate diagnosis of persistent vegetative state remains a major clinical challenge, and misdiagnosis is as high as 37–43% in some centers. Neither GCS nor the Coma Remission Scale can practically and effectively differentiate persistent vegetative state from MCS. Accordingly, specialized diagnostic measures have been provided to accomplish this goal. Revised Coma Recovery Scale (r-CRS) is an internationally-accepted scale which allows a clear differentiation between persistent vegetative state and MCS [9]. This instrument has become the diagnostic gold standard for this purpose as discussed in (www.coma.ulg.ac. be/medical/chronic.html). EEG-based techniques are also helpful in making an accurate diagnosis. Additionally, it has been found that PET can help clinicians to establish the correct diagnosis in around 85% of cases with disorders of consciousness [8].

14.1.2 Photobiomodulation Therapy for Disorders of Consciousness and Vegetative State in Humans

In 2015, Hesse et al. [10] evaluated for the first time the effects of transcranial PBM therapy in five traumatic brain injury (TBI) patients with severe disorders of consciousness. To assess the response to sensory stimuli, the r-CRS was administered at various time points including at baseline and post-PBM therapy. Besides, the Barthel Index and modified Rankin Scale were used to measure the activities of daily living and caregiver ratings, respectively. Transcranial PBM therapy was applied to the forehead using a 785 nm CW mode laser (10 mW/cm^2 for 10 min providing a scalp fluence of 6 J/cm^2) for 30 total daily sessions. Participants were examined from 21 days before the first PBM therapy, and then up to 70 days after the first laser treatment. According to the results, the scores of all participants showed some improvement from baseline to follow-up. For instance, PBM significantly improved nonverbal interaction and visual pursuit over the course of the treatment. Although the scores of r-CRS for all patients showed a significant improvement, changes in the Barthel Index and modified Rankin Scale were only minimal.

In 2016, Werner and colleagues [11] enrolled 14 patients with vegetative state/ unresponsive wakefulness syndrome, and 2 MCS patients, and randomly assigned them into transcranial PBM therapy and focused shock wave therapy groups. They used the same laser as Hesse et al. [10] and the treatment protocol consisted of transcranial PBM therapy using 785 nm laser (10 mW/cm² for 10 min providing a scalp fluence of 6 J/cm²), five times a week for four weeks. According to the r-CRS, FOUR scale (for testing eye movement, motor response to a pain stimulus or a verbal command, brain stem reflexes, and breathing), and SMART scale assessment (for testing five sensory abilities, taste, vision, hearing, tactile sense, and smell), the consciousness status of both groups was improved and continuously maintained, suggesting that these two neuromodulation modalities might be used to increase alertness and awareness in patients with chronic disorders of consciousness.

In a single case report, Nawashiro et al. [12] demonstrated the beneficial neurotherapeutic effects of transcranial LED PBM therapy in a 40-year-old male comatose patient who had been in a persistent vegetative state for 8 months. The patient received 146 transcranial PBM treatments using 850 nm NIR light (11.4 mW/cm², 20.5 J/cm², 30 min, twice a day) for 73 days. Transcranial PBM therapy was administered bilaterally to the forehead. Five days after beginning the PBM therapy (after 10 treatments), the patient started to move his left arm and hand to touch the tracheostomy site when suctioning was conducted. It is worthwhile to note that he had never shown any spontaneous movement before the PBM therapy. Although after 2 months of PBM therapy, the patient always showed spontaneous movement, he never obeyed commands. Nevertheless, no further improvement in the patient's neurological condition after the therapy was noticed. Concerning brain hemodynamics, the regional CBF in the viable left anterior frontal lobe also increased (by 20%) 30 min after the 146th LED treatment.

14.2 Multiple Sclerosis

14.2.1 The Problem of Multiple Sclerosis

Multiple sclerosis (MS) is a chronic immune-mediated disorder attacking the central nervous system (CNS), and is of inflammatory origin [13]. MS attacks the myelinated axons of the CNS and destroys them to a varying extent [14]. MS is the most common cause of disability in young and middle-aged individuals [15]. It is estimated that as many as 250,000–350,000 persons suffer from MS in the United States [13]. The most noticeable epidemiological property of MS is its seemingly uneven distribution over the globe. MS is quite common among white and high-income individuals of Nordic origin, who live in temperate zones [15]. Several factors play a role in the pathophysiology of MS, including genetics, geography-associated physical environment, lifestyle factors (smoking, exposure to ultraviolet radiation, body mass index, and vitamin D deficiency), and socioeconomic determinants [16].

The pathogenesis of MS involves neuroinflammation of both white and gray matter in the CNS, caused by focal infiltration of immune cells. Antigen-presenting cells (APCs) interact with T lymphocytes, which causes the differentiation of T cells into Th1, Th2, or Th17 phenotypes. The Th1 or Th17 subtypes promote neuroinflammation by the release of IL-17, IL-21, IL-22, and IL-26 (Th17) or interferon gamma (IFNγ) and tumor necrosis factor alpha (TNF-α) (Th1) [17]. Lymphotoxin (transforming growth factor beta TGF-β) and TNF-α released by B lymphocytes induce even more neuroinflammation in MS patients [18]. The role of CD8 + T cells (cytotoxic T cells) has also been confirmed in the pathogenesis of MS by the secretion of perforin (a cytolytic protein) [19].

Because MS can affect any area of the brain and the CNS, it can produce a wide variety of neurologic symptoms. The symptoms of MS can be generally divided into three distinct groups, primary, secondary, and tertiary symptoms. Primary symptoms are more common and include sensory disturbances, gait difficulties, vision problems, intestinal and urinary system dysfunction, cognitive and emotional dysfunction, vertigo, dysphagia, sexual problems, dysphagia, headaches, seizures, hearing loss, and breathing problems. Secondary symptoms include urinary tract infections, immobility, and inactivity. Tertiary symptoms are depression, social difficulties, and vocational complications [20]. The prevalence of tertiary symptoms and psychological effects in MS patients is quite high. A study showed that the prevalence of anxiety disorders was 21.9%, alcohol abuse was 14.8%, bipolar disorder was 5.83%, substance abuse was 2.5%, and psychosis was 4.3%. The commonest symptom was depression which was as high as 23.7% (95% CI: 17.4–30.0%) [21].

Unfortunately, there has been no definitive treatment for MS up until now. Immunosuppressive and immune-modulating agents are the cornerstone of treatment in clinical practice. Disease-modifying treatments (DMTs) have also been introduced to delay progression and decrease the attack frequency [22]. The approval of IFNβ and glatiramer acetate for the treatment of MS has revolutionized the treatment approach. Subsequently, monoclonal antibodies such as natalizumab were developed to further improve the results of MS treatment. The introduction of oral drugs, such as fingolimod, dimethyl fumarate, cladribine, and teriflunomide was another major step forward. Several newer anti-monoclonal antibodies, such as alemtuzumab and ocrelizumab have been developed which may further improve the treatment of MS patients [23]. Management of comorbidities and psychological support are also of major importance in MS patients. The long list of comorbidities that follow or accompany MS precludes it from being mentioned here. However, the utmost attempts should be made to prevent their occurrence or exacerbation in these emotionally and medically fragile patients [24].

14.2.2 Photobiomodulation Therapy for Multiple Sclerosis in Animal Studies

In 2012, Muili et al. [25] sought to examine for the first time the effects of PBM therapy in the experimental autoimmune encephalomyelitis (EAE) mouse model of MS. The EAE model of chronic disease was induced by immunization with myelin oligodendrocyte glycoprotein (MOG_{35-55}) peptide. For PBM therapy, mice received 670 nm LED light (5 J/cm^2) on the dorsal surface beginning on the day after immunization and continuing for 10 days. PBM therapy significantly decreased the production of pro-inflammatory mediators, including IFN-γ and TNF-α, as well as increased production of anti-inflammatory cytokines, namely, IL-4 and IL-10. These cytokine changes could be expected to have beneficial effects in the EAE model of MS, and therefore the immune modulation by 670 nm PBM therapy observed in their study, could be expected to improve clinical symptoms of EAE. It should be also mentioned that PBM therapy once-daily for 10 days starting on the day of onset of clinical signs significantly ameliorated acute episodes of EAE, whereas, extending the treatment sessions through the peak of disease (15 days post-immunization) eliminated the beneficial effects.

Because studies have shown that axonal loss and the progression of disability in EAE/MS is mediated in part by nitrosoxidative stress [26, 27], Muili et al. [28] examined the effect of LED PBM therapy on nitrosative stress in the MOG-induced EAE mouse model. MS was induced in female mice (6–8 weeks old) by immunization with 100 mg MOG_{35-55} peptide. PBM therapy consisted of once-daily irradiation using 670 nm light (28 mW/cm^2 for 3 min, providing 5 J/cm^2 on the dorsal surface). PBM therapy was performed with a double treatment protocol, consisting of once-daily LED treatment for 7 consecutive days beginning at EAE onset, then followed by a 7-day resting period, and then a subsequent 7-day LED treatment period. 670 nm PBM therapy significantly down-regulated expression of inducible nitric oxide synthase (iNOS), the main source of NO in EAE pathology, over the course of the disease. In addition, PBM up-regulated the expression of Bcl-2 anti-apoptosis gene, increased the Bcl-2/Bax ratio, and decreased apoptosis within the CNS. It is now believed that NO plays a key role in EAE progression, therefore the down-regulation of iNOS mRNA by red LED treatment could be expected to have a therapeutic benefit on the disease severity. To explore whether iNOS could play a role in the mitigation of EAE severity by PBM therapy, they induced EAE by MOG_{35-55} immunization in both iNOS knockout or wild-type (WT) mice. The results showed that PBM therapy was not able to ameliorate MOG-induced EAE in mice deficient in iNOS, suggesting that NO generated by iNOS was an essential element to the mechanism of neuroprotection by 670 nm LED PBM therapy [29]. In other words, this observation confirmed a role for the reversal of nitrosative stress in the improvement of MOG-induced EAE by PBM therapy.

In 2016, Gonçalves et al. [30] investigated the effect and underlying mechanism of action of PBM therapy using 660 nm far-red (10 J/cm^2) and 904 nm NIR (3 J/cm^2) lasers on the spinal cord of mice during the development of the EAE model of MS. The researchers induced EAE in female mice by immunization with MOG$_{35-55}$ peptides emulsified in complete Freund's adjuvant. Daily irradiation using both 660 and 904 nm lasers for 30 days consistently decreased the clinical EAE score and significantly delayed the disease onset, and prevented the loss of weight induced by immunization. They suggested that the beneficial effects of far-red/NIR PBM therapy were accompanied by the down-regulation of NO levels in the spinal cord and spleen (but not in the inguinal lymph nodes). Nevertheless, PBM could not suppress lipid peroxidation and restore antioxidant defenses (GSH levels) in either the spinal cord or inguinal lymph node tissues during EAE. Histological analysis also revealed that only PBM therapy using 660 nm laser could suppress neuroinflammation by decreasing inflammatory cells (especially lymphocytes) in the CNS, as well as inhibiting demyelination in the spinal cord. However, both lasers (AlGaInP, 660 nm; GaAs, 904 nm) significantly attenuated the production of pro-inflammatory cytokines (e.g., IL-17, IFN-γ, and IL-1β) during EAE.

14.2.3 Photobiomodulation Therapy for Multiple Sclerosis in Humans

In a recent randomized non-controlled clinical trial, Silva et al. [31] studied the effects of PBM therapy on the expression of IL-10 and nitrite levels in patients with MS. Fourteen patients diagnosed with Relapsing–Remitting MS with a score of up to 6.0 on the Expanded Disability Status Scale (EDSS) were enrolled in the study. PBM therapy was applied with an 808 nm laser with an irradiance of 0.8 W/cm^2 and fluence of 287 J/cm^2 for 6 min, twice a week for 12 weeks. Laser irradiations were performed either on the sublingual region (7 patients) or over the radial artery (7 patients). According to the authors, PBM therapy on the spinal column could not be performed due to the difficulty in palpating the vertebrae to locate the laser tip in obese patients. Peripheral blood was analyzed for serum levels of IL-10 and nitrite before and after the treatment. After 24 PBM therapy sessions, the IL-10 expression levels increased in both irradiation groups. On the other hand, nitrite levels were not significantly lower in either the radial artery group or the sublingual group. They suggested that further studies should be carried out with a larger sample and a control group, to assess the clinical performance of PBM therapy for the management of MS.

In two different studies from Tolentino's laboratory [32, 33], they evaluated the effects of PBM therapy on peripheral blood mononuclear cells and CD4+T cells isolated from MS patients. In the first study [32], isolated cells received PBM therapy in vitro, and the cytokine levels were measured in cell culture supernatants by ELISA assay. PBM using both 670 and 830 nm significantly increased IL-10 and decreased IFN-γ produced by peripheral blood mononuclear cells. Moreover, PBM with 670, 735, and 830 nm increased IL-10 levels, while 670 and 830 nm light (but not 735 nm) could decrease IFN-γ produced by CD4+T cells. PBM with 670 and 830 nm light decreased IFN-γ produced by peripheral blood mononuclear cells, whereas 830 nm increased IL-10 production by CD4+T cells isolated from healthy donors. These results indicate that depending on the light wavelength, PBM therapy could differentially modulate the immune response in MS patients and healthy donors, by increasing IL-10 and reducing IFN-γ. In a follow-up study using the same experimental setup, Tolentino et al. [33] also showed that PBM could decrease NO production by peripheral blood mononuclear cells from MS patients. Pulsed 640 nm (3 J/cm^2) and 830 nm (10 J/cm^2) light were the most effective PBM parameters for decreasing nitrite levels. The expression of IFN-γ was also correlated with nitrite levels using the above-mentioned parameters. It should be also noted that the beneficial effects of PBM therapy were directly affected by the drugs that patients were taking.

14.3 Drug Addiction

14.3.1 Photobiomodulation Therapy for Drug Addiction in Animal Studies

Chronic exposure to opiates normally leads to long-term behavioral and neurological changes producing mental and physical dependence [34]. The cessation of drug use in dependent individuals then results in a variety of symptoms ranging from mild to severe illness and a plethora of withdrawal signs such as anxiety, nausea, insomnia, muscle aches, etc. [35, 36]. For example, morphine withdrawal in a variety of laboratory animals has been suggested to be a valid model of drug addiction [37]. In this respect, studies have shown that morphine withdrawal is strongly linked to neurochemical and behavioral changes [38].

In 2009, Mirzaii-Dizgah and colleagues [39] investigated whether PBM therapy prior to naloxone injection could mitigate the withdrawal symptoms in morphine-dependent male Wistar rats. The dependent animals received a dose escalated subcutaneous (s.c.) injection of morphine sulfate (2 mL/kg) once-daily for 7 consecutive days starting with 6, then 16, 26, 36, 46, 56, and 66 mg/kg. The opioid withdrawal was precipitated by naloxone hydrochloride injection (3 mg/kg, s.c.) 1-day after the last morphine dose. Laser PBM therapy was applied 15 or 30 min before naloxone injection. The treatment protocol consisted of irradiation using 830 nm GaAlAs laser (power of 50 mW for 55 s, providing 12.5 J/cm^2) applied to the shaved scalp at the intersection between the interaural line and midline of the head. Naloxone administration considerably increased the total withdrawal score in morphine-dependent animals as compared to non-dependent animals. They used the total withdrawal score as an index of abstinence throughout their study. Results indicated that although the total withdrawal score was significantly lower in dependent rats who received PBM therapy 15 min before naloxone injection, PBM therapy 30 min prior to naloxone injection was not able to significantly improve the total withdrawal score as compared to control rats. Of note, 7 days of morphine sulfate administration also led to weight loss (6–8%) and occasional death (7%).

Similarly, in 2014, Ojaghi et al. [40] aimed to explore the potential involvement of the NO system in attenuating the effects of PBM therapy on naloxone-induced morphine withdrawal symptoms in a mouse model. Morphine sulfate (50, 50, and 75 mg/kg) was subcutaneously injected 3 times a day for 3 consecutive days. Also, an additional dose of morphine sulfate (50 mg/kg, s.c.) was injected on day 4 at 2 h prior to naloxone administration (2 mg/kg, s.c.). The PBM therapy protocol was the same as that described by Mirzaii-Dizgah et al. [39], using a single irradiation of 830 nm laser with a fluence of 12.5 J/cm^2. PBM therapy significantly decreased the "escape jump count" and stool weight, as typical withdrawal signs, in morphine-dependent naloxone-treated mice. It must be emphasized that they did not measure the cerebral NO levels in their study, nevertheless, because NO has been shown to have a pivotal role in opioid withdrawal syndrome, they proposed that PBM therapy may act partly via the NO pathway in attenuating morphine withdrawal signs.

14.4 Epilepsy

14.4.1 Photobiomodulation Therapy for Epilepsy in Animal Studies

Epilepsy is a serious and common neuropathological disorder affecting the quality of life of sufferers. According to the global estimation, almost 65 million people worldwide suffer from epilepsy, and more than 100,000 new cases develop each year [41]. Of note, the prevalence of epilepsy in developing countries is normally higher than in developed countries [42]. In epilepsy, the normal pattern of neuronal activity becomes disturbed, so that chronic seizures are initiated by abnormal paroxysmal electrical activity between neurons, and ultimately lead to irreversible damage to brain cells. It is believed that damage to inhibitory neuronal cells in the hippocampus during an epileptic attack can potentially change the balance of excitatory and inhibitory neurons and eventually causes hyperexcitability [43]. Hyperexcitability in turn can facilitate gliosis and subsequently results in mitochondrial dysfunction in the hippocampus and dentate gyrus neurons.

PBM therapy has been suggested as an effective treatment, which could overcome the shortcomings associated with anticonvulsant drugs. So far, two studies have explored the effects of transcranial PBM therapy using an in vivo rodent model of epilepsy [44, 45].

According to Ahmed et al. [46], 830 nm laser irradiation with a power of 90 mW showed a significant neurosuppressive effect on axonal conduction of cortical tissue in naïve rats, with a reduction in cortical glutamate, aspartate, and glutamine, and an increase in glycine. This preliminary finding from intact animals encouraged them to study the effects of daily laser PBM therapy in an epileptic rat model, which is characterized by neuronal hyperexcitability. In this regard, in 2009, they found that 830 nm laser irradiation with a power of 90 mW could potentially modulate the imbalance between neurotransmitters by regulation of GABA release and glutamate in the cortex and hippocampus of a pilocarpine-induced epilepsy model in rats. Specifically, they found that PBM therapy restored the increased cortical levels of glutamic acid, glutamine, glycine, and taurine as well as increased hippocampal levels of aspartate and glycine in pilocarpine-treated animals to near-normal values [44] (Table 14.1).

Table 14.1   Summary of in vivo studies on the effects of photobiomodulation therapy in other brain-related disorders

Study/year	Animal/model	Light source	Wavelengths	Irradiation parameters	Irradiation approach/sites	Findings
Mirzaii-Dizgah et al. [39]	Rat (naloxone-induced morphine withdrawal model)	Laser, ADVANCE, AMN50 (Australia)	830 nm	50 mW, 12.5 J/cm², 55 s, laser beam of 0.22 cm², immediate, 15 min, or 30 min before naloxone injection	Transcranially; at the cross-point between interaural line and midline of head	PBM applied immediately or 15 min prior to naloxone injection significantly attenuated the expression of withdrawal signs by decreasing total withdrawal score
Radwan et al. [44]	Rat (pilocarpine-induced epileptic model)	Laser, Lasermax, Inc. (Rochester, NY, USA)	830 nm	90 mW, 2.87 W/cm², 32.4 J, laser beam of 2 mm in diameter, 1 min per point, for 7 days	Transcranially; over the parietal region (perpendicularly at six points arranged symmetrically, three on each side of the sagittal suture of the skull)	Decreased the concentrations of glutamic acid, glutamine, glycine, and taurine in the cortex; decreased aspartate and glycine levels in the hippocampus; increased aspartate aminotransferase activity and decreased alanine aminotransferase activity and glucose content; and aspartate aminotransferase; decreased aspartate aminotransferase alanine aminotransferase activities and glucose content in the hippocampus
Moges et al. [56]	Mouse (SOD1 transgenic familial amyotrophic lateral sclerosis model)	Laser	810 nm	140 mW, 12 J/cm² per treatment site per day, spot size of 1.4 cm², 120 s, for 3 consecutive days every week	Transcranially; the primary motor cortex, the cervical enlargement of the spinal cord, and the lumbar enlargement of the spinal cord	No effect on the animal's survival; improved motor function as assessed by rotarod test in the early stage of the disease; reduced expression of the astrocyte marker, glial fibrilary acidic protein in the cervical and lumbar enlargements of the spinal cord; no effect on the number of motor neurons in the anterior horn of the lumbar enlargement

(continued)

Table 14.1 (continued)

Study/year	Animal/model	Light source	Wavelengths	Irradiation parameters	Irradiation approach/sites	Findings
Muili et al. [25]	Mouse (experimental autoimmune encephalomyelitis model of multiple sclerosis)	LEDs, Quantum Devices (Barneveld, WI, USA)	670 nm	2.1 W, 28 mW/cm^2, 5 J/cm^2, 3 min, once daily for 10 consecutive days starting 24 h post immunization	Positioned directly over the animal at a distance of 2 cm, covering the entire chamber and exposing the entire dorsal surface	Decreased disease severity; down-regulated proinflammatory cytokines (INF-γ, TNF-α) and up-regulated anti-inflammatory cytokines (IL-4 and IL-10)
Muili et al. [28]	Mouse (experimental autoimmune encephalomyelitis model of multiple sclerosis)	LEDs, Quantum Devices (Barneveld, WI, USA)	670 nm	2.1 W, 28 mW/cm^2, 5 J/cm^2, 3 min, once daily for 10 consecutive days starting 24 h post immunization	Positioned directly over the animal at a distance of 2 cm, covering the entire chamber and exposing the entire dorsal surface	Down-regulated iNOS gene expression in the spinal cord; up-regulated the Bcl-2 anti-apoptosis gene, an increased Bcl-2/Bax ratio, and reduced apoptosis within the spinal cord
Ojaghi et al. [40]	Rat (naloxone-induced morphine withdrawal model)	Laser, ADVANCE, AMN50 (Australia)	830 nm	50 mW, 12.5 J/cm^2, 55 s, laser beam of 0.22 cm^2, immediate, 15 min, or 30 min before naloxone injection	Transcranially; at the cross-point between interaural line and midline of head	Decreased escape jump count and stool weight, as typical withdrawal signs in morphine-dependent naloxone-treated animals
Khongrum and Wattanathorn [57]	Rat (Valproic Acid Model of Autism)	Laser, Xinland International Limited (Xi'an, Shaanxi, China)	405 nm	100 mW, 0.1 J/s, once daily from postnatal day 14 to day 40	Laser acupuncture; at HT7 acupoint on both the left and the right sides	Improved autistic-like behaviors; decreased MDA levels in the cortex, striatum, and hippocampus; increased glutathione peroxidase activity only in the striatum and hippocampus; no changes in SOD and catalase activities in the cortex, striatum, and hippocampus

(continued)

Table 14.1 (continued)

Study/year	Animal/model	Light source	Wavelengths	Irradiation parameters	Irradiation approach/sites	Findings
Gonçalves et al. [30]	Mouse (experimental autoimmune encephalomyelitis model of multiple sclerosis)	Lasers, Ibramed™ (Sao Paulo, Brazil)	660 or 904 nm	660 nm: 30 mW, CW, beam area of 0.06 cm^2, 10 J/cm^2, 20 s for each position for 30 days 904 nm: 70 W, 10 J/cm^2, pulsed regime (time of pulse 60 ns) and beam area of 0.10 cm^2, for 30 days	At 6 points located 0.5 cm distance between the points on the spinal cord	Reduced the clinical score of experimental autoimmune encephalomyelitis and delayed the disease onset, and prevented weight loss induced by immunization; down-regulated NO levels in the CNS; no effect on the lipid peroxidation and antioxidant defense; blocked neuroinflammation by decreasing inflammatory markers of IL-17, IFN-γ, and IL-1β and through a reduction of inflammatory cells in the CNS, especially lymphocytes, as well as preventing demyelination in the spinal cord
Mathangi et al. [47]	Rat (fluorescent light-induced brain damage model)	LEDs	670 nm	25 mW/cm^2, 9 J/cm^2, 6 min; 1, 15, or 30 days irradiations	Whole-body irradiation	Resulted in time-dependent variations in the enzyme activities of Na$^+$-K$^+$ ATPase and Ca^{2+} ATPase, with maximum response in 30 daily irradiation group; 15 and 30 daily irradiations increased CCO activity as compared to both non-treatment and once-daily treatment groups
Arias et al. [48]	Rat (minimal hepatic encephalopathy model)	LEDs, WARP 10, Quantum Devices (Barneveld, WI, USA)	670 nm	50 mW/cm^2, 9 J/cm^2, 3 min, irradiation area of 10 cm^2, once a day for 7 days	Transcranially; on the midline of the dorsal surface of the animals shaved head in the region between the eyes and ears	Improved spatial memory as assessed by MWM test, modulated cytochrome oxidase activity in various brain regions including frontal, septal, and temporal regions

(continued)

Table 14.1 (continued)

Study/year	Animal/model	Light source	Wavelengths	Irradiation parameters	Irradiation approach/sites	Findings
Romeo et al. [49]	Mouse (Fluorescent light-induced neurodegeneration model)	LEDs, Roithner Laser Technik GmbH (Austria)	710 nm	325 µW/cm^2, once daily for 3 months	Whole-body irradiation	No effect on the dopamine neurons in the substantia nigra; no effect on the contents of dopamine and its metabolites in the striatum; no effect on the mesencephalic cell viability; modulated the firing activity of extracellular-recorded dopamine neurons
Khongrum and Wattanathorn [50]	Rat (Valproic Acid Model of Autism)	Laser, Xinland International Limited (Xi'an, Shaanxi, China)	405 nm	100 mW, 0.1 J/s, once daily from postnatal day 14 to day 40	Laser acupuncture: at HT7 acupoint on both the left and the right sides	Decreased oxidative stress, IL-6 expression, and GABA-T activity but increased the expressions of GAD 65 kDa together with the density of Purkinje cells in the cerebellum
Ghanbari et al. [51]	Rat (methanol-induced retinal toxicity model)	LEDs	670 nm	28 mW/cm^2, 4 J/cm^2, 144 s, once daily for one week	Whole-body irradiation	Reduced retinal ganglion cell death; increased the number of BDNF positive cells in the visual cortex; decreased cell death (caspase 3 + cells) and reduced the NO levels, both in serum and brain tissue
Ghanbari et al. [52]	Rat (methanol-Induced neurotoxicity)	LEDs	670 nm	28 mW/cm^2, 4 J/cm^2, 144 s, once daily for one week	Whole-body irradiation	Improved memory function as increasing the latency period at 7, 14 and 28 days; decreased cell death (caspase 3 + cells) and cell edema at 7 and 28 days; increased the number of glial fibrillary acid protein astrocytes; increased the number of (BDNF + cells, but also circulating serum BDNF, at 7 and 28 days; increased the number of Ki-67 + cells and BDNF levels in the serum and hypothalamus

(continued)

Table 14.1 (continued)

Study/year	Animal/model	Light source	Wavelengths	Irradiation parameters	Irradiation approach/sites	Findings
Duarte et al. [53]	Mouse (cuprizone-induced demyelination model)	Laser, Photon Laser (DMC Equipamentos Ltda®)	808 nm	50 mW, 1.78 mW/cm², 36 J/cm², 0.028 cm² spot area, 20 s, 6 irradiations during 3 consecutive days	Transcranially: point equidistant between the eyes and ears of the animal	Improved motor coordination as assessed by rotarod test; attenuated demyelination and increased number of oligodendrocyte precursor cells; modulated microglial and astrocytes activation, and decreased toxicity by cuprizone
Salehpour et al. [54]	Mouse (sleep-deprivation-induced cognitive impairment)	Laser, Thor Photomedicine (Chesham, UK)	810 nm	200 mW, 4.75 W/cm², 8 J/cm² (at cortex), PW at 10 Hz with duty cycle of 88%, once daily for 3 days	Transcranially; covered the entire brain	Improved spatial and episodic-like memories; enhanced the hippocampal antioxidant status by decreasing MDA levels and increasing SOD and GPx activities and increasing TAC; increased number of active mitochondria and decreased ROS levels in hippocampus
Wang et al. [55]	Rat (acute myocardial infarction-induced ventricular arrhythmia)	LEDs, Conver-gence Technology (Wuhan, China)	610 nm	1.7 mW/cm², 2.0 J/cm², 3 h and 30 min	Transcranially; at the body surface of the hypothalamic paraventricular nucleus	Reduced the incidence of acute myocardial infarction-induced ventricular arrhythmia; reduced the left stellate ganglion neural activity; attenuated microglial activation and reduced IL-18, IL-1β and NGF expression in the peri-infarct myocardium

Abbreviations: ATPase—Adenosine triphosphatase; Bax—Bcl-2-associated X protein; BDNF—Brain-derived neurotropic factor; CCO—Cytochrome c oxidase; CNS—Central nerves system; CW—Continuous wave; GABA-T—γ-Aminobutyric acid-transaminase; GAD—Glutamic acid decarboxylase; GPx—Glutathione peroxidase; IL—Interleukin; INF-γ—Interferon-gamma; iNOS—Inducible nitric oxide synthase; LEDs—Light-emitting diodes; MDA—Malondialdehyde; MWM—Morris water maze; NGF—Nerve growth factor; NO—Nitric oxide; PBM—Photobiomodulation; PW—Pulsed wave; ROS—Reactive oxygen species; SOD—Superoxide dismutase; TAC—Total antioxidant capacity; TNF-α—Tumor necrosis factor-α

References

1. Tononi, G., and C. Koch. 2015. Consciousness: Here, there and everywhere? *Philosophical Transactions of the Royal Society B: Biological Sciences* 370 (1668): 20140167.
2. Dehaene, S., H. Lau, and S. Kouider. 2017. What is consciousness, and could machines have it? *Science* 358 (6362): 486–492.
3. Di Perri, C., et al. 2014. Functional neuroanatomy of disorders of consciousness. *Epilepsy & Behavior* 30: 28–32.
4. Edelman, G.M., J.A. Gally, and B.J. Baars. 2011. Biology of consciousness. *Frontiers in psychology* 2: 4.
5. Multi-Society Task Force on PVS. 1994. Medical aspects of the persistent vegetative state. *New England Journal of Medicine* 330 (21): 1499–1508.
6. Laureys, S., et al. 2010. Unresponsive wakefulness syndrome: A new name for the vegetative state or apallic syndrome. *BMC Medicine* 8 (1): 68.
7. Monti, M.M., S. Laureys, and A.M. Owen. 2010. The vegetative state. BMJ 341: c3765.
8. Bender, A., et al. 2015. Persistent vegetative state and minimally conscious state: A systematic review and meta-analysis of diagnostic procedures. *Deutsches Ärzteblatt International* 112 (14): 235.
9. Maurer-Karattup, P., et al. 2010. Diagnostik von Bewusstseinsstörungen anhand der deutschsprachigen Coma Recovery Scale-Revised (CRS-R). *Neurol Rehabil* 16 (5): 232–246.
10. Hesse, S., C. Werner, and M. Byhahn. 2015. Transcranial low-level laser therapy may improve alertness and awareness in traumatic brain injured subjects with severe disorders of consciousness: A case series. *International Archives of Medicine* 8.
11. Werner, C., M. Byhahn, and S. Hesse. 2016. Non-invasive brain stimulation to promote alertness and awareness in chronic patients with disorders of consciousness: Low-level, near-infrared laser stimulation versus focused shock wave therapy. *Restorative Neurology and Neuroscience* 34 (4): 561–569.
12. Nawashiro, H., et al. 2012. Focal increase in cerebral blood flow after treatment with near-infrared light to the forehead in a patient in a persistent vegetative state. *Photomedicine and Laser Surgery* 30 (4): 231–233.
13. Goldenberg, M.M. 2012. Multiple sclerosis review. *P & T: A Peer-Reviewed Journal for Formulary Management* 37 (3): 175–184.
14. Olek, M. 2011. Epidemiology, risk factors and clinical features of multiple sclerosis in adults. *Epidemiology and Clinical Features of Multiple Sclerosis in Adults* 31.
15. Koch-Henriksen, N., and P.S. Sørensen. 2010. The changing demographic pattern of multiple sclerosis epidemiology. *The Lancet Neurology* 9 (5): 520–532.
16. Kurtzke, J.F. 1965. Medical facilities and the prevalence of multiple sclerosis. *Acta Neurologica Scandinavica* 41 (5): 561–580.
17. Ouyang, W., J.K. Kolls, and Y. Zheng. 2008. The biological functions of T helper 17 cell effector cytokines in inflammation. *Immunity* 28 (4): 454–467.
18. Duddy, M., et al. 2007. Distinct effector cytokine profiles of memory and naive human B cell subsets and implication in multiple sclerosis. *The Journal of Immunology* 178 (10): 6092–6099.
19. Kasper, L.H. and J. Shoemaker. 2010. Multiple sclerosis immunology: the healthy immune system vs the MS immune system. *Neurology* 74 (1 Supplement 1): S2–S8.
20. Ghasemi, N., S. Razavi, and E. Nikzad. 2017. Multiple sclerosis: Pathogenesis, symptoms, diagnoses and cell-based therapy. *Cell Journal* 19 (1): 1–10.
21. Marrie, R.A., et al. 2015. The incidence and prevalence of psychiatric disorders in multiple sclerosis: A systematic review. *Multiple Sclerosis Journal* 21 (3): 305–317.

22. Gholamzad, M., et al. 2019. A comprehensive review on the treatment approaches of multiple sclerosis: Currently and in the future. *Inflammation Research* 68 (1): 25–38.
23. Tintore, M., A. Vidal-Jordana, and J. Sastre-Garriga. 2018. Treatment of multiple sclerosis— Success from bench to bedside. *Nature Reviews Neurology* 1.
24. Marrie, R.A., et al. 2015. A systematic review of the incidence and prevalence of comorbidity in multiple sclerosis: Overview. *Multiple Sclerosis Journal* 21 (3): 263–281.
25. Muili, K.A., et al. 2012. Amelioration of experimental autoimmune encephalomyelitis in C57BL/6 mice by photobiomodulation induced by 670 nm light. *Plos One* 7 (1): e30655.
26. Dutta, R., and B.D. Trapp. 2007. Pathogenesis of axonal and neuronal damage in multiple sclerosis. *Neurology* 68 (22 suppl 3): S22–S31.
27. Dutta, R., et al. 2006. Mitochondrial dysfunction as a cause of axonal degeneration in multiple sclerosis patients. *Annals of Neurology* 59 (3): 478–489.
28. Muili, K.A., et al. 2013. Photobiomodulation induced by 670 nm light ameliorates MOG35-55 induced EAE in female C57BL/6 mice: A role for remediation of nitrosative stress. *Plos One* 8 (6): e67358.
29. Tolentino, M., and J. Lyons. 2019. Photobiomodulation for multiple sclerosis in animal models. In *Photobiomodulation in the brain*, 241–251. Elsevier.
30. Gonçalves, E.D., et al. 2016. Low-level laser therapy ameliorates disease progression in a mouse model of multiple sclerosis. *Autoimmunity* 49 (2): 132–142.
31. Silva, T., et al. 2020. Effects of photobiomodulation on interleukin-10 and nitrites in individuals with relapsing-remitting multiple sclerosis–Randomized clinical trial. *Plos One* 15 (4): e0230551.
32. Tolentino, M., C.C. Cho, and J.-A. Lyons. 2019. Photobiomodulation therapy (PBMT) regulates the production of IL-10 and IFN-γ by peripheral blood mononuclear cells (PBMC) and CD4+ T cells isolated from subjects with multiple sclerosis (MS). *The American Association of Immunologists.*
33. Tolentino, M., C.C. Cho, and J.-A. Lyons. 2020. Photobiomodulation (PBM) regulates nitric oxide (NO) production by peripheral blood mononuclear cells (PBMC) isolated from multiple sclerosis (MS) patients.*The American Association of Immunologists.*
34. Elliott, K., et al. 1995. N-methyl-D-aspartate (NMDA) receptors, mu and kappa opioid tolerance, and perspectives on new analgesic drug development. *Neuropsychopharmacology* 13 (4): 347–356.
35. Zhu, H., and G.A. Barr. 2001. Opiate withdrawal during development: Are NMDA receptors indispensable? *Trends in Pharmacological Sciences* 22 (8): 404–408.
36. Craig, M.M., and D. Bajic. 2015. Long-term behavioral effects in a rat model of prolonged postnatal morphine exposure. *Behavioral Neuroscience* 129 (5): 643.
37. Koob, G.F., F. Bloom, and D. Kupfer. 2012. Animal models of drug addiction. In *Psychopharmacology: The fourth generation in progress/Bloom FE, Kupfer DJ—1995*, 345 p.
38. Crossland, J., and K. Ahmed. 1984. Brain acetylcholine during morphine withdrawal. *Neurochemical Research* 9 (3): 351–366.
39. Mirzaii-Dizgah, I., et al. 2009. Attenuation of morphine withdrawal signs by low level laser therapy in rats. *Behavioural Brain Research* 196 (2): 268–270.
40. Ojaghi, R., et al. 2014. Role of low-intensity laser therapy on naloxone-precipitated morphine withdrawal signs in mice: Is nitric oxide a possible candidate mediator? *Lasers in Medical Science* 29 (5): 1655–1659.
41. Ngugi, A.K., et al. 2010. Estimation of the burden of active and life-time epilepsy: A meta-analytic approach. *Epilepsia* 51 (5): 883–890.
42. Pedersen, M., et al. 2008. Cinnamamides from Piper capense with affinity to the benzodiazepine site on the GABAA receptor. *Planta Medica* 74 (09): PG34.

43. Sloviter, R.S., and D.W. Dempster. 1985. "Epileptic" brain damage is replicated qualitatively in the rat hippocampus by central injection of glutamate or aspartate but not by GABA or acetylcholine. *Brain Research Bulletin* 15 (1): 39–60.

44. Radwan, N.M., et al. 2009. Effect of infrared laser irradiation on amino acid neurotransmitters in an epileptic animal model induced by pilocarpine. *Photomedicine and Laser Surgery* 27 (3): 401–409.

45. Tsai, C.M., S.F. Chang, and H. Chang. 2020. Transcranial photobiomodulation attenuates pentylenetetrazole-induced status epilepticus in peripubertal rats. *Journal of Biophotonics* 13 (8): e202000095.

46. Ahmed, N.A.E.H., et al. 2008. Effect of three different intensities of infrared laser energy on the levels of amino acid neurotransmitters in the cortex and hippocampus of rat brain. *Photomedicine and Laser Surgery* 26 (5): 479–488.

47. Mathangi, D., and R. Shyamala. 2016. Effect of LED photobiomodulation on fluorescent light induced changes in cellular ATPases and Cytochrome c oxidase activity in Wistar rat. *Lasers in Medical Science* 31 (9): 1803–1809.

48. Arias, N., M. Méndez, and J.L. Arias. 2016. Low-light-level therapy as a treatment for minimal hepatic encephalopathy: Behavioural and brain assessment. *Lasers in Medical Science* 31 (8): 1717–1726.

49. Romeo, S., et al. 2017. Fluorescent light induces neurodegeneration in the rodent nigrostriatal system but near infrared LED light does not. *Brain Research* 1662: 87–101.

50. Khongrum, J., and J. Wattanathorn. 2017. Laser acupuncture at HT7 improves the cerebellar disorders in valproic acid-rat model of autism. *Journal of Acupuncture and Meridian Studies* 10 (4): 231–239.

51. Ghanbari, A., et al. 2017. Light-emitting diode (LED) therapy improves occipital cortex damage by decreasing apoptosis and increasing BDNF-expressing cells in methanol-induced toxicity in rats. *Biomedicine & Pharmacotherapy* 89: 1320–1330.

52. Ghanbari, A., et al. 2018. Light-emitting diode (LED) therapy attenuates neurotoxicity of methanol-induced memory impairment and apoptosis in the hippocampus. *CNS & Neurological Disorders-Drug Targets (Formerly Current Drug Targets-CNS & Neurological Disorders)* 17 (7): 528–538.

53. Duarte, K.C.N., et al. 2018. Low-level laser therapy modulates demyelination in mice. *Journal of Photochemistry and Photobiology B: Biology* 189: 55–65.

54. Salehpour, F., et al. 2018. Transcranial near-infrared photobiomodulation attenuates memory impairment and hippocampal oxidative stress in sleep-deprived mice. *Brain Research* 1682: 36–43.

55. Wang, S., et al. 2019. Noninvasive LED therapy: A novel approach for post-infarction ventricular arrhythmias and neuro-immune modulation. *Journal of Cardiovascular Electrophysiology*.

56. Moges, H., et al. 2009. Light therapy and supplementary Riboflavin in the SOD1 transgenic mouse model of familial amyotrophic lateral sclerosis (FALS). *Lasers in Surgery and Medicine: The Official Journal of the American Society for Laser Medicine and Surgery* 41 (1): 52–59.

57. Khongrum, J., and J. Wattanathorn. 2015. Laser acupuncture improves behavioral disorders and brain oxidative stress status in the valproic acid rat model of autism. *Journal of Acupuncture and Meridian Studies* 8 (4): 183–191.

Photobiomodulation for Brain Function in Healthy Young and Aging Adults

15.1 Cognitive Boosting in Healthy Young and Aged Adults

Cognitive impairment may accompany brain aging in otherwise healthy individuals. The decline may result from an alteration in cholinergic innervations and decreased acetyl-cholinergic tonus, triggering a number of downstream pathways including, oxidative stress, apoptosis, excitotoxicity, amyloid-β toxicity, neuroinflammation, and disturbances in neurotrophic factors. The net result of these changes is impaired memory, learning, and attention, often called age-induced cognitive impairment [1]. Nevertheless, a significant level of cognitive reserve continues to be preserved during aging [2]. An extensive body of literature has suggested the possibility of cognitive enrichment/boosting in young subjects, as well as improving cognitive capacity in later life [3]. These improvements can involve sensory, motor, cognitive, and social functions, as well as the reduction of stress [4]. Recent evidence suggests that lifestyle modifications such as cognitive exercise/training and social enrichment- or cognitive boosting- can improve cognitive function in elderly individuals [5, 6]. Moreover, the role of nutritional supplementation or "brain foods" has also been shown to be effective in the prevention of age-induced cognitive decline [7].

However, to understand the strategies to enhance cognitive performance in later life, we should examine the factors that may affect age-dependent cognitive impairment. According to a comprehensive review [8], it was proposed that education, activity, healthy blood pressure, and apolipoprotein E genotype are among the factors which can predict cognitive decline in later life; some are modifiable and others are non-modifiable. Accordingly, lifestyle changes and activity directed toward improving these factors in healthy individuals will increase their cognitive reserve as they get older. Both cognitive and physical training can also enhance cognitive function [9]. Life-long physical activity and cognitive enrichment have been shown to decrease the risk of age-related memory decline [10]. Several studies have shown that environmental enrichment using physical activity boosted hippocampus-dependent spatial memory consolidation in aged rodents, which is thought to be a direct effect of increases in the pre-synaptic marker protein synaptophysin, and cholinergic neurons in this region of the brain [11, 12]. Cognitive enrichment has also been found to have beneficial effects in later life. One study conducted on elderly participants aged between 65 and 94 assessed the effects of cognitive enrichment on domains, such as memory, reasoning, and speed of processing. When the training was continued for two years it showed a further enhancement in the cognitive performance [13].

It has been proposed that cognitive boosting, exercise, and nutrients can improve mitochondrial energy production, which plays a crucial role in synaptic plasticity and neuronal function. Indeed, ATP produced by mitochondria increases the production/release of brain-derived neurotrophic factor (BDNF) [14] and insulin-like growth factor 1 (IGF1), which then protects synapses against damage [15, 16]. IGF1 has also been shown to be involved in neurogenesis and memory consolidation [17, 18]. In addition, these interventions can increase the expression of silent information regulator 1 (SIRT1) which then contributes to a decrease in reactive oxygen species (ROS) and improved cognition by epigenetic regulation [16]. Exercise and brain foods together can enhance the expression of uMtCK, AMPK and UCP2 genes, which also improve brain plasticity [19].

15.2   Brain Photobiomodulation in Healthy Animals

15.2.1 Photobiomodulation for Healthy Young Animals

Several laboratory studies have reported the regulation of brain neurotransmitter levels after PBM therapy using both red and NIR light in healthy animals [20–25]. In 1982, Shu-Zhi and Li-Hua [21] inserted a fiber-optic into the intact rat caudate nucleus or frontal cortex, and compared the effects of 632.8 nm and 337.1 nm lasers on the levels of striatal neurotransmitter aminoacids. They showed that 632.8 nm laser irradiation to the caudate nucleus could decrease striatal dopamine, serotonin, aspartic acid, and glutamic acid, while it increased the striatal GABA content. However, 632.8 nm laser irradiation to the frontal area reduced serotonin and its metabolites, while it increased aspartic acid, glutamic acid, and GABA levels. Irradiation of the frontal area with 337.1 nm laser significantly decreased levels of norepinephrine, serotonin, and its metabolites, while it increased aspartic acid and GABA levels. Moreover, 337.1 nm laser irradiation to the caudate nucleus significantly decreased levels of dopamine, serotonin and its metabolites, while it increased only aspartic acid levels in the striatum.

Ahmed et al. [22] also investigated the effects of laser PBM on brain neurotransmitters in healthy rats. According to their data, low-power (90 mW) 830 nm laser irradiation showed a neurosuppressive effect on axonal conduction of cortical tissue, decreasing cortical glutamate, aspartate, and glutamine, while increasing glycine levels. It has also been shown that intracranial PBM using 840 nm pulsed laser into the rat intact brain could decrease glutamate and increase dopamine levels in the striatum [23].

Uozumi et al. carried out an in vivo study in healthy mice [26] using transcranial 808 nm laser PBM (10.6 W/cm^2). PBM was able to increase cortical nitric oxide (NO) levels (by 50%) immediately after starting the irradiation. PBM also gradually improved cerebral blood flow in the laser-exposed hemisphere (by 30%) and the opposite hemisphere (by 19%) 45 min after starting the irradiation. They suggested that the increase in transient cerebral blood flow after PBM most likely depended on nitric oxide synthase (NOS) activity and the levels of NO in the brain.

In 2012, Rojas et al. [27] conducted a series of experiments on healthy rats to investigate various neurobehavioral and neurochemical changes following transcranial 660 nm LED PBM therapy. In the first experiment, the in situ oxygen concentration in the prefrontal cortex (PFC) was assessed immediately following PBM therapy, and it was found that 660 nm light induced a dose-dependent increase in oxygen consumption by 5% after 1 J/cm^2 and 15.8% after 5 J/cm^2. In the second experiment, only PBM therapy using a fluence of 10.9 J/cm^2 (but not higher doses) significantly improved extinction memory as compared to control animals. Their third experiment indicated that PBM therapy could decrease fear renewal and prevent the re-emergence of extinguished conditioned fear responses. Altogether, behavioral data from the second and third experiments suggested a dose-dependent PBM increase in neural processes facilitating long-lasting extinction memory. In the fourth experiment, a scalp fluence of 10.9 J/cm^2 significantly up-regulated cytochrome c oxidase activity by 13.6% in the PFC, as compared to higher doses of 21.6 and 32.9 J/cm^2.

Basha et al. [28] investigated the effects of PBM therapy on the cellular enzyme activity of brain tissue in healthy rats. Whole-body LED pre-treatment was applied using 670 nm light with a skin fluence of 9 J/cm^2 daily for 1, 15, or 30 days irradiation. Na$^+$–K$^+$ ATPase, Ca^{2+} ATPase, and the levels of cytochrome c oxidase in the brain tissue were measured. LED PBM showed a significant increase in the enzyme activity of Na$^+$–K$^+$ ATPase and Ca^{2+} ATPase at all the time points assessed. In addition, PBM for 30 days resulted in a more than three-fold increase in cytochrome c oxidase levels as compared to the non-treated control group.

According to Tanaka et al. [29], both acute (1 day) and chronic (10 days) application of transcranial NIR PBM to the head of healthy rats also promoted hippocampal neurogenesis as measured by higher levels of BrdU-positive cells in the CA1 region.

15.2.2 Photobiomodulation for Healthy Aged Animals

Typically, in pre-clinical animal studies, rodents with an age of greater than 12 months are considered to be aged animals. Several studies on rodents have demonstrated that PBM can affect cognitive behavior in aged animals. For example, in 2008, Michalikova et al.

[30] showed that pre-conditioning with infrared light could improve cognitive function in 12-month-old female CD-1 mice. Specifically, animals received 6 min of whole body irradiation with 1072 nm light daily for 10 consecutive days. Treated animals made fewer memory errors in the 3D-maze task as compared to sham-treated aged animals, indicating improved working memory. Nevertheless, PBM did not produce a significant effect on the maze exploration activity or anxiety responses. In addition, in 2018, Salehpour et al. [31] tested a possible pro-cognitive effect of transcranial PBM therapy in intact aged mice. Eighteen months old BALB/c mice received 660 nm laser irradiation transcranially with a scalp fluence of 99.9 J/cm^2, daily for 2 weeks. Results from the Barnes maze task showed a significant improvement in spatial memory as indicated by a shorter latency time on the fourth day of the training session, and also the mice spent a longer time in the target quadrant as compared with age-control animals. ATP levels in the hippocampus of the laser-treated aged mice were also significantly higher than those in the age-control animals. However, no statistically significant difference in the locomotor activity in the open-field test was found among all the experimental groups.

El Massri et al. [32] investigated the effects of long-term PBM on the glial and neuronal organization in the striatum nucleus of aged C57BL/6 mice. Mice received PBM pre-conditioning with 670 nm light for 20 min once a day, starting at 5 months old up until 12 months of age. They observed a clear decrease in glial cell numbers, both astrocytes and microglia, in the striatum following 8 months of PBM in aged mice. However, the number of 2 different types of striatal interneurons (parvalbumin$^+$ and encephalopsin$^+$), together with the density of striatal dopaminergic terminals (and their midbrain cell bodies), did not show any significant difference after chronic treatment. Taken together, the authors suggested that long-term PBM could provide positive effects on the aged striatum by decreasing glial cell numbers with no deleterious effects on the striatum nucleus neurons and their terminations.

Besides the natural aging animal models discussed above, there is one experimental study on the effectiveness of PBM therapy in the enhancement of cognitive function in an animal model of artificial aging [33]. In this respect, a study conducted by Salehpour and colleagues [33] revealed a brain-boosting effect of red and NIR PBM therapy in D-galactose-treated BALB/c mice. Brain aging was modled in mice by daily administration of D-galactose (500 mg/kg/subcutaneous) for 6 weeks. For PBM therapy, 660 nm and 810 nm at 2 different fluences (4 and 8 J/cm^2) at 10-Hz pulsed wave mode were applied transcranially for 3 days per week. They found that either 660 nm or 810 nm at 8 J/cm^2 significantly improved D-galactose-impaired spatial and episodic memories. Likewise, a clear enhancement in indices of mitochondrial function (e.g., ATP, ROS, MMP, and cytochrome c oxidase activity) and beneficial changes in apoptotic markers (e.g., Bax, Bcl-2, and caspase-3) were observed following irradiation using either 660 or 810 nm at 8 J/cm^2. Nevertheless, PBM therapy at 4 J/cm^2 (both red and NIR wavelengths) did not provide beneficial effects on the neurobehavioral and molecular measures in this aging model. These findings suggested PBM therapy could be useful to improve learning and memory and attenuate cognitive deficits in aged rodents.

15.3 Photobiomodulation for the Brain in Healthy Human Volunteers

There is solid clinical evidence supporting the effects of transcranial PBM on improving cerebral blood flow in healthy adults [34–36]. As one example, Salgado et al. [37] showed that 4 weeks of transcranial LED PBM to the frontal and parietal areas could improve blood flow velocity of the left middle cerebral and the basilar arteries in healthy elderly women. The improved cerebral hemodynamics during or even after LED irradiation might be explained in part by the release of NO from neuronal cells [26]. It has also been suggested that PBM can act on superficial cerebral blood vessels (or subarachnoid vessels) and alter the elasticity of the cerebrovascular endothelium, at least in part by the release of NO [26].

Transcranial PBM has shown some promising beneficial results for cognitive enhancement in healthy individuals [38–41]. From the year 2013, several pieces of research conducted in Francisco Gonzalez-Lima and Hanli Liu's laboratories have proven that transcranial PBM at a specific NIR wavelength (1064 nm) can increase oxygen consumption and metabolic activity in the frontal cortex area, in turn leading to improved cognitive performance in several domains, such as memory, learning, attention, concentration, and executive function [35, 42]. For instance, in one placebo-controlled randomized study, they used transcranial PBM (1064 nm laser at 250 mW/cm^2 providing a scalp fluence of 60 J/cm^2) applied to the forehead of healthy young subjects to improve the performance of cognitive tasks related to the PFC, such as a psychomotor vigilance task (PVT), a delayed match-to-sample (DMS) memory task, and the positive and negative affect schedule (PANAS-X) to show an improved mood [38]. Another study in normal volunteers showed that 1064 nm transcranial laser irradiation could enhance performance in the Wisconsin Card Sorting Task (considered the gold standard test for executive function) [39].

According to Vargas et al. [43], five weekly sessions of 8 min irradiation using a collimated 1064 nm laser diode to the right PFC could improve cognitive functions and reaction time in healthy older adults. The improvements were observed both in terms of functional magnetic resonance imaging (fMRI) measures, and electroencephalographic power data. The alterations in electroencephalogram (EEG) power reported in their article, specifically higher delta amplitudes are controversial, because these features have been associated with increased dementia symptomology in some reports [44, 45], but not in others, especially if measuring symptoms in the earlier stages of dementia. A recent study conducted by Zomorrodi et al. [44] using a Vielight headset LED device and one intranasal applicator showed substantial initial increases in delta and theta power, with reduced slow-wave amplitudes in the fronto-central regions, and increased alpha, beta, and gamma (40-Hz) activity when active and sham treatments delivered one week apart were compared. It is worth noting that the global measure of connectivity used in their study demonstrated no significant change after transcranial PBM therapy, suggesting that connectivity abnormalities may be more directly improved using operant conditioning procedures, such as neurofeedback training based on real-time fMRI and LORETA EEG [46–48].

Chan et al. [49] carried out a randomized, sham-controlled study to explore the effects of transcranial LED PBM on frontal-related brain functions in older adults. The results showed significantly improved reaction time and category fluency after receiving PBM. Healthy adult cognitive functions are often linked to executive functions such as mental flexibility, discriminative capacity, and inhibitory control, as well as capacity. Research involving older individuals receiving transcranial PBM is essential to assess its clinical utility, as well as the underlying mechanisms of action and safety profiles.

Yao et al. [50] also tested whether and how transcranial NIR LED irradiation with varying frequencies could affect brain activity in healthy young subjects by analyzing the EEG signals during a single PBM therapy session. Forty healthy subjects (aged between 22 and 26 years) were randomly allocated into four groups. Each group received a 30-min 810 nm LED irradiation with 30.65 mW/cm^2 irradiance, 55.17 J/cm^2 fluence, and four different pulse frequencies (0-Hz, 5-Hz, 10-Hz, or 20-Hz) on the forehead. The remaining 10 participants formed the control group, who received a 30-min rest period without any light irradiation. The results showed that with increasing pulse frequency, the EEG gravity frequency increased more rapidly, indicating that brain activity was boosted, whereas the increase in the relative energy of C was larger, suggesting that PBM using higher frequencies (e.g., 10-Hz or 20-Hz) may have better effects on improving memory function. The findings also suggested that the main activation induced by PBM was in the frontal region, and the activation became more significant with an increased pulse frequency [50].

Finally, a recent systematic review and meta-analysis study by Salehpour et al. [35] aimed to address the question of whether transcranial PBM could improve cognitive function in healthy young and older adults. Their results showed a significant improvement in cognitive function in young healthy individuals. Indeed, transcranial PBM significantly improved attention-related outcomes without marked heterogeneity. The heterogeneity of the data obtained from the seven included studies in the memory subdomain was high, which could be explained either by the low number of published articles or the medium quality of the included data. Moreover, due to the limited number of studies carried out on older healthy adults, the data could not be adequately examined and analyzed for the older subjects. It is clear that further studies in this area are required to establish protocols and parameters to optimize the clinical results for various healthy subpopulations.

Table 15.1 Summary of in vivo studies on the effects of photobiomodulation therapy in intact (naïve) animals

Study/year	Animal/age	Light source	Wavelengths	Irradiation parameters	Irradiation approach/sites	Findings
Shen-Zeng et al. [51]	Rat	Laser, He-Ne or nitrogen	632.8 or 337.1 nm	632.8 nm laser: 20 mW, 3 min 337.1 nm laser: 1 mJ/pulse, pulse duration 5 nsec in 10 pulses/sec rhythm, 3 min	Transcranially; fiber optics implanted in frontal region or caudate nucleus	632.8 nm laser: *irradiation to the caudate nucleus:* decreased striatal dopamine, serotonin, aspartic acid, and glutamic acid, and increased striatal GABA levels *irradiation to the frontal region:* decreased serotonin and its metabolites and increased aspartic acid, glutamic acid, and GABA levels 337.1 nm laser: *irradiation to the caudate nucleus:* decreased dopamine and serotonin and its metabolites levels and increased only aspartic acid levels in striatal *irradiation to the frontal region:* decreased norepinephrine and serotonin and its metabolites levels, and in increased aspartic acid and GABA levels

(continued)

Table 15.1 (continued)

Study/year	Animal/age	Light source	Wavelengths	Irradiation parameters	Irradiation approach/sites	Findings
Shen-Zeng et al., [20]	Rat	Laser, He-Ne or nitrogen	632.8 or 337.1 nm	632.8 nm laser: 20 mW, 3 min 337.1 nm laser: 1 mJ/pulse, pulse duration 5 nsec in 10 pulses/sec rhythm, 3 min	Transcranially; fiber optics implanted in frontal region or caudate nucleus	Only 632.8 nm laser to the caudate nucleus increased its excitement, promoted general movement, and facilitated the formation of conditioned avoidance response, along with increased striatal concentrations of dopamine and norepinephrine
Igarashi and Inomata [52]	Neonatal rat	Laser, OhLase-3DI GaAlAs diode laser system, Japan Medical Laser Laboratory (Tokyo)	830 nm	60 mW, 1.9 W/cm2, 15 sec, twice per day from birth (day 1) to day 5	Transcranially; at two points located above the hippocampus	Decreased mean body weight; decreased the density of synaptic junctions of the neonatal rat hippocampus at day 20
Urciuoli et al. [53]	Rat	Laser, He-Ne medical laser supplied by NUOVA VI. TI. EMME Inc. (Turin, Italy)	632.8 nm	5 mW, 24 per spot	Transcranially; at 9 points symmetrically on the sinciput	Increased brain SOD activity
Rochkind [54]	Rat or dog	Laser, (Aerotech Inc.)	632.8 nm	16 mW, spot size 2 mm2, 30 J/cm2 for rats and 70 J/cm2 for dogs, for 21 days	Transcutaneously; on the closed operated wound	Prevented extensive glial scar formation (a limiting factor in CNS regeneration) between neural transplants and host brain; developed abundant capillaries

(continued)

Table 15.1 (continued)

Study/year	Animal/age	Light source	Wavelengths	Irradiation parameters	Irradiation approach/sites	Findings
Guillot Valls et al. [55]	Rat	Laser, Polytect 750 He-Ne laser	632.8 nm	5 mW, 5 min	Direct laser irradiation of the surgically-exposed pineal gland	Increased the medullary and cortical karyometric indices of the pineal body (with highest increase on the 3 days post-irradiation); decreased the karyometric on days 7 and 10 post-irradiation; increased metabolic activity on days 3 and 7, followed by a decrease in activity by day 10 with the appearance of numerous lipid droplets, pericanallicular dark cells and mesoglial cells
Zubkova and Vlikhailik [56]	Rat	Laser, Orion apparatus (Zhiva, Moscow, Russia)	890 nm	4.8 W pulse power, 10 min, PW at 10-Hz with 8-10 sec pulse duration	Transcranially; on the motor zone of the cortex	Activated DNA synthesis in brain cortex, skeletal muscle, and thymus tissues
Mochizuki-Oda et al. [57]	Rat	Laser, Panalas, 1000A; Matsushita Industrial Equipment Co. Ltd. (Osaka, Japan)	652 or 830 nm	4.8 W/cm2, 15 min	Direct laser irradiation of the parietal cortex on both sides (2 mm posterior and 1.5 mm lateral to the bregma)	830 nm laser: increased brain ATP content in the irradiated area, but no effect on the ADP levels 652 nm laser: no effect on either ATP or ADP contents

(continued)

Table 15.1 (continued)

Study/year	Animal/age	Light source	Wavelengths	Irradiation parameters	Irradiation approach/sites	Findings
Ilic et al. [58]	Rat	Laser, Photothera Inc. (San Diego, CA, USA)	808 nm	0.5–5.0 W; 7.5, 75, 375, or 750 mW/cm^2 with a corresponding fluencies of 0.9, 9, 45, and 90 J/cm^{2}2 min, CW or PW at 70-Hz	Transcranially	No long-term difference between laser-treated and control groups for both the scores from standard neurological tests and the histopathological examination up to 70 days post-irradiation; only resulted in adverse neurological effect in the CW mode group (750 mW/cm^2)
Ahmed et al., [22]	Rat	Laser, Laser Diode Driver [LDD 50]; LIMO Laser Systems (Dortmund, Germany); Lasermax, Inc. (Rochester, NY, USA)	808 or 830 nm	808 nm: 190 or 500 mW, 5.555 or 3.16 W/cm^2, 30 or 11.4 J/point 830 nm: 90 mW, 2.87 W/cm^2, 5.4 J/point	Transcranially	808 nm at 500 mW: decreased glutamate, aspartate, and taurine levels in the cortex; decreased GABA levels in the hippocampus 808 nm at 190 mW: increased aspartate and decreased glutamine and taurine levels in the cortex; decreased glutamate, glycine, and taurine levels, and increased glutamine and GABA levels in the hippocampus 830 nm at 90 mW: decreased glutamate, aspartate, and glutamine and increased glycine levels in the cortex; increased glutamate, aspartate, GABA, and taurine levels, and decreased glycine levels in the hippocampus

(continued)

Table 15.1 (continued)

Study/year	Animal/age	Light source	Wavelengths	Irradiation parameters	Irradiation approach/sites	Findings
Michalikova et al., [30]	Middle-age mouse	Laser	1072 nm	6 min/day for 10 days	Whole-body irradiation	Enhanced acquisition of working memory for spatial navigation; no effects on exploratory activity or anxiety responses
McCarthy et al. [59]	Rat	Laser, Photothera Inc. (San Diego, CA, USA)	808 nm	70 mW, 2230 mW/cm^2, 268 J/cm^2 (at scalp), 2 min, CW	Transcranially; to the parietal region right lateral to the sagittal crest	Resulted in no toxicologically effects for any hematologic parameters; resulted in no treatment-related abnormalities or induced neoplasia in brain and pituitary gland histopathology
Uozumi et al., [26]	Mouse	Laser, B&W Tek, Inc. (Newark, DE, USA)	808 nm	0.8, 1.6, or 3.2 W/cm^2, spot diameter of 3 mm, 45 min, one irradiation session, CW	Transcranially; at the left hemisphere, 2 mm posterior to and 3 mm left of the bregma F4)	At 1.6 W/cm^2: increased CBF in the irradiated and opposite hemisphere over the 45 min irradiation; increased cortical NO concentration; decreased neuronal apoptosis induced by transient cerebral ischemia in cerebral cortex and CA1 subfield of dorsal hippocampus

(continued)

Table 15.1 (continued)

Study/year	Animal/age	Light source	Wavelengths	Irradiation parameters	Irradiation approach/sites	Findings
Rojas et al., [27]	Rat	LEDs, LEDtronics, Inc. (Torrance, CA, USA)	660 nm	9 mW/cm^2, 1, 5, 5.4, 10.9, 21.6, or 32.9 J/cm^2; with corresponding duration of 111, 565, 600, 1200, 2400, or 4800 sec, respectively; CW	Transcranially; holding probe at the 1 cm from the dorsal head surface	Enhanced O$_2$ consumption in PFC (at 5.4 J/cm^2); improved extinction of fear-conditioned memories (at 10.9 J/cm^2); decreased renewal of conditioned-fear (at 5.4 J/cm^2); increased CCO activity in PFC (at 10.9 J/cm^2)
Gorshkova et al. [60]	Rat	laser	650 nm	20 mW/cm^2, 5 min	Irradiation of pial vessels	Laser-induced NO mainly affects major arteries and did not contribute to reactivity of small pial arteries and pre-cortical arterioles
Kuo et al., [23]	Rat	Laser	840 nm	2, 5, or 10 mW; 102, 255, or 510 W/cm^2, with corresponding doses of 0.12, 0.3, or 0.6 J, respectively; 1 min	Intracranially	At all three power levels: caused a clear decrease in glutamate concentration in a dose-dependent manner, with a maximal response at 5 mW. At only 10 mW: significantly increased secretion of dopamine levels
Zhang et al. [61]	C57BL/6 mouse young (4 weeks) or old (8-9 months old)	LEDs	670 nm	90 sec, for 5 days	Transcranially	In both young and old mice: increased expression levels of antioxidant superoxide dismutase-1

(continued)

Table 15.1 (continued)

Study/year	Animal/age	Light source	Wavelengths	Irradiation parameters	Irradiation approach/sites	Findings
Mintzopoulos et al. [62]	Dog	Laser, PhotoThera, Inc. (Carlsbad, CA, USA)	808 nm	4–10 mW/cm^2 on the dura mater, 0.48–1.2 J/cm^2 per site (total of 2 sites); spot diameter of 22 mm, 2 min, one session or 3×/week for 2 weeks, CW	Transcranially; at the anterior and posterior cranium midline locations	Improved cerebral bioenergetics via increase of PCr/β-NTP ratios and PCr level (at 2 weeks post-irradiation)
Salehpour et al., [33]	BALB/c mouse Aging model (D-galactose model (500 mg/kg/s.c. once daily for 6 weeks)	Laser	660 or 810 nm	200 mW, 4.75 W/cm^2, 4 or 8 J/cm^2 (at cortex), PW at 10-Hz with duty cycle of 88%, 3 days a week for 6 weeks	Transcranially; covered the entire brain	660 and 810 nm: improved spatial memory in Barnes maze test and episodic-like memory in WWWhich test (at 8 J/cm^2); increased number of vital mitochondria, ATP content, MMP and cytochrome c oxidase activity (at 8 J/cm^2); decreased ROS levels (at 4 and 8 J/cm^2); attenuated apoptosis via decrease of Bax/Bcl-2 and caspase-3 levels (at 8 J/cm^2)
Santos et al. [63]	C57BL/6 mouse	Laser, driver: CLD1010LP; laser LP405-MF80; Thorlabs	405 nm	300 mW/cm^2, 60 sec	Transcranially	Induced neurogenesis in the subventricular zone

(continued)

Table 15.1 (continued)

Study/year	Animal/age	Light source	Wavelengths	Irradiation parameters	Irradiation approach/sites	Findings
El Massri et al., [32]	C57BL/6 mouse old (12 months old)	LEDs	670 nm	20 min, once daily for 8 months	NR	Reduced glial cell number (both astrocytes and microglia) in the striatum; no effect on the number of striatal interneurons (parvalbumin$^+$ and encephalopsin$^+$); no effect on the density of striatal dopaminergic terminals
Salehpour et al., [31]	BALB/c mouse 18 (months old)	Laser	660 nm	200 mW, 6.66 W/cm^2, 99 J/cm^2, 15 sec, once daily for 2 weeks, CW	Transcranially; on the scalp at the bregma	Improved spatial memory in Barnes maze test; no effect on the open field test; increased hippocampal ATP content

Abbreviations: ADP, adenosine 5'-diphosphate; ATP, adenosine triphosphate; Bax, Bcl-2-associated X protein; β-NTP, b-nucleoside triphosphate; CCO, cytochrome c oxidase; CBF, cerebral blood flow; CNS, central nervous system; CW, continuous wave; DNA, deoxyribonucleic acid; GABA, gamma-aminobutyric acid; LEDs, light-emitting diodes; He-Ne, helium-neon; MMP, mitochondrial membrane potential; NO, nitric oxide; NR, not reported; PCr, phosphocreatine; PFC, prefrontal cortex; PW, pulsed wave; ROS, reactive oxygen species; SOD, superoxide dismutase; WWWhich, What-Where-Which task.

Table 15.2 Summary of clinical studies on the effects of photobiomodulation therapy in healthy young and aged brain

Study/year	Subjects (n)/age	Light source	Wavelengths	Irradiation parameters	Irradiation approach/sites	Findings
Litscher et al. [64]	Young volunteers (15) Mean = 25.0 years	Laser, Minilaser 2020F, Helbo-Medizintechnik (Gallspach, Austria)	685 nm	19 mW, 30 s per each acupoints, CW	Laser acupuncture; various acupoints (LI4, S36, B60, B65, B66, and B67)	No effect on the blood flow velocity in the posterior cerebral artery; increased the amplitudes of 40-Hz cerebral oscillations
Litscher et al. [65]	Young volunteers (18) Mean = 25.4 years	Laser, Minilaser 2020F, Helbo-Medizintechnik (Gallspach, Austria)	685 nm	30–40 mW, 4.6 kJ/cm², 20 min per each acupoints, CW	Laser acupuncture; various acupoints (LI4, S36, B60, and B67)	Increased mean blood flow velocity in the posterior cerebral artery measured by fTCD; decreased mean blood flow velocity in the middle cerebral artery; changed brain activity in the occipital and frontal gyrus assessed by fMRI
Siedentopf et al., 2005	Young volunteers (21) Age = 20 to 35 years old	Laser, Minilaser 2020F, Helbo-Medizintechnik (Gallspach, Austria)	670 nm	10 mW	Laser acupuncture; left and right acupoint of GB43	Resulted in ipsilateral brain activations within the thalamus, nucleus subthalamicus, nucleus ruber, the brainstem, and the Brodmann areas 40 and 22
Hsieh et al. [66]	Young volunteers (12) Mean = 27.9 years	Laser	808 nm	30 mW, CW or PW at 10-Hz with 50% duty cycle	Laser acupuncture; one acupoint (Kidney 1, K1)	**CW group:** activated the inferior parietal lobule, the primary somatosensory cortex, the precuneus of left parietal lobe, and medial and superior frontal gyrus of left frontal lobe; resulted in the cerebral hemodynamic responses **PW group:** activated the primary motor cortex and middle temporal gyrus of left hemisphere and bilateral cuneus; resulted in the cerebral hemodynamic responses

(continued)

Table 15.2 (continued)

Study/year	Subjects (n)/age	Light source	Wavelengths	Irradiation parameters	Irradiation approach/sites	Findings
Wu et al. [67]	Young volunteers (40) Mean = 20.55 years	Laser, 6 diodes, Advanced Chips & Products Corp. (Hillside, NJ, USA)	830 nm	7 mW per diode, 20 J/cm², 10 min, PW at 10-Hz with duty cycle of 50%	Laser acupuncture; 1 site, left palm	Increased amplitude power of alpha rhythms and theta waves mainly in posterior head regions; decreased amplitude power of beta activities in anterior head regions
Quah-Smith et al. [68]	Adult volunteers (16) Mean = 48.2 years	Laser, Moxla prototype fibreoptic infrared light laser Eurypheassa AB (Stockholm, Sweden)	808 nm	20 mW	Laser acupuncture; medial knee acupoint LR8	Resulted in significant activation in the left precuneus; resulted in some transient side effects such as tiredness and dizziness, vagueness, and nausea
Wu et al. [69]	Young volunteers (20) Mean = 21.0 years	Laser, 6 diodes, Advanced Chips & Products Corp. (Hillside, NJ, USA)	830 nm	7 mW per diode, 20 J/cm², 10 min, PW at 10-Hz with duty cycle of 50%	Laser acupuncture; 1 site, left palm with closed eyes	Resulted in markedly decrease in alpha band and less pronounced decrease in delta band of EEG; decreased beta band in the right occipital area
Raith et al. [70]	Term and preterm neonates (20) <28 weeks gestation were included	Laser, laser needle EG GmbH (Germany)	685 nm	10 mW, 1.5 J/cm² per acupoint, 5 min, CW	Laser acupuncture; right and left acupoint of LI 4 (Hegu)	Decreased regional cerebral oxygen saturation (rSO₂) accompanied by an increase in cerebral fractional tissue oxygen extraction (cFTOE); no effect on the heart rate and peripheral oxygen saturation values
Konstantinović et al. [71]	Young volunteers (14) Mean = 35.0 years	Laser, Endolaser 476, Enraf Nonius (Rotterdam, Netherlands)	905 nm	50 mW/cm², 3 J/cm² per site, 60 s, PW at 3000-Hz	Transcranially; 5 sites, over the primary motor cortex (M1) area, centered at the hot-spot for the FDI muscle	Transitory reduction of the excitability in the motor cortex

(continued)

Table 15.2 (continued)

Study/year	Subjects (n)/age	Light source	Wavelengths	Irradiation parameters	Irradiation approach/sites	Findings
Barrett and Gonzalez-Lima [72]	Young volunteers (40) Age = 18 to 35 years old	Laser, CG-5000, HD Laser Center (Dallas, TX, USA)	1064 nm	250 mW/cm², 60 J/cm², 4 min, one irradiation session, CW	Transcranially; 2 sites, unilateral (right frontal pole on 4 cm medial and lateral)	Improved reaction time in Psychomotor Vigilance Task and performance in a delayed match-to-sample memory task; appeared sustained positive emotional states 2 weeks post-irradiation
Chaieb et al. [73]	Young volunteers (55) Age = 18 to 35 years old	Laser, coupled with Weber Medical acupuncture	810 nm	500 mW/cm² at scalp via 4 laser needles, 10 min, one irradiation session, CW	Transcranially; 4 sites, over the primary motor cortex (M1) area	Decreased amplitude of motor-evoked-potentials; increased short interval cortical inhibition and decreased facilitation
Salgado et al. [37]	Elderly women volunteers (25) Mean = 72.3 years	LEDs	627 nm	0.2 W, 70 mW/cm², 10 J/cm², 30 s, twice per week for 4 weeks	Transcranially; 4 points on the frontal and parietal regions	Increase the blood flow velocity of the left middle cerebral (by 30%) and the basilar (by 25%) arteries
Blanco et al. [74]	Young volunteers (30) Mean = 20.4 years	Laser, Cell Gen Therapeutics, LLC (Dallas, TX, USA)	1064 nm	250 mW/cm², 60 J/cm², 8 min, one irradiation session, CW	Transcranially; 2 sites, (lower and upper portion of right lateral forehead at EEG map sites: Fp2, F4)	Improved executive function assessed by Wisconsin Card Sorting Task (WCST)
Shi et al. [75]	Young volunteers (20) Mean = 21.9 years	Laser, Laserneedle EG (Wehrden, Germany)	658 nm	40 mW, 5 min, CW	Laser acupuncture; various acupoints (GV26, LU11, SP1, PC7, BL62, GV16, ST6, CV24, PC8, GV23, and LI11)	Increased ReHo/ALFF value in the brain various regions including right inferior occipital gyrus, bilateral middle occipital gyrus, left inferior temporal gyrus, and bilateral postcentral gyrus; decreased ReHo/ALFF value in the cerebellum, left superior medial frontal gyrus, left precuneus, right middle temporal gyrus, right middle frontal gyrus and left superior parietal lobule

(continued)

Table 15.2 (continued)

Study/year	Subjects (n)/age	Light source	Wavelengths	Irradiation parameters	Irradiation approach/sites	Findings
Hwang et al. [41]	Young volunteers (60) Age = 18 to 30 years old	Laser, CG-5000, Cell Gen Therapeutics, LLC (Dallas, TX, USA)	1064 nm	250 mW/cm^2, 60 J/cm^2, 8 min, one irradiation session, CW	Transcranially; 2 sites, medial and lateral right forehead	Improved sustained attention in the Psychomotor Vigilance Task (PVT) and working memory in the delayed match-to-sample task (DMS)
Tian et al. [76]	Adult volunteers (12) Age = 18 to 40 years old	Laser, CG-5000, Cell Gen Therapeutics (Dallas, TX, USA)	1064 nm	250 mW/cm^2, 13.75 J/cm^2 per 1 min for 10 min, one irradiation session, CW	Transcranially; 2 sites, center of forehead (aimed at medial frontal lobes bilaterally) or right side of forehead (aimed at right lateral frontal lobe)	Increased oxygenated hemoglobin concentration and decreased deoxygenated hemoglobin concentration in both cerebral hemispheres over time during irradiation (10 min) and post-irradiation (6 min)
Blanco et al. [77]	Young volunteers (118) Age = 17 to 35 years old	Laser, CG-5000, Cell Gen Therapeutics, LLC (Dallas, TX, USA)	1064 nm	250 mW/cm^2, 60 J/cm^2, 8 min, one irradiation session, CW	Transcranially; 2 sites, lower and upper portion of right lateral forehead at EEG map sites: Fp2, F4, and F8 sites	Improved prefrontal rule-based learning; no significant effects on information-integration learning
Wang et al. [34]	Young volunteers (11) Mean = 31.0 years	Laser, CG-5000, Cell Gen Therapeutics, LLC (Dallas, TX, USA)	1064 nm	250 mW/cm^2, 13.75 J/cm^2 per 1 min for 8 min, one irradiation session, CW	Transcranially; 1 site, right forehead	Increased cerebral concentrations of oxidized CCO, oxygenated and total hemoglobin during and post-irradiation
Wang et al. [78]	Young volunteers (11) Mean = 31.0 years	Laser, CG-5000, Cell Gen Therapeutics, LLC (Dallas, TX, USA)	1064 nm	250 mW/cm^2, 13.75 J/cm^2 per 1 min for 8 min, one irradiation session, CW	Transcranially; 1 site, right forehead	Increased differential hemoglobin (HbD), total hemoglobin (HbT), oxyhemoglobin (HbO) as well as oxidized CCO concentrations

(continued)

Table 15.2 (continued)

Study/year	Subjects (n)/age	Light source	Wavelengths	Irradiation parameters	Irradiation approach/sites	Findings
Grover et al. [79]	Adult volunteers (31) Age = 16 to 65 years old	LEDs, LumiWave Infrared Light Therapy Device, BioCare Systems, Inc. (Parker, CO, USA)	903 nm	16.67 mW/cm^2, 20 J/cm^2 at skull, 20 min, CW	Transcranially; Multiple areas in the occipital, left temporal, and right temporal lobes above the ear line, as well as the frontal and parietal lobes	Improved reaction time in qEEG event-related response test
Nawashiro et al. [80]	Adult volunteers (5) Age = 32 to 54 years old	Laser, DioDent Micro 810, HOYA ConBio (Fremont, CA, USA)	810 nm	3.5 W, 518–599 mW, 204–236 mW/cm^2, 55–63 J/cm^2, 270 s	Transcutaneously; 2 sites, right and left lateral forehead (at EEG map site: F3 and F4)	Increased regional CBF as measured by blood-oxygen-level-dependent (BOLD) fMRI; the changes were dominant in the dorsolateral PFC just beneath the tip of the fiber and widespread brain regions such as ipsilateral parietal cortex
Moghadam et al. [81]	Young volunteers (39) Age = 18 to 24 years old	LEDs, multi-LED array source with 20 cells, Iranbargh (Tehran, Iran)	850 nm	285 mW/cm^2, 60 J/cm^2, 2.5 min, CW	Transcranially; 1 site, right frontal pole of the cortex (at EEG map site: Fp2)	Improved attentional performance in Level-1 of parametric Go/No-task
Vargas et al. [43]	Older adults (21) Age = 49 to 90 years old	Laser, CG-5000, Cell Gen Therapeutics, LLC (Dallas, TX, USA)	1064 nm	3.4 W, 250 mW/cm^2, 120 or 137.5 J/cm^2 per session, 240 s, CW, once a week for 5 weeks	Transcranially; 2 sites on the forehead at 4.2-cm diameter medial site and 4.2-cm diameter lateral site (at EEG map site: Fp2)	Improved reaction time and lapses in psychomotor vigilance task and correct responses in delayed match to sample task; increased resting-state EEG alpha, beta, and gamma power; promoted more efficient prefrontal blood-oxygen-level dependent-fMRI response
Litscher [82]	Adult volunteer (1)	Laser, prototype helmet (total of 360 diodes) from Weber Medical (Lauenförde, Germany)	810 nm	10 mW per diode, 20 min, one session	Transcranially; whole head irradiation with laser helmet	Increased regional oxygen saturation (rSO$_2$)

(continued)

Table 15.2 (continued)

Study/year	Subjects (n)/age	Light source	Wavelengths	Irradiation parameters	Irradiation approach/sites	Findings
Litscher [83]	Adult volunteer (1)	LEDs, prototype helmet (total of 256 diodes) from Suyzeko, Shenzhen Guangyang Zhongkang Technology Limited (China)	810 nm	60 mW per diode, 24 mW/cm², 15 min, one session	Transcranially; whole head irradiation with LEDs helmet	Increased the regional cerebral oxygen saturation during and after irradiation
Holmes et al. [84]	Young volunteers (34) Mean = 31.0 years ara>	Laser, CG-5000, Cell Gen Therapeutics, LLC (Dallas, TX, USA)	1064 nm	3.4 W, 250 mW/cm², 120 J/cm² per session; 8 min, CW, one session	Transcranially; 1 site, right frontal pole of the cortex (at EEG map site: Fp2)	Improved cognitive performance on the DMS; increased differential hemoglobin (HbD), total hemoglobin (HbT), oxyhemoglobin (HbO) concentrations on the anterior frontal cortex during performing cognitive tasks of PVT and DMS; resulted in a large increase in oxygenated hemoglobin in the anterior frontal region measured by the fNIRS
El Khoury et al. [85]	Young volunteers (24) Mean = 32.0 years	LEDs, alpha device, Vielight, Inc. (Toronto, Canada)	810 nm	100 mW, 100 mW/cm², 20 min, PW at 10-Hz with duty cycle of 50%, one session irradiation	Transcranially; multiple areas, bilateral mesial prefrontal cortex, precuneus/posterior cingulate cortex, angular gyrus (correspond to Fpz, Cz, T3 and T4 EEG points)	No effect on cerebral blood flow and global resting-state brain activity (task-negative); resulted in a clear reduction in evoked brain activity after finger-tapping (task-positive); putamen, primary somatosensory and parietal association cortex were most affected brain regions following PBM

(continued)

Table 15.2 (continued)

Study/year	Subjects (n)/age	Light source	Wavelengths	Irradiation parameters	Irradiation approach/sites	Findings
Zomorrodi et al. [44]	Older adults (20) Mean = 68.0 years	LEDs, Neuro Gamma device, Vielight, Inc. (Toronto, Canada)	810 nm	Transcranial: 75 and 100 mW, 75 and 100 mW/cm^2, 20 min, PW at 40-Hz with duty cycle of 50%, one active and one sham irradiations with one-week washout period between the two sessions Intranasal: 25 mW, 25 mW/cm^2, 20 min, PW at 40-Hz with duty cycle of 50%, one active and one sham irradiations with one-week washout period between the two sessions	Transcranially; multiple areas, bilateral mesial prefrontal cortex, precuneus/posterior cingulate cortex, angular gyrus (correspond to Fpz, Cz, T3 and T4 EEG points) Intranasally; left nose	Increased the power of the higher oscillatory frequencies of alpha, beta and gamma and decreased the power of the slower frequencies of delta and theta in subjects in resting state; affected integration and segregation of brain networks
Chan et al. [49]	Older adults (30) Mean = 66.2 years	LEDs, Model 1100; MedX Health, (Toronto, Canada)	633 + 870 nm	0.99 W, 44.4 mW/cm^2, 20 J/cm^2, 7.5 min, one session, CW	Transcranially; 3 sites on the left frontopolar at Fp1 and right frontopolar at Fp2 and Pz.	Improved neuropsychological functions such as action selection, inhibition ability, and mental flexibility
Fekri et al. [86]	Young volunteers (56) Age = 18 to 30 years old	Laser, Lumix 3 Plus Ultra model (Italy, Fisioline)	808 nm	628 mW, 200 mW/cm^2, 60 J/cm^2, 3.14 cm^2, 5 min, CW	Transcranially; 1 site, right or left primary motor cortex (at EEG map site: C3 or C4)	Irradiation only to the C4 site: improved motor performance as increased the number of finger taps in both right and left hands assessed by finger-tapping test (FTT)

(continued)

Table 15.2 (continued)

Study/year	Subjects (n)/age	Light source	Wavelengths	Irradiation parameters	Irradiation approach/sites	Findings
Jahan et al. [87]	Young volunteers (30) Mean = 21.2 years	LEDs	850 nm	0.4 W, 285 mW/cm^2, 60 J/cm^2, 1.4 cm^2, 2.5 min, CW	Transcranially; 1 site on the right frontopolar at Fp2	Decreased both relative and absolute power of delta band; decreased reaction time of the response in the attentional Go/No-Go task
Wang et al. [88]	Young volunteers (20) Mean = 26.8 years	Laser, CG-5000, Cell Gen Therapeutics, LLC (Dallas, TX, USA)	1064 nm	2.2 W, 162 mW/cm^2, 107 J/cm^2 per session, 11 min, CW, one session	Transcranially; 1 site on the right frontopolar at Fp2	Resulted in time-dependent, significant increases of EEG spectral powers at the alpha and beta bands at broad scalp regions, in particular across bilateral anterior and posterior sites

Abbreviations: ALFF, amplitude of low-frequency fluctuation; CBF, cerebral blood flow; CCO, cytochrome c oxidase; CW, continuous wave; DMS, delayed match-to-sample; EEG, electroencephalography; FDI, first dorsal interosseous; fMRI, functional magnetic resonance imaging; fTCD, functional multidirectional transcranial ultrasound Doppler sonography; LEDs, light-emitting diodes; PFC, prefrontal cortex; PVT, Psychomotor Vigilance Task; PW, pulsed wave; ReHo, regional homogeneity; qEEG, quantitative electroencephalography.

References

1. Majdi, A., et al. 2017. Revisiting nicotine's role in the ageing brain and cognitive impairment. *Reviews in the Neurosciences* 28 (7): 767–781.
2. Hertzog, C., et al. 2008. Enrichment effects on adult cognitive development: Can the functional capacity of older adults be preserved and enhanced? *Psychological science in the public interest* 9 (1): 1–65.
3. Milgram, N.W., et al. 2006. Neuroprotective effects of cognitive enrichment. *Ageing research reviews* 5 (3): 354–369.
4. Sztainberg, Y., and A. Chen. 2010. An environmental enrichment model for mice. *Nature protocols* 5 (9): 1535.
5. Buitenweg, J.I., et al. 2017. Cognitive flexibility training: A large-scale multimodal adaptive active-control intervention study in healthy older adults. *Frontiers in human neuroscience* 11: 529.
6. Kelly, M.E., et al. 2017. The impact of social activities, social networks, social support and social relationships on the cognitive functioning of healthy older adults: A systematic review. *Systematic reviews* 6 (1): 259.
7. Jannusch, K., et al. 2017. A Complex interplay of vitamin B1 and B6 metabolism with cognition, brain structure, and functional connectivity in older adults. *Frontiers in Neuroscience* 11: 596.
8. Anstey, K., and H. Christensen. 2000. Education, activity, health, blood pressure and apolipoprotein E as predictors of cognitive change in old age: A review. *Gerontology* 46 (3): 163–177.
9. Fischer, A., *Environmental enrichment as a method to improve cognitive function. What can we learn from animal models?* NeuroImage, 2016. **131**: p. 42–47.
10. Akbaraly, T., et al. 2009. Leisure activities and the risk of dementia in the elderly: Results from the Three-City Study. *Neurology* 73 (11): 854–861.
11. Frick, K.M., and S.M. Fernandez. 2003. Enrichment enhances spatial memory and increases synaptophysin levels in aged female mice. *Neurobiology of aging* 24 (4): 615–626.
12. Harati, H., et al. 2011. Attention and memory in aged rats: Impact of lifelong environmental enrichment. *Neurobiology of aging* 32 (4): 718–736.
13. Ball, K., et al. 2002. Effects of cognitive training interventions with older adults: A randomized controlled trial. *JAMA* 288 (18): 2271–2281.
14. Novkovic, T., T. Mittmann, and D. Manahan-Vaughan. 2015. BDNF contributes to the facilitation of hippocampal synaptic plasticity and learning enabled by environmental enrichment. *Hippocampus* 25 (1): 1–15.
15. Kramer, A.F., et al. 1999. Ageing, fitness and neurocognitive function. *Nature* 400 (6743): 418.
16. Laurin, D., et al. 2001. Physical activity and risk of cognitive impairment and dementia in elderly persons. *Archives of neurology* 58 (3): 498–504.
17. LLorens-Martín, M., I. Torres-Alemán, and J.L. Trejo, *Exercise modulates insulin-like growth factor 1-dependent and-independent effects on adult hippocampal neurogenesis and behaviour.* Molecular and Cellular Neuroscience, 2010. **44**(2): p. 109–117.
18. Trejo, J.L., M. Llorens-Martin, and I. Torres-Alemán. 2008. The effects of exercise on spatial learning and anxiety-like behavior are mediated by an IGF-I-dependent mechanism related to hippocampal neurogenesis. *Molecular and Cellular Neuroscience* 37 (2): 402–411.
19. Ding, Q., et al. 2006. Exercise affects energy metabolism and neural plasticity-related proteins in the hippocampus as revealed by proteomic analysis. *European Journal of Neuroscience* 24 (5): 1265–1276.

20. Shu-Zhi, L., and W. Li-Hua. 1983. Effects of laser guided by optic fiber into rat brain on conditioned avoidance response and brain chemistry. *Lasers in Surgery and Medicine* 2 (3): 231–239.
21. Shu-Zhi, L., and W. Li-Hua. 1982. Effects of a low power laser beam guided by optic fiber on rat brain striatal monoamines and amino acids. *Neuroscience Letters* 32 (2): 203–208.
22. Ahmed, N.A.E.H., et al. 2008. Effect of three different intensities of infrared laser energy on the levels of amino acid neurotransmitters in the cortex and hippocampus of rat brain. *Photomedicine and Laser surgery* 26 (5): 479–488.
23. Kuo, J.-R., et al. 2015. Deep brain light stimulation effects on glutamate and dopamine concentration. *Biomedical optics express* 6 (1): 23–31.
24. Romeo, S., et al. 2017. Fluorescent light induces neurodegeneration in the rodent nigrostriatal system but near infrared LED light does not. *Brain research* 1662: 87–101.
25. Lombard, A., et al., *Neurotransmitter content and enzyme activity variations in rat brain following in vivo He-Ne laser irradiation.* Proceedings, Round Table on Basic and Applied Research in Photobiology and Photomedicine, 1990: p. 10–11.
26. Uozumi, Y., et al. 2010. Targeted increase in cerebral blood flow by transcranial near-infrared laser irradiation. *Lasers in surgery and medicine* 42 (6): 566–576.
27. Rojas, J.C., A.K. Bruchey, and F. Gonzalez-Lima. 2012. Low-level light therapy improves cortical metabolic capacity and memory retention. *Journal of Alzheimer's Disease* 32 (3): 741–752.
28. Mathangi, D., and R. Shyamala. 2016. Effect of LED photobiomodulation on fluorescent light induced changes in cellular ATPases and Cytochrome c oxidase activity in Wistar rat. *Lasers in medical science* 31 (9): 1803–1809.
29. Tanaka, Y., et al. 2011. Infrared radiation has potential antidepressant and anxiolytic effects in animal model of depression and anxiety. *Brain stimulation* 4 (2): 71–76.
30. Michalikova, S., et al. 2008. Emotional responses and memory performance of middle-aged CD1 mice in a 3D maze: Effects of low infrared light. *Neurobiology of learning and memory* 89 (4): 480–488.
31. Salehpour, F., et al. 2018. A protocol for transcranial photobiomodulation therapy in mice. *JoVE (Journal of Visualized Experiments)* 141: e59076.
32. El Massri, N., et al. 2018. Photobiomodulation reduces gliosis in the basal ganglia of aged mice. *Neurobiology of aging* 66: 131–137.
33. Salehpour, F., et al. 2017. Transcranial low-level laser therapy improves brain mitochondrial function and cognitive impairment in D-galactose–induced aging mice. *Neurobiology of aging* 58: 140–150.
34. Wang, X., et al. 2017. Up-regulation of cerebral cytochrome-c-oxidase and hemodynamics by transcranial infrared laser stimulation: A broadband near-infrared spectroscopy study. *Journal of Cerebral Blood Flow & Metabolism* 37 (12): 3789–3802.
35. Salehpour, F., et al. 2019. Transcranial Photobiomodulation Improves Cognitive Performance in Young Healthy Adults: A Systematic Review and Meta-Analysis. *Photobiomodulation, photomedicine, and laser surgery* 37 (10): 635–643.
36. Salgado, A.S.I., et al. 2019. Cerebral blood flow in the elderly: Impact of photobiomodulation. In *Photobiomodulation in the Brain*, 473–477. Elsevier.
37. Salgado, A.S., et al. 2015. The effects of transcranial LED therapy (TCLT) on cerebral blood flow in the elderly women. *Lasers in medical science* 30 (1): 339–346.
38. Barrett, D.W., and F. Gonzalez-Lima. 2013. Transcranial infrared laser stimulation produces beneficial cognitive and emotional effects in humans. *Neuroscience* 230: 13–23.
39. Blanco, N.J., W.T. Maddox, and F. Gonzalez-Lima. 2017. Improving executive function using transcranial infrared laser stimulation. *Journal of neuropsychology* 11 (1): 14–25.

40. Gonzalez-Lima, F., and D.W. Barrett. 2014. Augmentation of cognitive brain functions with transcranial lasers. *Frontiers in systems neuroscience* 8: 36.
41. Hwang, J., D.M. Castelli, and F. Gonzalez-Lima. 2016. Cognitive enhancement by transcranial laser stimulation and acute aerobic exercise. *Lasers in medical science* 31 (6): 1151–1160.
42. Gutiérrez-Menéndez, A., et al., *Photobiomodulation as a promising new tool in the management of psychological disorders: a systematic review.* Neuroscience & Biobehavioral Reviews, 2020.
43. Vargas, E., et al. 2017. Beneficial neurocognitive effects of transcranial laser in older adults. *Lasers in medical science* 32 (5): 1153–1162.
44. Zomorrodi, R., et al. 2019. Pulsed near infrared transcranial and intranasal photobiomodulation significantly modulates neural oscillations: A pilot exploratory study. *Scientific reports* 9 (1): 1–11.
45. Bonanni, E., et al. 2012. Differences in EEG delta frequency characteristics and patterns in slow-wave sleep between dementia patients and controls: A pilot study. *Journal of Clinical Neurophysiology* 29 (1): 50–54.
46. Berman, M.H., and J.A. Frederick. 2009. Efficacy of neurofeedback for executive and memory function in dementia. *Alzheimer's & Dementia* 4 (5): e8.
47. Ruiz, S., et al. 2013. Acquired self-control of insula cortex modulates emotion recognition and brain network connectivity in schizophrenia. *Human brain mapping* 34 (1): 200–212.
48. Scheinost, D., et al. 2014. Resting state functional connectivity predicts neurofeedback response. *Frontiers in Behavioral Neuroscience* 8: 338.
49. Chan, A.S., et al. 2019. Photobiomodulation improves the frontal cognitive function of older adults. *International Journal of geriatric psychiatry* 34 (2): 369–377.
50. Yao, L., et al., *Effects of stimulating frequency of NIR LEDs light irradiation on forehead as quantified by EEG measurements.* Journal of Innovative Optical Health Sciences, 2020: p. 2050025.
51. Shen-Zeng, X.-J., S. Lin, and L. Wang. 1982. Effects of a low power laser beam guided by optic fiber on rat brain striatal monoamines and amino acids. *Neuroscience Letters* 32: 203–208.
52. Igarashi, H., and K. Inomata. 1991. Effects of low-power gallium aluminium arsenide diode laser irradiation on the development of synapses in the neonatal rat hippocampus. *Cells Tissues Organs* 140 (2): 150-155.
53. Urciuoli, R., P.M. Rolfo, and V. Rossetti. 1991. SUPEROXIDE DISMUTASE (SOD) LEVELS IN RAT BRAIN AFTER HeNe LASER IRRADIATION COMPARED TO INTRAVENOUS ADMINISTRATION OF SOD SOLUTION: AN EXPERIMENTAL MODEL FOR THE PREVENTION OF CEREBRAL VASOSPASM AFTER SUBARACHNOID HAEMORRHAGE. *LASER THERAPY*, 3 (4): 183–186.
54. Rochkind, S. 1992. Central nervous system transplantation benefitted by low-power laser irradiation. *Lasers in medical science* 7 (1-4): 143-151.
55. Valls, G., T. Hernández Gil de Tejada, and F. Martínez Soriano. 1995. A morphometric and statistical study of the effects of soft laser (He-Ne) irradiation on the pineal gland. *Histology and histopathology* 1995.
56. Zubkova, S. and L. Mikhailik. 1995. Effect of pulsed infrared laser radiation on DNA synthesis in tissues of intact rats and during strenuous physical exercise. *Bulletin of Experimental Biology and Medicine* 119 (6): 602-604.
57. Mochizuki-Oda, N., et al. 2002. Effects of near-infra-red laser irradiation on adenosine triphosphate and adenosine diphosphate contents of rat brain tissue. *Neuroscience Letters* 323 (3): 207–210.
58. Ilic, S., et al. 2006. Effects of power densities, continuous and pulse frequencies, and number of sessions of low-level laser therapy on intact rat brain. *Photomedicine and Laser Therapy* 24 (4): 458–466.

59. McCarthy, T.J., et al. 2010. Long-term safety of single and multiple infrared transcranial laser treatments in Sprague-Dawley rats. *Photomedicine and Laser Surgery* 28 (5): 663–667.
60. Gorshkova, O., V. Shuvaeva, and D. Dvoretsky. 2013. Role of nitric oxide in responses of pial arterial vessels to low-intensity red laser irradiation. *Bulletin of experimental biology and medicine* 155 (5): 598.
61. Zhang, L., et al. 2015. Near Infrared (NIr) Light Increases Expression of a Marker of Mitochondrial Function in the Mouse Vestibular Sensory Epithelium. *JoVE (Journal of Visualized Experiments)* (97): e52265.
62. Mintzopoulos, D., et al. 2017. Effects of near-infrared light on cerebral bioenergetics measured with phosphorus magnetic resonance spectroscopy. *Photomedicine and Laser Surgery* 35 (8): 395–400.
63. Santos, T., et al. 2017. Blue light potentiates neurogenesis induced by retinoic acid-loaded responsive nanoparticles. *Acta Biomaterialia* 59: 293–302.
64. Litscher, G., L. Wang, and M. Wiesner-Zechmeister. 2000. Specific effects of laserpuncture on the cerebral circulation. *Lasers in Medical Science* 15 (1): 57–62.
65. Litscher, G., D. Rachbauer, S. Ropele, L. Wang, D. Schikora, F. Fazekas, and F. Ebner. 2004. Acupuncture using laser needles modulates brain function: first evidence from functional transcranial Doppler sonography and functional magnetic resonance imaging. *Lasers in Medical Science* 19 (1): 6–11.
66. Hsieh, C.-W., et al. 2011. Different brain network activations induced by modulation and non-modulation laser acupuncture. *Evidence-Based Complementary and Alternative Medicine*.
67. Wu, J.-H., et al. 2012. Effect of low-level laser stimulation on EEG. *Evidence-Based Complementary and Alternative Medicine*.
68. Quah-Smith, I., et al. 2013. Differential brain effects of laser and needle acupuncture at LR8 using functional MRI. *Acupuncture in Medicine* 31 (3): 282–289.
69. Wu, J.-H. and Y.-C. Chang. 2013. Effect of low-level laser stimulation on EEG power in normal subjects with closed eyes. *Evidence-Based Complementary and Alternative Medicine*.
70. Raith, W., et al. 2013. Near-infrared spectroscopy for objectifying cerebral effects of laser acupuncture in term and preterm neonates. *Evidence-Based Complementary and Alternative Medicine*.
71. Konstantinović, L.M., et al. 2013. Transcranial application of near-infrared low-level laser can modulate cortical excitability. *Lasers in Surgery and Medicine* 45 (10): 648–653.
72. Barrett, D.W. and F. Gonzalez-Lima. 2013. Transcranial infrared laser stimulation produces beneficial cognitive and emotional effects in humans. *Neuroscience* 230: 13–23.
73. Chaieb, L., et al. 2015. Neuroplastic effects of transcranial near-infrared stimulation (tNIRS) on the motor cortex. *Frontiers in behavioral neuroscience* 9: 147.
74. Blanco, N.J., W.T. Maddox, and F. Gonzalez-Lima. 2015. Improving executive function using transcranial infrared laser stimulation. *Journal of neuropsychology* 11 (1): 14–25.
75. Lv, J., et al. 2016. The brain effects of laser acupuncture at thirteen ghost acupoints in healthy individuals: A resting-state functional MRI investigation. *Computerized Medical Imaging and Graphics* 54: 48–54.
76. Tian, F., et al. 2016. Transcranial laser stimulation improves human cerebral oxygenation. *Lasers in surgery and medicine* 48 (4): 343–349.
77. Blanco, N.J., C.L. Saucedo, and F. Gonzalez-Lima. 2017. Transcranial infrared laser stimulation improves rule-based, but not information-integration, category learning in humans. *Neurobiology of learning and memory* 139: 69–75.
78. Wang, X., et al. 2017. Impact of heat on metabolic and hemodynamic changes in transcranial infrared laser stimulation measured by broadband near-infrared spectroscopy. *Neurophotonics* 5 (1): 011004.

79. Grover Jr, F., J. Weston, and M. Weston. 2017. Acute effects of near infrared light therapy on brain state in healthy subjects as quantified by qEEG measures. *Photomedicine and laser surgery* 35 (3): 136–141.

80. Nawashiro, H., et al. 2017. Blood-oxygen-level-dependent (BOLD) functional magnetic resonance imaging (fMRI) during transcranial near-infrared laser irradiation. *Brain Stimulation: Basic, Translational, and Clinical Research in Neuromodulation* 10 (6): 1136–1138.

81. Moghadam, H.S., et al. 2017. Beneficial effects of transcranial Light Emitting Diode (LED) therapy on attentional performance: an experimental design. *Iranian Red Crescent Medical Journal* 19 (5).

82. Litscher, G. 2018. *Transcranial laser stimulation research—A new helmet and first data from near infrared spectroscopy.* Multidisciplinary Digital Publishing Institute.

83. Litscher, G. 2019. *Brain photobiomodulation—preliminary results from regional cerebral oximetry and thermal imaging.* Multidisciplinary Digital Publishing Institute.

84. Holmes, E., et al. 2019. Cognitive enhancement by transcranial photobiomodulation augments cerebrovascular oxygenation of the prefrontal cortex. *Frontiers in neuroscience* 13: 1129.

85. El Khoury, H., J. Mitrofanis, and L.A. Henderson. 2019. Exploring the effects of near infrared light on resting and evoked brain activity in humans using magnetic resonance imaging. *Neuroscience.*

86. Fekri, A., et al. 2019. Short term effects of transcranial near-infrared photobiomodulation on motor performance in healthy human subjects: An experimental study. *Journal of Lasers in Medical Sciences* 10 (4): 317–323.

87. Jahan, A., et al. 2019. Transcranial near-infrared photobiomodulation could modulate brain electrophysiological features and attentional performance in healthy young adults. *Lasers in medical science* 1–8.

88. Wang, X., et al. 2019. Transcranial photobiomodulation with 1064-nm laser modulates brain electroencephalogram rhythms. *Neurophotonics* 6 (2): 025013.

Conclusions and Future Directions 16

The loss of one's mental faculties is a major health concern for the majority of the world's population. Whether this arises from dementia, Alzheimer's disease, stroke, TBI, psychiatric disorders, or simple aging, it will eventually affect everybody at some stage in their lifetime. Therefore brain specialists always look for effective, inexpensive, and sustainable treatment modalities. Because real-world mental activities require complex cognitive processes, such as decision-making, sustained attention, and executive function, novel modalities for improving brain function are in great demand. Therefore, we believe that brain PBM therapy will become a promising alternative or complementary strategy for neurorehabilitation in the coming years. The practitioners in this field are in need of information about the dosage, treatment regimens, preferred irradiation sites, power of devices, time of treatment, and any possible side effects of this approach. For example, the application of pulse mode irradiation for patients with migraine may result in adverse effects, or PBM in bipolar patients in the manic phase could result in more brain metabolic activity. On the other hand, knowledge about how light interacts with cellular and molecular targets, such as mitochondria, DNA, ion channels, and various light-sensitive receptors is important in the selection of wavelengths and fluences for effective therapy. Moreover, to allow treatment planning in terms of uniform delivered fluence on the cortex, magnetic resonance imaging (MRI) can play a key role in the exact measurement of the skull and scalp thickness, as these values are different from individual to individual. In order to improve brain functions, such as learning and memory, concentration, problem solving, and other high-level cognitive functions in healthy individuals (e.g., surgeons, military forces, athletes, students, elderly subjects, etc.), brain PBM could provide more natural cognitive enhancement instead of brain-boosting drugs.

In regard to intranasal PBM (i-PBM) therapy, as our thorough review of the literature on this topic suggests, the lack of any approved indication for its clinical use is not surprising, considering the limited scientific evidence available. Nonetheless, i-PBM products are currently available in the market for the enhancement of mental function in healthy individuals. The devices are very convenient and user-friendly, and are easily applied in the nose as portable home-based devices for self-administration of red and near-infrared (NIR) light. Compared to transcranial PBM helmet devices on the market, they are relatively inexpensive. Particularly, these are ideal for red and NIR light users who wish to benefit from the systemic effects of photobiomodulation to maintain their current healthy state. Research is still in its preliminary stages to elucidate the mechanisms of the neurotherapeutic effects seen with nostril-based i-PBM. The recent surge in public awareness of the need to mitigate the risk of neurodegenerative diseases and to reduce the burden of neuropsychiatric disorders, has led to a wider desire to improve and preserve brain health. This is the reason why intra-nostril i-PBM is favored by many individuals. It is noteworthy that while several systemic beneficial effects have been reported in the literature, no evidence has yet been produced to support improved mental well-being or prevention of mental disorders. Since i-PBM is easy and straightforward to use, it could be an ideal first-line treatment for brain disorders. The clinical use is off-label and should be used only when the patient is resistant to, refuses, or does not tolerate the available approved treatments. Based on dosimetry studies, one limitation of intra-nostril i-PBM is its inability to deliver sufficient light to the brain, and hence does not allow direct brain neuromodulation. Some authors have proposed indwelling i-PBM devices, according to current advances in technology. Currently, novel i-PBM interventions such as indwelling devices in the sphenoid sinus or in the olfactory submucosa appear to be unrealistic at the moment. Surgical procedures are needed to implant LEDs at the back of the roof of the nose, and these are limited by the minuscule, curved, and rigid nature of the intranasal cavity. An implanted indwelling device is not possible with current surgical equipment unless the bony structures are surgically drilled or crushed. The nasal bone structures are extremely thin and the cavity is highly vascularised, which presents a risk of iatrogenic hemorrhage, olfactory nerve damage, cerebrospinal fluid (CSF) leakage, and even damage to brain structures like the pituitary gland. Hence, it is reasonable to regard these deep i-PBM indwelling devices as a clinical utopia. Deep-nose i-PBM was proposed to allow for a much higher light fluence delivered to the ventromedial prefrontal cortex (vmPFC) and orbitofrontal prefrontal (vmOFC), and even the limbic system. In contrast, mid-nose submucosal i-PBM devices and frontal sinus i-PBM devices are reasonably practical, and their benefits could surpass intra-nostril and transcranial PBM devices to some extent. Most notably, these indwelling devices will need septoplasty-equivalent minimally-invasive otorhinolaryngology (ENT) surgery, so they could be implanted during an outpatient-day procedure or as an elective surgery. Although the delivery of NIR light to the vmPFC and vmOFC would be equivalent to an external transcranial PBM light source on the forehead skin, i-PBM indwelling devices are more promising. Some limitations of transcranial PBM have been identified in its off-label use in neuropsychiatric and neurorehabilitation PBM clinics. Firstly, transcranial PBM is time-consuming, as it necessitates either an in-office visit for laser light delivery or self-administration with LED devices. Even the latter requires at least 60 min a day to achieve a neurotherapeutic dose because of the low irradiance. The inevitable time-consuming nature of transcranial PBM hinders its use in functioning healthy individuals. Indwelling mid-nose or frontal sinus i-PBM devices could overcome this constraint due to their automatic pre-set programs administered without interfering with routine life. Secondly, for patients with mental or physical impairments, transcranial PBM applied at home requires support from family members or other caregivers for the positioning and the control of light delivery. In these patients with neurodevelopmental disorders or dementia, the need for the involvement of caregivers is avoided by indwelling i-PBM devices. Thirdly, an innovative approach using smart neuromodulator devices is likely to be developed, where it would be possible to "read" the state of the brain and "write" into the brain using light, with the appropriate stimulation to help brain function. In accordance with these advances, transcranial PBM is the modality of choice among closed-loop devices, which incorporate both brain activity sensors and red/NIR stimulation to the brain. Moreover, i-PBM indwelling devices have another advantage as they allow for real-time, automated light delivery where transcranial PBM may be impractical or requires conscious effort of patients. This may include critical conditions such as an unexpected craving for drugs of abuse, self-harm, or suicidal ideation. It should be noted that regardless of all the benefits of i-PBM, it still remains to be demonstrated whether the two modalities have equivalent therapeutic effects on brain disorders. Some of the proposed NIR mechanisms of action is only relevant to transcranial PBM [1]. To sum up, although intra-nostril i-PBM is presently available for boosting brain function in healthy individuals, there is still no proof of its efficacy in achieving and/or preserving brain health. However, the novel neurotherapeutic modalities of submucosal or frontal indwelling i-PBM are not available on the market, nor is there sufficient scientific evidence to support them. Deep-nose i-PBM, i.e., under the cribriform plate or within the sphenoid sinus, is still only a theoretical approach without established technology and with uncertain clinical efficacy, which may be tested in the future.

It is likely that the application of PBM will be combined with other neurorehabilitation approaches. Although there are only a few studies investigating the potential effects of combining neuro-rehab approaches with PBM in brain disorders, combination therapy with drugs or supplements could be considered. These could include metabolic modulators (e.g., methylene blue, coenzyme Q_{10}, pyrroloquinoline quinone [PQQ], hydroxyl or adenosylcobalamin, N-acetyl-l-cysteine [NAC], magnesium threonate, curcumin, pregnenolone, docosahexaenoic acid [DHA], riboflavin, L-carnitine), antidepressants (e.g., fluoxetine citalopram, and lithium), or neurotherapeutic agents (e.g., γ-Secretase and tissue plasminogen activator [tPA]). Moreover, additional device-based neuromodulation methods (e.g., transcranial magnetic stimulation [TMS], magnetic seizure therapy [MST], vagus nerve stimulation [VNS], transcranial direct current stimulation [tDCS], deep brain stimulation [DBS], low-intensity focused ultrasound pulsation, or electroconvulsive therapy [ECT]) could be combined with PBM. Stem cell therapy, as well as cognitive improvement methods (e.g., aerobic exercise, environmental enrichment), could be combined with PBM for patients who suffer from trauma, stroke, Alzheimer's disease (AD), Parkinson's disease (PD), as well as psychiatric disorders [2–12]. In addition, a combination of neurofeedback and PBM has been shown to be promising for the safe and effective renormalization of neural connectivity and to halt the progression of neurodegeneration [13]. Nevertheless, much more work is required to define the optimal protocols for combining other modalities with PBM.

Research studies on PBM have compared different operation modes (pulsed wave [PW] vs. continuous wave [CW]), different treatment protocols (combination therapy vs. monotherapy), different light delivery approaches (brain irradiation vs. remote irradiation), different treatment protocols (acute irradiation vs. chronic irradiation), different physical parameters (comparing various wavelengths, and light sources such as LEDs or lasers). Moreover, different diagnostic tools (functional MRI [fMRI], positron emission tomography [PET], single-photon emission computed tomography [SPECT], and functional near-infrared spectroscopy [NIRS]) can be used to monitor the effects of PBM, and eventually to design reliable treatment protocols and guidelines [14].

Numerous LED PBM devices have received Food and Drug Administration (FDA) clearance for sale over the counter worldwide. They are considered safe for general health and wellness applications, provided the manufacturers do not claim specific therapeutic efficacy. Nevertheless, these devices are a low-cost option for home treatment of brain conditions, under proper medical supervision. If PBM were to be approved as a safe and effective modality for neurorehabilitation, these devices would be well suited for adoption among the expanding range of possible therapeutic options for brain disorders.

Although the potential of PBM as a novel treatment for neurodegenerative diseases remains unconfirmed, a variety of studies have demonstrated the efficacy of PDM for epilepsy, Alzheimer's disease, Parkinson's disease and other neurodegenerative diseases. PBM may become an effective approach for these disorders in the future, but several points must be considered regarding its application. Firstly, PBM should if possible, be administered during the early phase of disease development. Some studies assessing PBM for the treatment of neurodegenerative diseases have been conducted during the later phase of progression, and have shown that, like other therapies, PBM cannot rescue neurons already undergoing degeneration from apoptosis, or restore them to normal healthy conditions. Taken together, these findings indicate that PBM should be used to inhibit neuronal degeneration before it really starts, and to prevent apoptosis from occurring. Secondly, one major advantage of PBM is its relative lack of side effects, which suggests that it may be used as an adjunctive therapy combined with current effective treatments. For example, pharmacological side effects can be reduced by lowering the dose of the drug, and the application of PBM could reduce the required drug dose. Thirdly, it is important to consider how to apply PBM to the relevant brain lesion or region using a helmet with LEDs, or by surgically implanting a fiber optic into the brain. The current extent of research and development for the clinical application of PBM therapy remains insufficient. However, the therapeutic possibilities and efficacy of PBM for the treatment of neurodegenerative diseases are clearly evident, and warrant the further attention of researchers and clinicians.

References

1. Herisson, F., et al. 2018. Direct vascular channels connect skull bone marrow and the brain surface enabling myeloid cell migration. *Nature neuroscience* 21 (9): 1209.
2. Fukuzaki, Y., et al. 2013. 532 nm low-power laser irradiation recovers γ-secretase inhibitor-mediated cell growth suppression and promotes cell proliferation via Akt signaling. *PLoS ONE* 8 (8): e70737.
3. Dong, T., et al. 2015. Low-level light in combination with metabolic modulators for effective therapy of injured brain. *Journal of Cerebral Blood Flow & Metabolism* 35 (9): 1435–1444.
4. Peplow, P.V. 2015. Neuroimmunomodulatory effects of transcranial laser therapy combined with intravenous tPA administration for acute cerebral ischemic injury. *Neural regeneration research* 10 (8): 1186.
5. Lapchak, P.A., and P.D. Boitano. 2016. A novel method to promote behavioral improvement and enhance mitochondrial function following an embolic stroke. *Brain research* 1646: 125–131.
6. Lapchak, P.A., et al. 2008. Safety profile of transcranial near-infrared laser therapy administered in combination with thrombolytic therapy to embolized rabbits. *Stroke* 39 (11): 3073–3078.
7. Hwang, J., D.M. Castelli, and F. Gonzalez-Lima. 2016. Cognitive enhancement by transcranial laser stimulation and acute aerobic exercise. *Lasers in medical science* 31 (6): 1151–1160.
8. Meynaghizadeh-Zargar, R., et al., *Effects of transcranial photobiomodulation and methylene blue on biochemical and behavioral profiles in mice stress model.* Lasers in medical science, 2019: p. 1–12.
9. Salehpour, F., et al. 2019. Near-infrared photobiomodulation combined with coenzyme Q10 for depression in a mouse model of restraint stress: Reduction in oxidative stress, neuroinflammation, and apoptosis. *Brain research bulletin* 144: 213–222.
10. Salehpour, F., et al., *Photobiomodulation and Coenzyme Q10 Treatments Attenuate Cognitive Impairment Associated with Model of Transient Global Brain Ischemia in Artificially Aged Mice.* Frontiers in cellular neuroscience, 2019. **13**.
11. Cassano, P., *Photomedicine and Pharmaceuticals: A Brain New Deal*, 2019, Mary Ann Liebert, Inc., publishers 140 Huguenot Street, 3rd Floor New ….
12. Gonzalez-Lima, F., and A. Auchter. 2015. Protection against neurodegeneration with low-dose methylene blue and near-infrared light. *Frontiers in cellular neuroscience* 9: 179.
13. Berman, M.H., and T.W. Nichols. 2019. Treatment of Neurodegeneration: Integrating Photobiomodulation and Neurofeedback in Alzheimer's Dementia and Parkinson's: A Review. *Photobiomodulation, photomedicine, and laser surgery* 37 (10): 623–634.
14. Zein, R., W. Selting, and M.R. Hamblin. 2018. Review of light parameters and photobiomodulation efficacy: Dive into complexity. *Journal of biomedical optics* 23 (12): 120901.